Treating Dairy Cows Naturally
Thoughts & Strategies

Hubert J. Karreman, V.M.D.

Acres U.S.A. LLC
PO Box 351
Viroqua, Wisconsin 54665 U.S.A.

Treating Dairy Cows Naturally
Thoughts & Strategies

Copyright © 2004, 2007 Hubert J. Karreman, V.M.D.

Acres U.S.A. LLC
PO Box 351
Viroqua, Wisconsin 54665 U.S.A.
(512) 892-4400 • fax (512) 892-4448
info@acresusa.com • www.acresusa.com

Printed in the United States of America

Publisher's Cataloging-in-Publication

Karreman, Hubert, J. 1962-
Treating dairy cows naturally: thoughts and strategies / Hubert J. Karreman.
Austin, TX, ACRES U.S.A., 2007
    xx, 412 pp., 25 cm.
    Includes Index
    Includes Bibliography
    2nd edition
    ISBN 978-1-60173-000-8 (trade)

    1. Dairy cattle – Diseases – Homeopathic treatment. 2. Dairy Cattle – Diseases – Prevention. 3. Dairy farms – Management. 4. Holistic veterinary medicine – United States. I. Karreman, Hubert, 1962- II. Title.

SF961.   636.2089

A HANDBOOK FOR

ORGANIC & SUSTAINABLE FARMERS

TREATING DAIRY & COWS NATURALLY

THOUGHTS & STRATEGIES

HUBERT J. KARREMAN

V.M.D.

Disclaimer

The contents of this book represent views on traditional and contemporary treatments and compile clinically useful information but are not intended or offered to supplant the advice or service of a trained professional veterinarian that has an established veterinary-client-patient-realtionship.

For Becky — a gift from God
and for Emily, our little creation who loves her
moooooo cows!

Acknowledgements

This book would not have been made possible without the input, inspiration, help and assistance of some very important people.

First, I would like to thank Dr. Joe McCahon for being my mentor when I finished veterinary school and was still quite "green" to some of the realities of veterinary practice. I remember three occasions where he revealed to me the art of being a front line practitioner: a prolapsed uterus, a right-sided torsion of the abomasum and a fetotomy. The prolapse was at midnight, the torsion was on a Saturday afternoon and the fetotomy was during the evening on a weeknight. Such is the life of a practitioner. We first met at a meeting for veterinary homeopathy in 1991, he a seasoned practitioner and I a veterinary student, both interested to further our learning in the alternative medical realm. Over the years we have maintained a wonderful professional relationship by discussing cases through countless phone calls and maintaining a healthy friendship all along. I am very grateful for his wisdom and expertise.

Another person I would like to thank is David Griffiths at Seven Stars Farm in Kimberton, PA. My life was rearranged for the better because of the Biodynamic practices that I learned while being herdsman on his farm. The care and affection that I could express for the cows in the herd will last with me forever. He and his wife, Edie, produce the best yogurt that I have ever tasted — Seven Stars Biodynamic Yogurt. Although I did not grow up on a farm, Seven Stars is my "home farm," which I still visit whenever time allows. David also accepted the responsibility to be a reviewer of the text for this book. Thank you, David.

Part of what changed my view about dairy cows while being a herdsman at Seven Stars Farm was being introduced to homeopathic remedies for various health conditions. These little remedies worked so well that it dawned on me to go to veterinary school to learn the "real thing." That chain of events certainly steered me onto my career path of now using whatever method is appropriate for a given situation. Our homeopathic veterinarian was Dr. Ed Sheaffer; he is simply the best teacher of farm animal homeopathic medicine there is.

I would also like to thank the following reviewers of this text who brought to light various issues, each stemming from their unique perspectives. First, Dr. Jill Beech, Professor of Medicine at New Bolton Center, University of Pennsylvania School of Veterinary Medicine. Jill took painstaking notes and added minute corrections to everything from the general ideas presented in the text to subtle grammatical changes. She is a friend of ecological agriculture and always stimulates me to think about the ethical implications of farming practices. Jerry Brunetti of Agri-Dynamics Inc., who is without doubt the world's preeminent "professor of holistic farm management," added many important insights and concerns. Jerry's long standing passion and concern for healthy agro-ecosystems is unsurpassed. Kelly Shea, Director of Organic Agriculture at Horizon Organic Dairy, Inc., continues to provide me with valuable insights regarding healthcare issues for organic livestock on the regulatory front. Kelly's love for dairy cows and their well-being is obvious to anyone who meets her for even a few minutes. I would also like to thank my long time friend, Dr. Mark Hodgson, who has always "ribbed me" about alternative veterinary treatments — hopefully this book will have explained to him exactly how I practice. His wife, Patty (an equine practitioner who is certified in acupuncture and chiropractic) and I will continue to try to teach Mark about alternative therapies.

I would also like to express my thanks to two veterinary colleagues who hold very different views on treatments for animals yet who both sincerely aim to do what is best for the animals in their care. Dr. Susan Wynn pointed me in the direction of the Eclectic practitioners after my constant questioning of how to arrive at the correct medicinal concentrations of natural treatments. Susan is one of the primary leaders of the complementary and alternative veterinary medicine (CAVM) movement in the United States. Dr. David Ramey, an outspoken proponent of evidence-based medicine, enlightened me to the term "hormesis" — the realm of low concentration compounds that are still within material reality. I met David while we were both sitting on the American Veterinary Medical Association's "Taskforce on Complementary and Alternative Medicine." I thank them both for helping me to better define my clinical approaches in medicine.

Finally, I wish to thank all the wonderful dairy farmers with whom I work. They are the folks that have allowed me to actually use complementary and alternative therapies on their cows (with the certified organic farmers being very open to trying out new methods). They are the people who ultimately decide which treatments are worth repeating over time. My daily interaction with their animals helps to continually expand my use of all kinds of therapies — using whatever is best for each cow that needs treatment.

Mission Statement
of Penn Dutch Cow Care

My goal as a veterinarian is to promote environmentally sound methods of dairy farming. I will dedicate my professional knowledge to help farmers use water, soil, pasture and manure resources in ways that create a biologically healthy landscape for dairy cattle and society. In medicine and reproduction, I will primarily use natural approaches instead of only synthetic antibiotic and hormonal treatments. In emergency situations, I will use both conventional and non-conventional treatments as needed for the well-being of the animal. I will endeavor to coordinate local grassroots efforts with state and national programs to promote sustainable agriculture. I will remain up-to-date regarding prohibited and allowed substances for certified organic clients. Additionally, in order to expand my knowledge and deliver new information to clients, I will also attend meetings and classes oriented toward achieving the goals stated herein.

Hubert J. Karreman, V.M.D.
December 1999

Veterinarian's Oath

Being admitted to the profession of veterinary medicine, I solemnly affirm to use my scientific knowledge and skills for the benefit of society through the protection of animal health, the relief of animal suffering, the conservation of animal resources, the promotion of public health, and the advancement of medical knowledge.

I will practice my profession conscientiously, with dignity, and in keeping with the principles of veterinary medical ethics.

I accept as a lifelong obligation the continual improvement of my professional knowledge and competence.

Contents

Cows on pasture with farm in background.

Introduction

The title of this book, *Treating Dairy Cows Naturally,* may seem rather generic for some readers. However, when I use the phrase, I am referring to the way we treat dairy cows in all aspects. This means thinking about how the cows are impacted by the overall landscape they live in — the conservation and health of soils which create the pastures from which they graze, the crops that are fed to them in the winter, the waterways that flow through the farm, the barnyard environment as well as the actual treatments that are used if and when they need medicine.

This book originally started as a small booklet with natural treatment recipes for farmers to use. Time has gone on, and as organic dairy farming has grown, many issues within the organic sphere have come to the fore. In addition, my own perspective as a veterinarian of natural treatments enlarged when I was introduced to the Eclectic school of therapeutics and its pharmacology, which is based predominantly on using medicinal plants. This book is intended to give various approaches to working with dairy cows in a more natural way; it is presented in a more thorough context than just the simple and quick recipes as originally envisioned. Hopefully, I will create a solid foundation from which the reader can build a working knowledge of holistic management and natural medicines, as they pertain to dairy cows. The information and insights I have to offer is that of a clinical practitioner. I am called to treat cows for a variety of medical conditions on a daily basis — acute illnesses, chronic conditions, as well as life threatening emergencies on about 85 certified organic and 30 conventional farms. Therefore my approach to the reader is to describe what thoughts I have regarding treating livestock, what kind of discussions I have with the farmers and also the treatments that we end up utilizing. I gladly share many of the recipes or formulas that I use so readers will become adept at using natural treatments and hopefully excited to learn more about these kinds of treatments. Although I write from my direct experiences "out in the trenches" with my client farmers, this book is especially for those dairy farmers and veterinarians who are at a distance and contact me, but with whom I cannot meet or personally examine their cows. Another reason for the existence of this book is that there are already so many books on the natural treatments of cats, dogs and horses that I thought it was high time for dairy cows to be given their own forum.

There may be a few recipes (from the old veterinary texts) that relate to sheep, pigs and horses, but the recipes in this book focus predominantly on dairy cows. Information provided is meant to be incorporated by farmers into an overall holistic livestock management plan for their farm. It is also for veterinary practitioners who are open to ideas based partly on scientific knowledge, partly on theoretical hypotheses and balanced by clinical concern for the cow — in effect, the art of veterinary practice. The information contained in this book is a mixture of technical details as well as common sense approaches. Hopefully the explanations presented regarding natural treatments and how I justify using them will help to bridge farmers and veterinarians into the realm of applied pharmacognosy — a realm of medicine that combines real amounts of natural materials (especially plants as crude drugs) administered for their known clinical use. By researching older veterinary textbooks, I have come to realize that there are many plant and mineral remedies buried just beneath the surface of time and these need to be recovered, revived and applied to modern dairy cows — especially organic dairy cows that are not allowed to be given most synthetic treatments. This work is geared to such an effort — to blend the old with the new, to combine modern scientific understanding with quaint laid-aside treatments, and to use practical on-the-farm procedures to enhance the effectiveness of natural treatments. I don't pretend to have all the answers, but by utilizing holistic approaches to common problems, the cases truly needing antibiotics and hormones will hopefully be few and far between. Other parties interested in natural types of approaches to managing dairy cows may also be interested in the topics discussed. Those parties may include organic certifying agencies, organizations dedicated to promoting ecological dairy farming and also consumers of natural dairy products. I do not expect anyone to read this book straight through to the end; instead, I envision it to be a work that can be quickly referred to while also being thought-provoking at other times. Natural treatments work best if a truly caring and seriously dedicated cow-conscious person is in charge of the herd. Therefore, it depends not as much on what type of farming system is in place, but more on the person who is the primary care giver for the cows.

Hubert J. Karreman, V.M.D.
Bartville, PA

PART 1
THOUGHTS

Long and winding road.

Chapter 1

WHAT IS A DAIRY FARM?

What exactly *is* a dairy farm? The question seems rather basic. But is it? Answering this question may give us an idea as to how society currently views agriculture. We can then look at what a truly healthy dairy farm is, arrive at what a healthy dairy cow would be, and see how best to keep her that way.

There are two parallel tracks in agriculture, each following a qualitatively different path from the other, but having the same goal — to produce food for society. Farmers can jump tracks. It may be a conventional "traditional" family farm willing to either process milk into a finished "value-added" niche market product, or when the farmer consciously shifts thinking, transitioning into grazing and then certified organic production. Conversely, it can be a traditional family farm that undergoes major expansion into a total confinement situation. Although total confinement farms pump out large quantities of milk, they only receive the lower commodity blend price for the milk, and there are hidden costs to always keeping cows inside on concrete.

Dairy cows are living, breathing, sentient creatures created to *eat grass* and turn it into milk. Animals of the family Bovidae have four stomachs for a specific reason. They, along with other ruminants, are here on earth to fill an extremely important ecological niche — to digest plant materials that we humans cannot. We are the beneficiaries of this by being able to consume and assimilate that biologically nutritious substance called milk.

I would forecast many, if not most, remaining family farms (as consumers perceive them today) will be Amish or Mennonite in the years ahead, and that other family farms will become graziers +/- organic. Or, on the other hand, some family farms will expand towards large, total confinement operations. Actually, it's already happening. Fortunately, many farmers are considering the grazing +/- organic route as they feel that this is a profitable and healthy way to take care of their animals. Healthy cows and a healthy environment make for a healthy business and also are rewarded by consumers who will pay a premium for products from such farms.

So, just what is a *healthy* dairy farm? Obviously, it is highly doubtful a total confinement operation would be, unless only the bottom line net profit is looked at to be a reflection of health. This is actually how an agricultural economist may view farm health, much like a physician may analyze a patient by way of a urinalysis or liver profile and declaring the person "well" based solely on a single external exam finding. Some may believe that cows simply producing maximum amounts of milk are what makes for healthy cows and farms. I believe, however, that we need to look at the internal functioning of the system as a whole, interwoven, living, biological entity — the agro-ecological interaction of cows, soil, water, crops and conservation practices — as well as the human enjoyment of participating in it all. We need to reflect on all these components *as a whole* in order to grasp what a truly healthy dairy farm is. Members of modern society at times become aware of farming but usually form opinions only based on perceived regional environmental effects. Rural residents may be more attuned, but usually think no further than the general farm landscape scenery as they drive by it.

Viewed in this way, a healthy dairy farm may likely be one that is at least "traditional" — one that allows the cows outside in nice weather and practices soil conservation with the visually pretty, contoured cropland. For consumers driving by, this is a pleasing view. Most likely it is an organic or ecological farm that strongly utilizes conservation principles, enhancing cow well-being on a daily basis in conjunction with land health on a perennial basis by intensively grazing pastures and never using herbicides, insecticides and fungicides. Both cow and land have been directly tied to each other forever. Total confinement farming tries to divorce this most basic concept that has been with civilization since the first recorded keeping of domesticated livestock. In agriculture we humans are in control - but hopefully it is control *with compassion and respect* for the creatures to express their natural behaviors of walking on the land and grazing. I mention cows specifically "on a daily basis" because we can intrinsically connect with them simply by living each day and sharing somewhat similar reactions to our immediate environment and daily needs to eat, rest and be comfortable. On grazing farms, farmers interact closely with each of the cows at milking times and when walking them back and forth to pasture. As a herdsman, life revolves around the cows, and they are completely dependent upon your attention, care and compassion. Over time, it is rather easy for a person sufficiently sensitive to immerse their entire consciousness into that of the herd and the individual animals that create it. Being with the cows daily can become a long-term meditation. Like the look of cows peacefully chewing their cud, life around dairy cows can be a deeply enriching experience.

As for the sense of the land and its health, this may be more clearly understood in one's heart on a seasonal basis. As the snow lays quietly on the fields, as the meadows turn green and the creeks rise, as lush pasture plants of springtime change with the heat of summer, as crickets and lightning bugs frame summertime and as dust is kicked up during the annual crop harvest in the autumn, we come to understand the interplay among the different layers of life in the landscape. We may reflect back over a certain season and see changes in the plants and health of animals even within a particular week. Without a doubt, true stewards of the land vividly sense that everything on the farm is interconnected — woods, wildlife, hedgerows, birds, insects, meadows, pastureland, cows, crops, soils, water and air. It is clear that all sources and types of life have their place on the farm, continually enhancing each other, since everything is completely inter-dependent. Individuals on organic farms who go about their daily work with true awareness of their surroundings know this to the core. Think for a moment of birds flying among trees that come down to perch on shrubs, only again to fly back to the trees and how they connect our senses from pasture level to the sky and back again. Then closer to our ears is the hum of the insect sphere closely aligned with the herbaceous plants. Similarly, the scent of the air near the ground with the fragrance of wild flowers, grasses and weeds differs from the air we breathe if we climb tall trees and take notice of the scent of the bark and green leaves. If we observe the birds flying in their arboreal realm we may realize their life-sphere, that they are to the trees as insects are to the low lying green shrubbery, and how at ground level all terrestrial life eventually merges with the Earth's soil through manure deposition and its slow breakdown, yielding humus over time.

Consider the earth where plant rootlets reach out into the vast soil, where they disappear into a fine mat, creating their own group identity within the subterranean biosphere. If we dig up healthy soil, especially in the early morning or near dusk when dew is present, an invigorating burst of pure moist earth strikes us at a most visceral level, invigorating something quite instinctual within us. Then breaking a clump open and observing it, we see the channels that worms have left behind, the very fine root hairs blending into the soil, tiny insects crawling about, as well as the soil's own blocky, platey or crumbly physical structure. Hard science, with its staunch objectivity, would separate out all these aspects and define them apart from each other. As truly aware individuals we can allow the whole clump of earth to resonate with something nearly indefinable in our being — something instinctual that harkens us back to a simpler mode of living, a clearer way. This fertile portion of the earth, whose scent invokes in us a sense of being fully alive, enables us to feel connected to all life on earth.

Cattle on many levels, but especially dairy cows, seem to be the embodiment of all that is connected to the land, and that which comes forth from the land to us as people. Sitting quietly near a group of grazing cows is a peaceful, resonant activity to which people may be easily drawn. In their agro-ecological niche, dairy cows are animals that effortlessly change grass to milk, that renewable resource from which humanity enjoys a multitude of products. In an anthropocentric way, they are the symbol of all that is good and pure in man's dealing with domesticated livestock. From the standpoint of soils, cows' inherent vitality nourishes the land with their enriching manure. From an artistic view, they are symbols of tranquility, especially when watching a group that is grazing in a lush meadow with a stream running by and surrounded by woods. They give us a sense of peace as they show their contentment chewing their cud, often times with their eyes half closed. While outside, they soak up the sunshine and breezes, and contribute to the health of the countryside — simply by just being there. Viewed in such ways, dairy cows are central to life itself. A reverence for dairy cows is a reverence for life itself.

Yet we also need to keep in mind that with "green" agriculture there is more explicit responsibility on the part of the farmer to his/her consumer. With community supported agriculture (CSA), the community is invited to partake in various aspects of the garden life. As a whole, organic farmers usually enjoy visitors and the interest that those visitors have in their farm and are usually quite candid with them. I remember well how often people would stop in the barn at Seven Stars Farm, especially during milking time, to visit and connect with the cows and calves. I think the reason is actually rather simple. Grazing cows have freedom all day long to walk about, graze fresh grass, lay down in the sun or shade and then "touch base" at milking times. It is a concept somewhat akin to people being out all day, and then coming home to gather with family. It creates a sense of place, of home. This is what I believe intrinsically draws people to the realm and reality of dairy cows on family farms. When interested folks do not readily have access to dairy farms because of where they live, if given the choice at the store shelf, they will actively support farms that practice traditional methods of letting cows walk upon the landscape. Of course the dairy processor has to be market-savvy to promote their ecological product through good and honest public relations for the "removed" consumer to be attracted to the milk carton or dairy product. The most removed people, those in the city, may only be able to make a connection to cows by buying commercialized objects like posters and stuffed animals — and well-marketed dairy products. However, if consumers are given tours of total confinement farms and traditional farms, there is little doubt in my mind that consumers will choose the traditional farms, preferring milk from cows that walk upon the land eating fresh grass from pastures rather than from cows kept forever indoors to eat only silage at a feedbunk. One thing is certain, people *like* cows out in the pasture; it reminds them of their own connection to nature.

Farmers and promoters of "green" ecological farming also need to keep in mind what these visitors see when they drop by. Just because a farm is managed ecologically does not necessarily mean that consumers will always be pleased with what are actually quite normal practices on farms. A very large gap or "disconnect" exists between the typical consumer and farmer. This can make for awkwardness between what a farmer considers routine and what a consumer can easily accept. We must remember that with the greater connection we strive to create in a more environmentally friendly agriculture, and especially because of the premiums paid for products produced in this manner, the greater the need for a well-developed sense of compassion and respect in relation to the animals under the farmer's care. By combining a true compassion for all life on the farm along with the farmer's own clear sense of place within the farm's agro-ecology, a healthy dairy farm will be obvious to all who stop by and visit.

Some consumers may still have the perception that organic farms are not as cleanly managed as high tech confinement farms. However, with the introduction of the USDA Certified Organic official seal, any lingering fears regarding farm hygiene by consumers should be minimal, since there are strict guidelines regarding what it takes to be USDA Certified Organic. However, just as there are sloppy confinement operations, there certainly can be sloppy organic operations. *There is no room for either.* In addition, I believe the organic farmer has the greater duty to maintain a higher standard of hygiene due to the consumer paying more for the product. If consumers want to pay bare bottom prices for conventionally raised food, their input into how farms are managed should be minimal. As a dairy veterinarian, I keep concepts of humane treatment of animals and hygiene close at hand when asked to help formulate livestock healthcare standards for current and future organic regulations. Healthy cows and a healthy farm are what visitors will notice and remember after being on the farm. This can be a challenge to any dairy farmer, but doing the right preventative steps in a biologically compatible manner will strengthen demand and create profits while enhancing the local agro-ecology.

Seasons on a Farm

Early spring.

Summer.

Summer morning.

Autumn.

Winter.

SUMMARY OF FARM TYPES

TOTAL CONFINEMENT

Farmer's perspective	**Result**
1) Maximize return on capital investment	1) High density of animals housed together inside
2) Maximize milk per cow	2) Round-the-clock milking/Routine use of rbST injections
3) Minimize days open	3) Routine use of hormone injections for estrus synchronization
4) Analyze cows by computer	4) Cows are numbered and in groups
5) Maximize mechanization	5) Heavy capitalization costs for machinery/ Increased safety concerns for laborers
6) Hire cheapest labor	6) Laborer interest not same as farmer interest
7) Contract out field work and manure management	7) Decreased self-reliance

Net result for society: A cheap product with hidden costs

TRADITIONAL/CONVENTIONAL

Farmer's perspective	**Result**
1) Personal pride in herd and farm	1) Potential of "value-added" product
2) Continuation of the family farm	2) Long hours, but near family
3) Enjoys cow and crop work	3) Self-reliant
4) Becomes attached to certain cows	4) Cows are viewed as individual creatures
5) Uses pasture in a limited fashion	5) Cows get to go outside and exercise on soil/paddocks
6) Long-term commitment to farm	6) Inherent incentive for conservation practices

Net result for society: An essentially healthy production system

ECOLOGICAL/ORGANIC/GREEN

Farmer's perspective	**Result**
1) Personal enjoyment with herd	1) Family *likes* to be involved
2) True sense of place, home	2) Attuned to nature
3) Deep satisfaction from being with cows and land	3) Spiritual growth
4) Conscious effort to improve the farm's agro-ecology	4) Biodiversity
5) Insight into farm as its own whole organism	5) Self-sufficient as possible
6) Cow manure critical for soil biology	6) Cows are central to the vitality of the farm
7) Attains Certified Organic status	7) Premium received for product

Net result for society: An ecologically healthy production system with no hidden costs

Chapter 2
Two Different Views of Cow Management

Let us now look more closely at the two very distinct and contrasting farming strategies — those farmers who believe in the latest technological advances and those farmers who believe in more basic approaches and minimal technology. These break out roughly along the lines of "confinement" versus "green" types of farming. As mentioned previously, it is similar to two sets of train tracks, both have the same destination (food production), but they originate from two different philosophies. Both require a high level of management expertise and efficiency to be successful in their differing spheres. The weather patterns of a year will often dictate who is faring better, especially when considering the overall reliance on home-raised crops versus purchased feeds. Ideally, a farm is a self-sufficient biological entity. Family farmers generally strive to maximize this concept; the difference comes in the approach to the prime reason of business, the dairy cow. Is she a creature that we confine, control and manipulate to produce profit by forcing her genetic potential through synthetic inputs and biotech? Or is she a creature that effortlessly produces profit by allowing her to fulfill her biological niche as a grazing animal? It should be pointed out that making lots of milk doesn't necessarily mean making lots of money. The constant knowledge of income over feed costs will either delight or frustrate a dairy producer. And this is where a critical difference arises between grazing and total confinement farming systems. But in the end, it all really comes down to how well the farm is managed. This means the whole farm: animals, barns, labor, crops, fields, machinery, soil health, conservation, record keeping and financial analysis.

Confinement Systems

In general, total confinement farming tends to have large groups of cows kept inside continuously with no access to outdoors. Formulated and thoroughly mixed rations are mechanically fed to the cows with virtually any feed ingredient that is not specifically illegal. These rations are theoretically balanced on paper. This ration is called a "Total Mixed Ration" (TMR) and, as the ingredients can be tightly manipulated, it is the "main course." Conventional nutritionists favor this type of feed system over more traditional individual sequential feeding of hay-silage-grain due to the uniformity of diet. Indeed, a cow is by nature nearly the ultimate "creature of habit," and production drops can occur if the cow's life changes in any way relative to her normal routine. A large part of the TMR is based on ensiled forages, purchased grains and fortified with added vitamins and minerals. Other feed commodities, such as citrus pulp, cottonseed, tallow and other industrial by-products that are essentially foreign to the cow's digestive tract can be added at will (usually reflecting what is cheapest that month). TMR's can work really well for milk production, yet if a mistake is made in calculations or in adding ingredients, digestive and metabolic problems can quickly surface as confinement cows totally rely on the mix that is fed to them. TMR's are not bad in and of themselves, especially if used in the barn to top off a ration on farms where the cows are outside grazing most of the day. On many farms, the cows are allowed out to a nearby earthen area where they can exercise and forage for any "dessert" that they might find (usually a variety of weeds). However, there are an increasing amount of farms where the cows *never* experience bare earth or even a weedy area; rather, they spend their entire life on concrete rotating between walkways, the milking parlor, feed bunks and stalls. In this respect, modern dairy farming is rapidly becoming like poultry and hog farming in that animals are forever kept within a building with mechanical ventilation and lighting until the day they are sent to market for slaughter.

In these production units, managers often rely on routine injections with hormones to regulate the reproductive cycle or increase milk production. The animals may also be at increased risk to a host of environmentally related diseases due to high densities of animals housed in close proximity on concrete. Intense vaccination schedules are usually practiced, although this protection can be overcome by sheer concentrations of animals and persistently infected carriers (as in the case of BVD and *Lepto hardjo bovis).* With the advent of rbST (rBGH/"bovine growth hormone"), some dairies do not even bother breeding back animals. Instead they will extend the lactation of the cow indefinitely with rbST until computerized economic analyses indicate it is time to "ship her" for slaughter. With intensive systems, the cows are managed to produce the largest quantities of milk per cow possible in the shortest period of time. Ironically, small family farmers benefit from this as prices paid for newborn heifer calves have increased dramatically because confinement farms become dependent on purchasing animals to maintain their milk production. (Small family farms traditionally have extra animals to sell since they tend to maintain animals in the herd longer.)

Everything a capital intense confinement farm does is measured on a "per cow" basis. Economies of scale come into play, and more cows housed in a building make for more efficient use of inputs (machinery, land, borrowed money, etc.). Each cow is responsible for the production unit's fixed costs and variable costs. The cow is seen more as merely an economic entity rather than a living, breathing creature with biological needs specific to being a ruminant. It is quite common, even during the planning of an expansion to have more cows in an area than there are stalls, on the assumption some cows will be eating, drinking or walking around while some are resting in their stalls. In other words, if a barn will have 1000 stalls, it is likely that there will be 10% more than that, or 1100 cows in an area that only provides 1000 stalls. This is actually quite commonly done on modern expansion herds. It then becomes a question not of how the animals are healthiest, but how the animals can produce the most with the least obvious illness. Unfortunately, the mode of feeding these confined animals is such that they are often borderline acidotic if not suffering from actual rumen acidosis (due to maximal grain feeding) which gives rise to more insidious problems like laminitis, changes in the pH of the digestive tract that allow opportunistic pathogens to proliferate, liver abscesses and decreased immune system function. In my opinion, one of the main causes of health problems in all kinds of cattle in the United States is subacute rumen acidosis (SARA). This is brought on by minimizing the amount of hay fed while feeding as much grain as possible. On such farms, cows are being fed like a monogastric pig, not as a ruminant cow needs to be. The pursuit of profit is a two-edged sword when using each cow (as a production unit) to bear the burden of capital investments. Cheap milk is not cheap in terms of cow health.

Grazing Systems

In grazing systems, the whole farm is geared to the local ecology of native grasses/legumes, and cows are usually fed sequentially with the individual components of hay, silage, grains, minerals and vitamins when in the barn (although TMR's are fairly common). In order to keep their diets balanced (at least the smart graziers strive to maintain a balanced nutritional plane for their cows), the "main course" comes from grazing pre-determined strips of pasture while their "dessert" is fed in the barn to complete their ration. This can cause fits for a conventionally trained nutritionist, as the inputs to the cow are constantly changing because of grasses and legumes growing at different rates throughout the growing season. It is conceivable that fresh grasses have more active constituents and higher biological quality than ensiled and highly processed feeds. Unfortunately, all the bioactive compounds and biological qualities of fresh plants, in terms of the activities of constituent phytochemicals, would be nearly impossible to measure (except possibly by a few dedicated researchers). Many graziers no longer even listen to conventional nutritionists since the two groups do not share a common language when feeding cows. Although this may be understandable, it should always be remembered that dairy cows, no matter what stage of lactation or pregnancy, do have well known, basic metabolic requirements. If the bare basics are ignored due to a rebellious mind-set of the grazier, the cows will show problems at some point, whether it is poor conception ability or more obscure problems of micro-nutrient mineral imbalances. For cows to be healthiest and to determine what needs to be corrected if a problem arises, it is *critical* to test soil and forages to know what is being delivered to the cows (both quality *and* quantity).

Intensively grazed cows are either milked in tie stalls or in parlors, depending on the size of the herd. With grazing systems, the cows are managed not for the largest possible quantities of milk but for absolute least cost of producing milk. This is accomplished by replacing the input of purchased feeds and by growing luxuriant pasture stands as well as maximizing the animals' inherent ability to graze and harvest their own feed. Production per cow usually drops with less purchased input feed but often herds are expanded to keep the bulk tank full. Animals have a chance to live longer since they are not pushed as hard for maximal milk production, although older age can bring on a host of different health problems (cancer, arthritis, etc.). Economic analyses are usually on a "per acre" basis in grazing herds, rather than on the traditional "per cow" basis. This encourages grazing farmers to invest in their soil and land-based resources and in so-doing alleviates some of the pressures on each cow to bear the burden of all the capital investments (which are considerably less for grass based dairies).

Grazing Compared to Confinement

As has been briefly touched upon, there *is* a fundamental difference between analyzing farms on a "per acre" versus a "per cow" basis. When using the "per cow" approach, there is intense pressure on each animal to be profitable so as to pay owner and employee salaries, pay down debt (whether machinery or buildings) — let alone the cow having to "pay her way" for her own feed intake. Using the "per acre" approach, as is increasingly done on intensively grazed farms, the focus of efficiency is placed on land productivity. This in effect places the cow as a "pass through" organism which more directly converts the land's natural products into a marketable product. On the "per acre" type of analysis, there is emphasis on getting the most value out of the land (through milk), which means carefully managing soil fertility and the growth of pasture species. This is done by using correct stocking rates, adjusting for time of the season (fast growth or slow growth), frequently adjusting size of strips to be grazed and testing the quality and quantity of pasture periodically. To graze cows well takes intensive management just as total confinement of cows takes intensive management. The difference is in how we control the cow and feeding. Do we rely on the cow's natural ability to graze, or do we feed her only what we say she will eat? Certainly there are metabolic requirements for a cow to maintain body condition, grow (if first calf heifer), lactate and be re-bred. There is certainly a balance to be struck for the good of the cow and the profit of the farmer. In the "per acre" analysis, we strive for making most of the annual milk from a pasture-fed diet, whereas in a "per cow" analysis we strive for making most of the annual milk from a strictly controlled feeding system, often with a high percentage of bought ingredients.

It is in the continual harvesting, hauling, feeding of cows and subsequent barn clean-out that depreciates mechanical equipment at a faster rate with intensive confinement farming than with intensively grazed farms. And who is going to pay for the replacement equipment? The cows, of course. It is no wonder that cows need to be pushed for maximum production everyday if pressure is placed on them to continually pay for needed repairs or to update machinery due to heavy usage, as well as payment for labor, management and farm debt. It is a kind of circular logic. And although labor, management and farm debt are costs common to both farming types, heavy mechanization has by far the most costs associated with it due to initial purchase price, monthly payments, repairs and depreciation (and potential accidents).

Pushing for maximal milk production certainly has its benefit in creating lots of milk for an increasing world population. However, there are hidden costs, as well as explicit costs, to maximizing production. First and foremost, it should be pointed out that it is always assumed there will be cheap sources of fuel and electricity. Total confinement farms are very dependent on a steady supply of cheap fuel and electricity. Land in the USA is plentiful. That means that a confinement facility can set-up pretty much anywhere, rapidly extract from the soil all it's inherent fertility, create concentrated areas of manure, tap into the area's groundwater resources (and also affect it) and then leave the area if desired.

As for the cows in a total confinement operation, there is much more input needed to keep the cows producing and reproducing. This is seen by way of increased nutritional requirements and rBST to force the cow to produce even more milk (which in turn causes them to need to be fed more yet). Hormonal drugs like gonadorelin (GnRH) and prostaglandin (PGF2) which are allowed by the FDA to treat cysts and uterine infections, now are heavily relied on to synchronize breeding of many cows at once. In addition to the questionable use of drugs which are approved for therapy but are being used for purely production purposes, it is particularly disquieting to see how many synchronization programs have spun off the original one. The original one is called OvSynch. But now there is CoSynch, Heat Synch and Re-Synch, which are all supposed to increase the chance of OvSych to work. Next we'll hear of the latest, newest synchronization program — Kitchen Synch! When cows are exposed to such "management tools," their endocrine systems may become dependent on, or upset by, these artificial inputs and general health may ultimately suffer due to artificial manipulation of the adrenal and pituitary glands. When a cow that is being pushed in such ways becomes sick, it potentially takes more powerful medicines to make her better, if possible. Being a practicing veterinarian, I am no longer amazed at how farmers simply accept all the injections that their cows need daily. It is as though managing with crutches such as hormones and antibiotics seems more appealing to some farm managers than trying to be free of such substances. For many farmers, even those with only 50 cows, giving hormone and antibiotic shots has become almost as routine as feeding the cows. There is no questioning involved, just simple acceptance that to farm profitably, synthetic production drugs are a fact of life.

Certainly in large and small grazing herds the cows are managed to produce as well (what is a dairy farmer in business for if not to produce milk?). But I think grazing herds are more or less "coaxed" to produce by strict pasture management rather than pushed outright by various synthetic means while confined in the barns. Also, I believe that when cows are taking in fresh grasses and exercising by foraging, their bodies are in a state of good biological equilibrium. Thus their immune systems are able to achieve full functionality so that various treatment options are available when they do need medical attention. Granted, all cows are born with a genetic blueprint of their own unique immune system just as they also have for their own genetic potential to make milk. Perhaps the key difference between confinement and grazing systems is based upon the exploitation of these two pre-determined factors. That is, in grazing herds, cows generally maintain a vigorous healthful existence, owing to optimally functioning immune systems because of grazing fresh plant materials much of the year. They produce milk effortlessly, in a sense. In confinement herds, the genetic potential for maximal milk production is achieved through strict manipulation of what is offered to the cows in the form of fine-tuned TMR ration. This means pushing for milk regardless of the stress on the digestive tract, thereby leaving the cow in a potentially precarious metabolic state. The old adage "you are only what you eat" is true whether a human being or an animal. Regardless of what kind of farming style a farmer practices, I know that every farmer likes a vigorous cow full of vitality that lives a good *long* life and produces above average levels of milk. In the end it comes down to the basic question of short-term extremely high production or long term consistent production (but not as high). The cow will tell you by the age she attains how well she has been cared for.

Perhaps owners of enormous total confinement operations play the odds and are willing to allow for "acceptable" levels of disease while in the pursuit of profit. As an operation adds on more cows, each animal loses some individual economic value and thus "acceptable levels" of certain problems may become "tolerable." Losing one animal from a large herd of 5000 cows is not as relatively big a loss as it is to a farmer who has a more traditional size herd of 100-500 cows (or for a farmer in a developing country who has only 5 cows).

A group of cows, no matter how many, is considered "a herd." The issue becomes whether one looks at the farm enterprise primarily from a whole herd perspective or as individuals that go to make up the herd. If one only sees a whole herd as is emphasized in production medicine, then each individual animal *as a biological organism* has a relatively decreased value to the farmer, and its economic rank is the sole motivating force for a farmer to pay any further attention to it. This is often accomplished by monitoring the animals' milk production or feed intake by way of a computerized review at the end of a day. Contrasting this is the situation where one watches each individual animal within a herd for the slightest hint of something possibly wrong and spends the extra few moments observing her in order to nip a problem in the bud. There is inherently more value and benefit to careful, direct hands-on observation of an animal in its environment than only by reviewing computer records to alert for any possible problems. In my opinion, the optimal way to stay on top of cow health is to keep a watchful eye on them individually throughout the day and make use of records to analyze herd performance over set intervals. Computers are useful but should not replace hands-on knowledge of the cows in a herd. To be the best possible herdsman, then, probably means not to partake in running lots of machinery to harvest feed for the cows and computer monitoring but rather by spending time with the cows and observing them while grazing and milking. In short, grazing a herd of cows takes intensive management just as confinement operations require; it just comes down to a choice of reliance on personal observation and intuition versus impersonal reliance on mechanization and computers.

Chapter 3
Grazing Basics

All cows should be allowed to graze, at least to some extent. By nature, cows are herbivores and need to chew grasses, fresh and dried (hay). Modern dairy farming demands hi-tech diets for genetically hi-tech cows. It is true that rations need to be balanced to meet the cows' needs for biological maintenance, growth, milk production and reproduction. Almost all farms have a reserved area for the cows to be exercised. An area of only 10 acres, divided up into one acre paddocks and seeded with nutritious grasses suited to your climate is enough to offer 40-50 cows at least some fresh herbivorous intake in addition to their official ration fed in the barn. Although commercial laboratory analysis can assess the standard nutritional value of pasture species, there are many other beneficial plant compounds not analyzed and whose effects are not yet understood. The myriad of phytochemicals contained in plants, in addition to the other easily quantifiable components, act synergistically to give an overall character to the plant. Many compounds that are not biologically understood can be perhaps helpful to the cow in maintaining her vitality and internal equilibrium. This is simply due to the cow being an herbivore and being able to process plant materials to the fullest extent. Animals can sometimes be observed to nibble on technically non-nutritious plants at times and this *may* be an instinctual way of seeking out medicinally helpful substances that are hidden in a plant's make-up (see "species memory-response" in Chapter 9). It may even be wise to purposely plant various medicinal herbs (*e.g.*

arnica, dandelion, celandine, gentian, fennel, etc.) in the pasture or along its borders in the hedgerows from which the cows can freely choose if they feel the urge to do so. This parallels feeding bicarbonate on the curb of a tie-stall from which the cow can lick if desired or the practice of free-choice mineral feeders, which works on the same principle. Certainly, most cows will only graze on the intended grasses instead of the "weeds" that may be in a pasture. It may be wise then to actually dedicate a small paddock that is purposely seeded with garlic, purple coneflower, dandelions, chicory, amaranths, fennel, celandine and the like to actually mow and harvest to store for sporadic feeding if need be. Mowing such a pasture before it goes to seed would be wise so other pastures don't become loaded with unintended plant species. Planning of such a paddock will take some thought as the parts of medicinal herbs (root, vegetation, flowers, and seeds) contain the active constituents in differing

amounts. Therefore, planting all herbs with the vegetation known to contain medicinal compounds would be wise so as to make hay (or graze it) whereas plants with seeds or roots containing medicinal compounds need to be harvested at later times of the year.

I will leave details as to what species and how much of certain seed mixes to sow for regular pasture use in your area up to you and your local agronomist. (The Barenbrug Company does much research on pasture species and is well respected among many graziers.) I just want to give some extra tips from a health perspective regarding pasturing.

Cow level view of tall sorghum grass with hedgerow and trees in background.

Try to move the cows to fresh strips every 12-24 hours depending on the growth rate of the grass. On matured pastures that have developed a perennial stand, try not to graze the grasses to less than 3 inches. If you do, rate of regrowth will be slower and the quality of pasture may suffer as opportunistic weedy species may invade. In the spring time you will move your cows through the paddocks quite rapidly and even have some paddocks needing to make hay. Then the lush spring growth will slow down and you will need to be careful not to overgraze paddocks.

The hottest, driest months are often a challenge. With periodic rainfall, even cool season grasses will re-grow and provide needed nutrition if not previously grazed too close to the ground. But, as insurance against drought conditions, it's probably best to have a few paddocks of brown midrib sorghum Sudangrass. As a herdsman, I had favorable results with using sorghum Sudangrass for the two hottest months of the summer. It's a tropical grass that does well in the heat of summer in the northeast USA. The newer varieties (brown midrib, BMR) can provide as much nutritional value as other more desirable pasture species. If it gets ahead of you because of its rapid growth, it can be cut and made into baleage with nutritional values equivalent to corn silage, yet giving much better fiber. But don't graze sudan grass before it's about 18 inches high as the prussic acid content may lead to toxic amounts of cyanide. Also don't graze it right after a frost as cyanide content could be higher. I generally found the cows graze it down about 10 inches in 12 hours. Try to "stagger plant" it over a couple weeks'

time, so it will grow and not be ready all at once since it grows rapidly in the heat. The seed can't be drilled in any deeper than ½ inch and the soil needs to be 60 F.

In extremely hot and humid conditions, it may be best to bring the cows inside during the hottest part of the day, but only if you have decent ventilation in the barn. The cows will have to be fed then, which may seem to defeat the purpose of grazing. However, if the welfare of the cows is to be kept in mind, keep-

Lush pasture — what the cows notice. White clover has high energy content.

ing them cool becomes a priority. Remember, most commercial dairy breeds were originally from cool, northern climates and do best in 40-60 F. Heat waves will eventually pass and the cows can be back out on pasture 24 hours a day (except milking times). Alternatively, you can force the cows to stay out on pasture during the hottest time of day, even without shade, if you have an adequate and steady supply of fresh water available at all times. Shade, even if artificial, will be beneficial. If electing to keep the cows outside like this, you may get the best grazing intake if you move the cows every two hours to fresh grass. This is labor intensive, but necessary if you wish to maintain a decent production level during the hottest times. Some animals may get sunburn, especially those white areas of the hair coat. Irrigation of pastures can be done as well. This requires capital expenditure but will enable a farmer to weather the worst droughts in a reasonable fashion. Pond or creek water is fine. Farms that routinely use irrigation for crops (as done in the arid west) should also use

Cows strip grazing.

irrigation water to give continual lush pasture. Some operations also irrigate from a lagoon where the milk house water and barnyard run-off are collected. If this is the case, don't put the cows out on the paddocks for a few weeks so pathogens can die due to exposure from sun and wind.

In general, pastures are excellent sources of protein but not as high in energy. The exception is when high energy clovers are growing abundantly — and cows love to eat clover. Alternatively, if there is lots of sunshine and active photosynthesis, the carbohydrate (energy) content of the pasture will be higher. This is perhaps why hay grown in the arid west with lots of sunshine has generally high relative feed values (RFV). However, since pastures commonly have lower energy than protein, it may be important to supplement the cows' pasture intake with energy in the form of grains or molasses. Molasses may help to keep down the excessive levels of ammonia that build up in the rumen due to all the soluble protein being consumed out on pasture. There are graziers who boast about feeding almost nothing in the barn at all during grazing season. This is an extreme position and I can't agree with it from a veterinary perspective as cows tend to lose excessive body condition and are in negative energy balance sometimes for months on end. I do believe there is rationale for feeding some hay, grain and even some corn silage in the barn during the grazing season. (There should never be more than 15-20 pounds corn silage as fed regardless the time of the year.) For one, a cow's rumen needs to be set-up with hay before being put out to pasture to avoid bloat (especially in the cooler months of grazing legumes). This is straightforward enough. The second reason is that it takes biological energy for the excessive ammonia produced in the rumen (due to digestion of massive amounts of fresh plant material) to cross the rumen wall and be transformed in the liver to urea which then needs to be eliminated as urine. The third reason is that cows tend to lose some body condition during the grazing season, although they look to be in good health with nice shiny coats. With excessive loss of body condition it is more likely that reproductive problems may develop, and cows may not breed back in a timely manner. If grazing, use of a bull or two would be best to eliminate heat detection for articial insemination. By the time grazing season effectively ends sometime in late November or early December (in southeast Pennsylvania), cows need to already be pregnant for a few months to calve again by the next pasture season. Poor body condition tends to favor infertility as cows are in a negative energy balance. Therefore it is important to consider feeding energy sources to keep body condition good. Proper mineral nutrition is critical as well, and all forages need to be tested for the standard constituents. Ideally, no mineral supplements should be needed if the soil health is optimal to grow nutritious crops for herbivorous consumption. However, regional soil deficiencies must be addressed or the crops consumed will obviously need to be supplemented.

Since I am not a trained dairy nutritionist but do have a strong interest in trying to keep grazing animals essentially healthy by way of balanced rations, I would like to present a set of different rations from an excellent book by Frank Morrison, *Feeds and Feeding*. The particular edition quoted was published in 1949 — when grazing was an integral part of most U.S. dairy farms. Only in recent years has the grazing craze found a renaissance, but mainly from folks who are sick and tired of hearing about the latest high-tech feeding strategies from their local land grant university. Since the renaissance in grazing is often based on seat-of-the-pants trial and error (with the cows taking the brunt of the errors), I think it would be interesting to show how grazing cows were fed back when grazing was a mainstream part of dairy farming. I am not saying that these tables should substitute for proper nutrition advice on the farm. However, if a farmer reading this is unfortunately in an area where there are no grazing friendly nutritionists, these tables may help.

The best source of nutrition consulting, in my opinion, is from Tom Weaver and associates of K.O.W. Consulting Associates in Wisconsin. K.O.W. Consulting is staunchly in favor of basing rations on very healthy amounts of forages to maximize rumen health (and therefore cow health and long-term productivity). As a veterinarian who cares deeply about dairy cows and their true ecological niche, I can only applaud this kind of thinking. They do everything to avoid rumen acidosis. Although

Cows strip grazing.

mentioned earlier, I will repeat that I believe that rumen acidosis is the root cause of many modern dairy cow health problems — especially rumen health, digestion, and laminitis. Too many nutrition programs push the cow to maximize milk while seeming to forget that the cow's rumen functions best with a much larger percentage of the diet being forage based as opposed to grain/concentrate based. K.O.W. gets great production from cows while keeping them exceptionally healthy — all with minimum to no corn silage and nearly zero abomasal displacements. Although I do believe in feeding corn silage (no more than 10-20 pounds a day), it is probably one of the most expensive crops to grow when considering all the time and energy invested in cultivating, spraying, harvesting and feeding it out. Body condition on cows fed by K.O.W. Consulting couldn't be better, even as grazing cows. For excellent grazing information in the form of a periodical, subscribe to *GRAZE* magazine. This will give timely information on a variety of topics to the grass fed dairy farmer. See the Contact Resources at the end of this book for contact information. Additionally, Gary Zimmer of MidWestern Bio-Ag and Jeff Mattocks of Fertrell Company do excellent work with grazing farmers and organic herds.

Notice that in the following example rations from previous years, there was *never corn and barley in the same ration*. That is because they probably knew it was probably too hot a ration for the rumen. Oats are excellent to help regulate the gut. It was an integral part of standard rations of yesteryear. Unfortunately, it is not used enough in modern rations. Graziers would do well to use it as it can help keep the manure from becoming too soupy when on lush pasture (especially pasture with clovers). Of course feeding the proper amount of dry hay for effective fiber is required in order to keep the fiber mat in the rumen functioning well. Otherwise, many of the nutrients the cow takes in pass through the intestinal area before her system gets a chance to actually absorb and make use of them.

Modern rations almost always contain sodium bicarbonate as a buffer to keep rumen pH from becoming too acidic. Notice that none of the rations listed use it. Could it be that they fed cows in ways which did not create rumen acidosis? Also, most modern rations add in trace minerals, vitamins and other micronutrients. Could it be that the crops grown these days don't deliver as much in the way of minerals because the soils have been depleted over the years? In order to ensure proper salt, vitamin and micronutrient content in the cows' diet, add in K.O.W. Consulting TMSK salt and VTM vitamin mineral packs if these nutrients are lacking in pasture swards or harvested feeds.

Farm Animals Benefit from Pastures, Sunshine & Fresh Air

Morrison's Rations for Dairy Cows

A. Mixtures containing approximately **12% protein**

- For cows in milk which are fed good alfalfa, soybean, or cowpea hay (at least 1 lb. daily per 100 lbs. live weight) with corn silage, sorghum silage, corn fodder, sorghum fodder, or roots. When alfalfa, soybean or cowpea hay is fed as the only roughage, a mixture of farm grain supplies sufficient protein for all, except unusually high-producing cows.
- For cows in milk which are fed red clover hay as the only roughage.
- For cows in milk which are on excellent pasture.
- For dry cows, when at least 1/3 the roughage (on the dry basis) is legumes.
- For heifers over 6 months old, when 1/2 the roughage (on the dry basis) is alfalfa, soybean or cowpea hay.

	Feed	Lbs.
1	ground corn	1160
	ground oats	500
	wheat bran	200
	soybean oil meal	120
	(or cottonseed meal)	
	salt	20
	Total	2000
	Dig. protein	9.8%
	TDN	75.4%
2	ground corn	1130
	ground oats	500
	wheat bran	200
	linseed meal	150
	salt	20
	Total	2000
	Dig. protein	9.8%
	TDN	75.3%
3	ground barley	1030
	(or wheat)	
	ground oats	700
	wheat bran	250
	salt	20
	Total	2000
	Dig. protein	10.2%
	TDN	73.0%

	Feed	Lbs.
4	ground corn	1000
	ground oats	640
	wheat bran	200
	corn gluten feed	70
	(or dried distillers' grain)	
	soybean oil meal	70
	salt	20
	Total	2000
	Dig. protein	9.8%
	TDN	74.6%
5	ground kafir*	1400
	ground oats	330
	wheat bran	200
	cottonseed meal	50
	(or soybean oil meal)	
	salt	20
	Total	2000
	Dig. protein	9.9%
	TDN	77.2%

kafir is a type of grain sorghum

	Feed	Lbs.
6	corn-and-cob meal	1310
	wheat bran	500
	cottonseed meal	85
	(or soybean oil meal)	
	linseed meal	85
	salt	20
	Total	2000
	Dig. prot.	9.6%
	TDN	71.0%

B. Mixture for fitting dry cows and for freshening cows — **16% protein**

(When little or no legume roughage is available)

1

Feed	Lbs.
ground corn	760
ground oats	600
wheat bran	500
linseed meal	100
bone meal	20
salt	20
Total	2000
Dig. protein	13.3%
TDN	71.8%

2

Feed	Lbs.
ground corn	590
ground oats	500
wheat bran	500
linseed meal	360
phosphorus	20
ground limestone	10
salt	20
Total	2000
Dig. protein	13.3%
TDN	71.8%

C. Mixtures containing approximately **14% protein**

- For cows in milk which are fed red clover hay (at least 1 lb. daily per 100 lbs. live weight) and corn or sorghum silage or corn or sorghum fodder, when protein supplements are expensive.
- For cows in milk which are on *very good* pasture.
- For dry cows when only $1/4$ of the roughage (on the dry basis) is legume.
- For heifers over 6 months old, when $1/2$ the roughage (on the dry basis) is clover hay.

1

Feed	Lbs.
ground corn	1055
ground oats	500
wheat bran	200
soybean oil meal (or cottonseed meal)	225
salt	20
Total	2000
Dig. protein	11.4%
TDN	75.3%

2

Feed	Lbs.
ground corn	980
ground oats	500
wheat bran	200
linseed meal	30
salt	20
Total	2000
Dig. protein	11.6%
TDN	75.1%

C. Mixtures containing approximately **14% protein** (cont.)

	Feed	Lbs.
3	ground barley	1090
	ground oats	600
	wheat bran	200
	soybean oil meal	90
	(or cottonseed meal)	
	salt	20
	Total	2000
	Dig. protein	11.3%
	TDN	73.6%

	Feed	Lbs.
4	ground kafir	1275
	ground oats	330
	wheat bran	200
	cottonseed meal	175
	salt	20
	Total	2000
	Dig. protein	11.4%
	TDN	76.5%

	Feed	Lbs.
5	ground corn	955
	ground oats	500
	wheat bran	225
	corn gluten feed	150
	soybean oil meal	150
	salt	20
	Total	2000
	Dig. protein	11.5%
	TDN	74.9%

	Feed	Lbs.
6	corn/cob meal	1220
	wheat bran	450
	cottonseed meal	155
	(or soybean oil meal)	
	linseed meal	155
	salt	20
	Total	2000
	Dig. protein	11.4%
	TDN	76.5%

D. Mixtures containing approximately **16% protein**

- For cows in milk which are fed good red clover or alsike clover hay (at least 1 lb. daily per 100 lbs. live weight) with corn silage, sorghum silage, corn fodder, sorghum fodder, or roots.
- For cows in milk which are fed good mixed clover and grass hay (containing at least 30% clover) and corn or sorghum silage or corn or sorghum fodder, when protein supplements are unusually expensive.
- For cows in milk which are on *good* pasture.
- For dry cows which are fed little or no legume roughage.

	Feed	Lbs.
1	ground corn	945
	ground oats	500
	wheat bran	200
	soybean oil meal	335
	(or cottonseed meal)	
	salt	20
	Total	2000
	Dig. protein	13.1%
	TDN	75.2%

	Feed	Lbs.
2	ground corn	820
	ground oats	500
	wheat bran	220
	linseed meal	440
	salt	20
	Total	2000
	Dig. protein	13.3%
	TDN	74.7%

D. Mixtures containing approximately **16% protein** (cont.)

	Feed	Lbs.		Feed	Lbs.
3	ground barley	1000	**4**	ground kafir	1180
	ground oats	560		ground oats	300
	wheat bran	200		wheat bran	200
	soybean oil meal	220		cottonseed meal	300
	(or cottonseed meal)			(or soybean oil meal)	
	salt	20		salt	20
	Total	2000		Total	2000
	Dig. protein	13.1%		Dig. protein	12.9%
	TDN	73.8%		TDN	76.0%
5	ground corn	830	**6**	corn/cob meal	1220
	ground oats	500		wheat bran	450
	wheat bran	200		cottonseed meal	155
	corn gluten feed	225		(or soybean oil meal)	
	soybean oil meal	225		linseed meal	155
	(or cottonseed meal)			salt	20
	salt	20		Total	2000
	Total	2000		Dig. protein	11.4%
	Dig. protein	13.1%		TDN	76.5%
	TDN	74.9%			

E. Mixtures containing approximately **18% protein**

- For cows in milk which are fed mixed clover-and-timothy hay or other mixed clover-and-grass hay containing at least 30% clover (at least 1 lb. of hay daily per 100 lbs. live weight), this hay being fed with corn or sorghum silage, corn or sorghum fodder, or roots.
- For cows in milk which are on fair pasture.
- For heifers over 6 months old, when only about $1/4$ the roughage (on the dry basis) is legume.

	Feed	Lbs.		Feed	Lbs.
1	ground corn	685	**2**	ground corn	540
	ground oats	500		ground oats	500
	wheat bran	200		wheat bran	200
	soybean oil meal	270		linseed meal	340
	distillers dried			corn gluten feed	400
	corn grains	325		salt	20
	salt	20		Total	2000
	Total	2000		Dig. protein	15.1%
	Dig. protein	14.4%		TDN	74.2%
	TDN	75.6%			

E. Mixtures containing approximately **18% protein** (cont.)

	Feed	Lbs.
3	ground barley	940
	ground oats	500
	wheat bran	200
	soybean oil meal	340
	(or cottonseed meal)	
	salt	20
	Total	2000
	Dig. protein	14.7%
	TDN	74.1%

	Feed	Lbs.
4	ground kafir	1045
	ground oats	300
	wheat bran	200
	cottonseed meal	435
	(or soybean oil meal)	
	salt	20
	Total	2000
	Dig. protein	14.5%
	TDN	75.2%

	Feed	Lbs.
5	ground corn	680
	ground oats	500
	wheat bran	200
	corn gluten feed	300
	soybean oil meal	300
	(or cottonseed meal)	
	salt	20
	Total	2000
	Dig. protein	14.8%
	TDN	74.6%

	Feed	Lbs.
6	corn/cob meal	970
	wheat bran	450
	cottonseed meal	435
	(or soybean oil meal)	
	linseed meal	280
	salt	20
	Total	2000
	Dig. protein	14.6%
	TDN	71.3%

F. Mixtures containing approximately **20% protein**
(Add 20 lbs. ground limestone per ton if roughage is grown on soil very deficient in calcium.)

- For cows in milk which are fed mixed legume-and-grass hay containing less than 30% legumes, this hay being fed with corn or sorghum silage, corn or sorghum fodder, or roots.
- For cows in milk which are fed non-legume roughage of good quality and which are producing sufficient milk so that they require at least 8-10 lbs. of concentrate or grain mixture.
- For cows in milk which are on poor pasture.
- For heifers over 6 months old which are fed no legume roughage.

	Feed	Lbs.		Feed	Lbs.
1	ground corn	630	**2**	ground corn	525
	ground oats	400		ground oats	300
	wheat bran	200		wheat bran	200
	soybean oil meal	350		linseed meal	455
	distillers dried			corn gluten feed	400
	corn grains	400		salt	20
	salt	20			
				Total	2000
	Total	2000		Dig. protein	17.0%
	Dig. protein	16.0%		TDN	74.8%
	TDN	76.2%			
3	ground barley	915	**4**	ground kafir	915
	ground oats	400		ground oats	300
	wheat bran	200		wheat bran	200
	soybean oil meal	465		cottonseed meal	565
	(or cottonseed meal)			(or soybean oil meal)	
	salt	20		salt	20
	Total	2000		Total	2000
	Dig. protein	16.5%		Dig. protein	16.1%
	TDN	74.5%		TDN	74.5%
5	ground corn	430	**6**	corn/cob meal	870
	ground oats	300		wheat bran	400
	wheat bran	200		cottonseed meal	335
	corn gluten feed	450		(or soybean oil meal)	
	soybean oil meal	400		linseed meal	355
	(or cottonseed meal)			salt	20
	salt	20			
				Total	2000
	Total	2000		Dig. protein	16.3%
	Dig. protein	16.6%		TDN	71.5%
	TDN	72.6%			

G. Mixtures containing approximately **24% protein**
(Add 20 lbs. of ground limestone per ton if roughage is grown on soil very deficient in calcium.)

- For cows in milk which are fed good quality non-legume roughage and which are not producing sufficient milk to require so much as 8 lbs. of concentrate or grain mixture.
- For cows in milk which are fed only non-legume roughage of fair to poor quality.

1

Feed	Lbs.
ground corn	380
ground oats	300
wheat bran	200
soybean oil meal	500
distillers dried corn grains	600
salt	20
Total	2000
Dig. protein	19.5%
TDN	76.8%

2

Feed	Lbs.
ground corn	380
ground oats	200
wheat bran	200
linseed meal	500
corn gluten feed	500
soybean oil meal	200
salt	20
Total	2000
Dig. protein	20.5%
TDN	75.1%

3

Feed	Lbs.
ground barley	580
ground oats	400
wheat bran	200
soybean oil meal (or cottonseed meal)	500
linseed meal	300
salt	20
Total	2000
Dig. protein	20.1%
TDN	74.4%

4

Feed	Lbs.
ground kafir	680
ground oats	300
wheat bran	200
cottonseed meal	500
soybean oil meal	800
salt	20
Total	2000
Dig. protein	19.6%
TDN	74.4%

5

Feed	Lbs.
ground corn	220
ground oats	200
wheat bran	200
corn gluten feed	600
soybean oil meal (or cottonseed meal)	560
cane molasses	200
salt	20
Total	2000
Dig. protein	20.0%
TDN	72.7%

6

Feed	Lbs.
corn/cob meal	620
wheat bran	400
cottonseed meal (or soybean oil meal)	500
linseed meal	460
salt	20
Total	2000
Dig. protein	19.7%
TDN	71.5%

Grain Feeding Table
for Cows on Pasture
Total pounds of grain or concentrate to feed

| Quality of Pasture | | | Percentage of Fat in Milk | | | | | | |
| Excellent | Good | Fair | | | | | | | |
[MILK PRODUCED DAILY]			3.0%	3.5%	4.0%	4.5%	5.0%	5.5%	6.0%
22	13	***	***	***	***	***	***	***	1.2
24	15	***	***	***	***	***	1.2	2.0	2.3
26	17	***	***	***	***	1.9	2.2	3.1	3.5
28	19	10	***	***	1.6	2.8	3.2	4.2	4.6
30	21	12	1.5	2.0	2.4	3.8	4.2	5.3	5.7
32	23	14	2.3	2.8	3.3	4.7	5.2	6.3	6.8
34	25	16	3.0	3.6	4.2	5.6	6.2	7.4	8.0
36	27	18	3.7	4.4	5.0	6.5	7.2	8.4	9.1
38	29	20	4.5	5.2	5.9	7.5	8.2	9.5	10.2
40	31	22	5.2	6.0	6.8	8.4	9.2	10.5	11.3
42	33	24	6.0	6.8	7.6	9.3	10.2	11.6	12.5
44	35	26	6.7	7.6	8.5	10.3	11.2	12.7	13.6
46	37	28	7.4	8.4	9.3	11.2	12.2	13.7	14.7
48	39	30	8.2	9.2	10.2	12.1	13.2	14.8	15.8
50	41	32	8.9	10.0	11.1	13.1	14.2	15.8	17.0
52	43	34	9.6	10.8	11.9	14.0	15.1	16.9	18.1
54	45	36	10.4	11.6	12.8	14.9	16.1	18.0	19.2
56	47	38	11.1	12.4	13.7	15.9	17.1	19.0	20.3
58	49	40	11.8	13.2	14.5	16.8	18.1	20.1	21.5
60	51	42	12.6	14.0	15.4	17.7	19.1	21.1	22.6
62	53	44	13.3	14.8	16.3	18.7	20.1	22.2	23.7
64	55	46	14.1	15.6	17.1	19.6	21.1	23.3	***
66	57	48	14.8	16.4	18.0	20.5	22.1	***	***
68	59	50	15.5	17.2	18.9	21.4	***	***	***
70	61	52	16.3	18.0	19.7	22.4	***	***	***
72	63	54	17.0	18.8	20.6	23.3	***	***	***
74	65	56	17.7	19.6	21.4	***	***	***	***

*Morrison, Frank. *Morrison's Feed and Feeding*. The Morrison Publishing Company, Ithaca, NY. 1949 (pg.1188).

Author's note – most graziers will find these grain feeding rates to be unacceptably high. Do not push grazing cows with too much grain – even grazing cows can experience rumen acidosis if not enough fiber is fed. Try to find a comfortable balance from within this table. Many graziers that have cows maintaining good body condition will feed 10-12 lbs. of grain per day to their highest producers. For the graziers who want to feed very little to no grain at all, remember that cows still must take in a certain amount of *energy* or carbohydrates daily (not just protein and fiber). If not, grazing cows will become skinny quickly and likely experience more subtle problems over a long time period. In addition, I find that cows which become sick on farms that place severe restrictions on grain feeding usually are down cows and so weak that nothing will help return them to a healthy state. There is a balance to be struck when feeding grain. If needed, add in molasses to the hay in order to give energy when using a low grain diet.

Clipping or dragging
pastures reduces parasites

What should you do once the cows have just finished being in a paddock? There are a few options: do nothing, clip it to uniform height, or let other animals graze it further e.g. horses, sheep and/or goats. As a veterinarian and former herdsman with grazing experience, I recommend clipping (or dragging with tines). This can be done with a small mower, up to a couple days after the paddock was grazed (so that a few paddocks get done at once and save some labor). Clipping pastures spreads out the manure that cows drop when grazing. This spreading has a two-fold benefit. First, from a health standpoint, exposing parasitic stomach worm larva to direct sunlight and drying will kill them before they can become infectious to the cows when they are put back in the paddock. In addition, the pasture will more quickly regenerate through splattered manure, whereas it takes much longer to sprout through an intact mound of manure. This means the cows will have more square feet of grazing area when returned to the paddock. In a small grazing nutrition study I did during veterinary school, I observed that cows would not graze closely near an intact manure pile. I then measured how close cows will graze up to intact manure and found there to be a 7 foot diameter "zone of repugnance" wherein cows will not graze if given a choice. This is a simple observation but also has implications: you need to gradually increase the size of paddocks throughout the season if not clipping or risk creating increased levels of internal parasites by forcing cows to graze more closely to intact manure than they otherwise would. And *do not* have young animals follow cows after the cows have been in pasture. This is because the worm larva that the cows leave behind in their manure will infect the young stock quickly. Mature cows can live in balance with intestinal worms, young cattle cannot. They don't have a mature immune system yet and can become fatally infested if grazed on paddocks where mature cows have recently been. When dragging or clipping manure, you will see grass blades come up quickly through the dried out, flattened manure splatter. Compare this to the long-lived internal moisture (and

A manure pie left intact after 3 weeks time. Open it up and it will be moist and contain parasites. No vegetaion can grow through it.

thus parasites) when you let manure paddies stand intact. Another way to break up cow pies is by letting chickens or hogs follow through the pastures.

If forcing cows to graze more closely to intact manure in subsequent runs through a paddock, a farmer might consider deworming his cows periodically in order that the cows are not robbed of nutrients by the parasitic worms (see treatment section for suggestions). Although certified organic farmers are allowed to *occasionally* use conventional dewormers for emergencies in livestock (ivermectin) they should *definitely* first consider clipping as a means of enhancing animal health as well as pasture re-growth. Cows should be checked before the pasture season for worm burden, in order to set-up a base line of

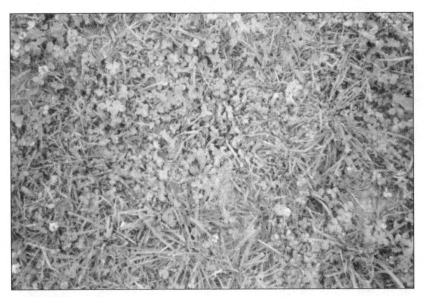

An area where a manure pie was splattered out soon after the paddock was grazed — 3 weeks later there are only crumbly remains and vegetation growing up through it.

information. Periodically during the grazing season they should be checked again. As a herdsman during one particular season, I found that using diatomaceous earth, DE, in addition to clipping pastures, kept worm burdens to a minimum (almost non-existent). Diatomaceous earth is sea sediment with a lot of silica in it that acts somewhat like micro-glass shards upon the parasites when in contact. I would strongly recommend the use of DE to begin with when worm burdens are light. I believe it enhances a *balance* between cow and parasite load. Conventional dewormers do this as well but with a large kill-off at first, then allowing the animal to try to come to equilibrium with the next assault of parasites. The main difference between use of diatomaceous earth and conventional wormers, in my opinion, is that DE is a natural compound that works synergistically with the cows' normal digestive processes. Use 10-20 lbs. of DE per ton of grain mix and add liquid molasses to keep all the dust down from the DE. Contrasted to that is the typical wormer which is a blast of synthetic chemical that does a very effective job but probably upsets a cow's digestive process during its action and may leave questionable residues in the manure that the cow excretes. The residues may hinder the normal organisms in the pasture ecology from breaking down the manure.

The parasites that do the most damage to ruminant health are the common stomach worms. These worms can hibernate in the true stomach (abomasum) lining all winter long. They can sense when it is the right time to emerge and come back to life. Perhaps it is the green grass being re-introduced into the stomach, perhaps it's the different enzymes at work to digest fresh grasses, perhaps the different internal pH or different microbial activity in the rumen — no one knows for sure actually. But, it is well known that young stock are most susceptible to infestations and true health problems such as anemia (due to the worms sucking blood from the animal), weight loss and decreased immune capabilities. It is basic parasite biology. Calves born in the late summer will become infected and carry the hibernating worms through the winter, then the worms will emerge with lethal strength in the spring.

Although worming adult cattle is done more for *economic* benefit than for health concerns, worming young stock so they don't seed pastures with their worm-infested manure is a wise move. As mentioned earlier, and worth repeating, don't follow cows with young stock through your pastures. Adult cows can live in balance with worms in general due to their mature immune system, but they will still potentially shed eggs and larva which young stock can easily pick up if grazing the paddocks cows just went through. (Shedding of Johne's disease by adult cattle also occurs and young stock are more susceptible to infection due to their immature immune system capabilities.) Allow at least 21 days until you put animals back in the same paddock. Eggs and larva deposited with manure need to be re-ingested within 21 days, or they will die in general. If there is no obvious infestation or signs of worm burden in your animals, it is still a good management practice to check the manure under the microscope occasionally — especially in young stock.

Streambank Fencing, Water Quality & Conservation Practices

Inexpensive water systems can be installed so cows don't drink directly from streams. Streambank fencing with controlled crossings (using hog slats as a partially submerged bridge) allows access to various paddocks and keeps direct contact with the stream to a minimum. Hog slats also help decrease puncture wounds caused by stones on the creek bottom. The real benefit of this type of conservation practice is the vegetation, songbirds and other wildlife that rapidly return when the stream banks are given a chance to re-grow by not being trampled. This adds a significant dimension of life to the farm, increasing its diversity, its vitality and overall beauty to both farmer and non-farmer. Streambank fencing also protects water quality.

Frank Lucas, a USDA soil conservationist in Lancaster, PA always likes to point out that "there's nothing good in the stream for the cows, and there's nothing good the cows put into the stream." In areas dense with dairy farms along creeks that drain fertile farmland, it is too common to see dairy animals of various ages wading through the water depositing urine and manure directly into the waterway. With many farms along the main channel, the farms down-stream receive the brunt of this "organic" pollution. Congregating areas

Submerged stones can cause lameness.

Perfect conservation — cows grazing in the sunshine, stream bank fencing with nicely vegetated stream banks, trees protected.

and feed troughs should be at least 100 feet from any nearby drainage ditches or streams in order for the soils to absorb and neutralize the residues excreted by cattle. Fortunately, at least in my practice area, there are less and less cows seen in waterways as most farmers are well aware of the health issues concerning such practices.

Water, in a very real sense, is perhaps the most important nutrient in a cow's diet — at least as far as quality and quantity are concerned. Cows drink on average between 10 and 30 gallons of water per day, depending on which part of lactation, body size and season. In addition to water intake, only about 62% of the feed taken in is digested — with 75-100 pounds of wet manure and 50-70 pounds of urine excreted by a single cow *each day.* Keep this waste material away from your animals! Also think about getting a water-meter to make sure appropriate quantities of water can be delivered to the cows during peak times as well as to measure how much they are drinking outside. Simple black PVC 1" diameter hose is all you need as water lines out to the troughs in the pasture. They will quickly become covered over by grass and therefore direct sunlight won't heat up the water in the hoses too badly. Shallow 50-100 gallon water tubs are indispensable. These shallow tubs are easy to dump, clean out algae and move to another area. I used to use a few wads of tall grasses to scrub off any algae. The water won't heat up either if using shallow tubs since refilling will be more frequent than in the big tubs. Larger containers holding a few hundred gallons of water are even better, if they are re-filled daily. Your cows are your income — they deserve lots of good, clean water.

Without good, clean water, and lots of it, your cows will not produce to their potential. If not provided, cows *will* search for water — in ditches, puddles and streams. Cows *never* "get used to" bad water. Standing water can harbor diseases. Puddles in barnyards are sources of concentrated manure and urine. Salmonella, coliforms and leptospires thrive in puddles and saturated manure. Clinical signs include

Standing water in barnyard — potential source of contaminated water for livestock.

diarrhea, fever, mastitis and abortions. Saturated manure in areas where cattle congregate is ripe with disease potential. Warm and wet conditions can hurt hooves and also cause mastitis. Muck and puddles will soften hooves and the hairline to such an extent that bacteria can enter, and gravel will easily puncture the soles giving abscesses, or footrot and strawberry heel will occur . This may be exacerbated, in my opinion, if the cows are being fed a "hot" ration high in grain, weakening the cow's immune system due to subacute rumen acidosis (SARA). Leaky, heavy producers will have environmental coliforms or *strep non-ag* enter the teat canal. Stressed animals (just fresh or high producing cows) are at increased risk. Slow moving water carries diseases with it — this is true of pasture run-off as well as permanent streams, especially in the summer. Cows standing in streams are also exposed to problems happening upstream. Cows urinating and dropping manure into streams create problems for your neighbors downstream. We are all, in a sense, "downstream."

In a study with Johns Hopkins University, I collected manure samples from strategic locations where cattle congregated or calved-in from local farms along the Pequea Creek (a tributary of the Susquehanna River). The study found that a majority of the samples had *Cryptosporidia* present in the manure. *Cryptosporidia* can live for long times in waterways and may affect public water supplies downstream. Moist bedding in the stall or lying as a group under a tree also gives rise to this. One of the main conclusions from the study was that stream bank fencing *was* a critical factor in reducing loading of waterways with *Cryptosporidia*. (Grazyk).

Obviously, the source of water is particularly important. Well water often is best because it is a protected source compared to stagnant or slow moving water. Sometimes, water treatment is needed for well water if coliforms are present or nitrates are high. If water quality is a problem, due to coliform bacteria, nitrates or other contaminants consider using metered-in hydrogen peroxide, an ultra-violet light, sand filters or other neutralizing devices. Many people put peroxide systems in, and the infused hydrogen peroxide can be beneficial. Probably more effective is an ultraviolet light where all water has to pass by. If nitrates are present, various types of salt filters are beneficial. When using homeopathic remedies in a water trough, peroxide-treated water is probably OK if it is far back in the water system. But make sure that if a peroxide system is being used in the barn for the cows' water bowls that the peroxide is not being used up before it reaches the last water bowl. Hydrogen peroxide

Poorly drained flood plain soil — can harbor pathogenic bacteria. Also a breeding ground for mosquitoes (the vector for West Nile virus).

is a great antiseptic, but the moment it contacts organic substances, it becomes used up as it does its job (see effervescence/bubbling). It is probably wise to do an in-depth water analysis once every few years to check for heavy metals and other compounds.

Situations that Negatively Affect Cow Health or Water Quality

These cows have no choice but to swim to get to the pasture on the other side.

A bad situation for both cows and water.

Situations that Negatively Affect
Cow Health or Water Quality (cont.)

Submerged stones can cause lameness.

Cow drinking from unprotected water source — potential source of contaminated water.

Terrible walking conditions.

Uneven walking surface in barnyard can create hoof problems.

Manure deposited directly into stream.

A mare urinates in the stream while another horse is about to drink.

Poor water source.

Conservation Practices that Enhance Cow Health & Water Quality

Conservation — improved laneway under construction.

Conservation — stock water tub.

Conservation — spring water development.

Cows on nicely improved lane-way.

Conservation — controlled streambank fencing & wildlife.

Conservation — controlled stream crossing.

Well-vegetated waterway with recognition.

Conservation Practices that Enhance
Cow Health & Water Quality (cont.)

Well-vegetated waterway — this will slow down velocity of water and reduce erosion. Notice controlled crossing area.

Underwater crossing (hog slats) — protects cows' hooves from sharp, submerged stones.

Perfect conservation — controlled stream crossing (notice underwater slats), ducks populating area, cows grazing.

Same location with pending storm.

Same location during a September hurricane. Notice the controlled flow of water due to excellent vegetation.

Conservation — bird box provided.

Conservation Practices that Enhance
Cow Health & Water Quality (cont.)

Barnyard — providing clean water, feed and grooved pavement.

Cemented barnyard for good water shedding with grooved surface for traction.

Chapter 4
ORGANIC DAIRY FARMING

An outgrowth of the pasture-based system is that of becoming "certified organic." There are already many dairy farms that have become certified organic and many more either considering the idea or undergoing active transition. In the Lancaster County area of Pennsylvania where I practice, there is now a "critical mass" of this kind of farmer — from only a tiny handful just a few years ago in the mid-1990's. Although organic dairy farming is still a minority faction within the dairy industry, there is certainly a lot more interest in it by various agricultural professionals who work with farmers. At any given field day that spotlights organic farming, one will no doubt see at least a few professors from land grant universities, Extension personnel, agricultural lenders, feed mill representatives, machinery representatives and local farm store owners in attendance, in addition to the many seasoned and novice organic farmers. Organic dairy farming is big business. It is no longer a radical fringe that attracts only the most innovative and risk-taking farmers. In addition, when alternative-minded professionals are all in the local area and can provide seed, soil, animal nutrition and veterinary service, the conditions are ripe to fully support the organic dairy sector. This may help local farmers that are "sitting on the fence" and nervous about taking "the plunge" since it is necessary to have sympathetic professionals helping with the various and inevitable problems of soils, crops and animals that can occur.

The premiums associated with organic milk production are certainly quite enticing. Typically a farmer in the Northeast will receive a consistent US$20 per hundred pounds (cwt) of milk produced. This is in contrast to the roller-coaster prices paid for conventional milk that typically averages about US $12-14 over the long run. Although the organic premiums are substantial, there are increased costs of bought feed that has to be certified organic as well as strict rules that farmers must play by to remain certified. In general, an organic dairy farmer will net more in revenue compared to a conventional dairy farmer. But, it will not be the simple difference between organic and conventional milk prices due to increased organic feed prices. Just as more research is needed in alternative medicines for livestock, there also needs to be more economic data on organic livestock farming operations. If a farmer is simply in it due to the economic premiums, s/he will no doubt have a hard time of it due to the hurdles involved in managing a farm in an organic way. There is no universal prescription on how to farm organically, but there are definite restrictions regarding prohibited substances. It is probably safe to say that a person has to really have their heart and mind set on becoming certified organic. Then the rules aren't so burdensome and will probably make sense, at least most of the time. Many farmers with whom I work are already ecologically motivated to some degree, so transitioning to certified organic is not all that difficult a thinking shift for them.

What drives the certified organic market? Without a doubt, it is a highly consumer-driven commodity. This is true of course in any market, but even more so in the organic dairy market when considering the rules pertaining to the management of soils, crops and animals. The typical consumer of organic products either has: 1) extra income and is in a socio-economic status which allows him/her to buy premium products, 2) a deep philosophical conviction regarding the earth and its natural resources, 3) a health condition which necessitates the consumption of foods potentially untainted by synthetic herbicides, fungicides, insecticides, antibiotics, hormones, etc., and/or 4) a religious reason. As issues of the environment and animal welfare (freedom of movement and to exhibit natural behaviors) come more and more under the scrutiny of consumers, mainstream market chain stores are now carrying whole sections of organically and ecologically produced products. And this simply re-emphasizes that the market is totally consumer driven. If the demand didn't exist, the stores would have no interest in giving their limited shelf space for such items.

People often wonder why prices of organic products are higher than conventional products if, as generally perceived, fewer inputs are needed to produce them (i.e., no cost of expensive synthetic herbicides, insecticides and fungicides in organic systems). There are four basic reasons for this situation, of which only the first may change and provide for less expensive organic goods on store shelves. This first reason is simply supply and demand. In general, there are not enough suppliers of organic products, and the demand outstrips supplies, therefore the price goes higher until the store manager sees a leveling off of demand for a certain product. That is what is called fair market price. The second reason is somewhat intertwined with the first and that is increased price due purely to greed. If a store manager can sell something at a higher price, why make it cheaper? Store managers are in business to generate profit, so why not go for the most, especially if your shoppers are not complaining? (To be fair, some farmers can also be somewhat guilty of charging exceedingly high prices when selling products right at the farm — but perhaps that can be forgiven due to the length of time that they've had to take commodity pricing for the "cheap" food they've previously produced for society.) The third reason has to do with labor costs. Simply put, organic farming tends to be more labor intensive. If a farmer is restricted from spraying herbicides and the like, someone has to cultivate the weeds between the crop rows. Cultivating is a slow and tedious process and integral to organic farming. Human labor must be paid for. In the case of large organic produce farms that exist in the western parts of the United States, the migrant laborers may still work long hours and for not very high wages, but at least on organic farms they do not come into contact with herbicides and pesticides as they do on conventional farms. The last reason why organic food is more expensive on the store shelf is a bit more obscure but probably most startling. It started with the huge increases in crop yields (due to newly developed fertilizers after World War II) making cheaper the feed used to fatten livestock. Livestock farming increased and became more efficient in part due to the pharmaceutical companies helping to solve some of the problems associated with increased densities of poultry, hogs and cattle. All the while, livestock farmers kept applying relatively cheap synthetic nitrogen, phosphorus and potassium fertilizer to the soil for their crops. They diligently listened to the latest research that the universities provided, but they also kept applying one of the by-products of their livestock operation — the manure. Over the years of bountiful crops and increasing livestock density, nothing seemed wrong. But at some point, the problem of groundwater pollution came to the surface. Certain aquifers began to have unacceptably high levels of nitrates due to over fertilization for decades. The "cheap food" which people had become so accustomed to actually had a real cost, but it was *hidden and delayed*. The cost is nearly incalculable when trying to figure out how to clean up an underground aquifer (or surface waters leading to large bodies of water). So, once again, why are organic products more expensive? One reason is definitely at the retail level. But the other reason is that there are no hidden costs, as there are with mass produced "cheap" food.

Therefore, when you buy organic, you are paying for its true cost of production with no environmental costs down the road. A "new" hidden cost of mass production will probably be crops that are genetically engineered as they release their pollen with foreign DNA to other crops in the area. Who knows what the consequences of this will be on the gene pool of that which exists in nature.

For better or worse, the government has historically supported a "cheap food" policy and has aggressively pursued its maintenance. This is mainly due to macro-economic ramifications: it allows consumers to spend a higher percent of the "sacred dollar" on durable goods like cars, appliances, houses, etc., which in turn keeps people employed to make more of such things. Within the U.S., consumers pay the least for food as a percent of their dollar compared to any other country. The U.S. is a very consumer-dominated society and has a high standard of living — *but at what cost?* Do we want to be dependent on foreign oil as a cheap energy source and allow our groundwater to be endangered by cheap synthetic petroleum-based fertilizers and pesticides? Do we want genetic manipulation of crops with unforeseen consequences, just as groundwater pollution reared its ugly head over decades? Do we want huge barns with no windows so we can confine high densities of animals under highly controlled artificial conditions? Do we want animals that are continuously confined to walking on concrete slats or wire cages? Do we want streams that carry such a heavy load of agricultural run-off that we cannot swim in them and that they are no longer able to maintain healthy conditions for birds and fish that have lived there for centuries? These questions will continue to come to the foreground if the "cheap food" policies continue unabated. If multi-national corporations have their way, the *de facto* official government policy of cheap food will make the United States dependent on other developing countries for continual sources of food. It is an anthropological axiom that civilizations perish when they can no longer feed themselves. We need not prove this again.

It is up to the consumers to decide these kinds of issues in a free-market economy. Educating consumers about conventional and organic farming practices is a necessity in order to sustain both a producing earth and a living earth. The government will not do it since many former executives of huge corporations often comprise the scientific advisory councils of government entities. It is a revolving door with many former government officials becoming executives in the corporations that they themselves previously regulated. This again emphasizes the need for the consumer to vote with his or her dollar for the food production system that best reflects his or her values regarding the earth and its limited resources. If consumers "speak" loud enough with their wallets, huge corporations actually will change and dictate to their producers and growers the ways in which to produce food. For instance, if consumers don't want genetically engineered potatoes as the basis for their french fries, then huge corporations will dictate to their buyers to buy only non-genetically engineered potatoes. In Europe, some of the fast food chains are now even using organic ingredients due to consumer demand. Scientific organizations may not like this as it divorces science from the decision making basis. Such scientific organizations must realize that huge corporations exist in "real-time" and are beholden every moment to the consumers' mighty dollar and consumer perceptions, not to scientific dictates and long, drawn out academic discussions. Even governments cannot necessarily influence business choices in a free market economy. Thus, education of consumers by advertisers (or activists) will likely influence the public opinion well into the future.

What I find pleasing as a livestock veterinarian in working with many certified organic dairy farms are the rules requiring cows to receive a portion of their nutritional requirements from managed pastures. Cows grazed on pastures of fresh feeds have less inflammatory conditions and are therefore healthier than cows confined to concrete indoors and eating mainly ensiled feeds. I also particularly like the rule that tails cannot be "docked" (chopped off) as many conventional dairies do routinely now. There is evidence accumulating that tail docking gives absolutely *no* benefit to milk quality or the cow — the only benefit is that the dairy farmer isn't swatted by the tail. Freedom of movement with cows grazing well-managed paddocks is paramount to the animals having a healthy immune system. The simple act of allowing ruminants to graze as designed by nature gives them healthy exercise and provides vibrant fresh feed. It is also what consumers think of when they think of dairy cows. The consumers also don't want cows to be routinely given antibiotics and hormones. They don't care that grazed cows don't make as much milk as cows on confinement farms; they simply want a clean product that's from "happy cows." What the conventional dairy industry tries to advertise as a great natural and healthful product is truly accomplished, in my opinion, mainly by the certified organic sector.

From another perspective, a somewhat lower milk production level due to grazing by the conventional dairy industry could be good — for both farmer income and cow health. But the government has a history of sending mixed signals to the dairy sector, which has in the long run kept increasing supplies. There are, at times, initiatives taken by the conventional dairy sector to boost farmer income by reducing milk production. One particular program, the Cooperative Working Together (CWT) of 2003, only received enough overall support nationally to reach the 80% participation level to trigger the original plan but did garner enough support for a weaker plan. In addition, certain sectors of the industry put pressure on the dairy co-ops participating in the CWT plan to not ban the use of certain bio-technology products which get routinely injected into cows in order to increase milk production. For instance, in 2003, total U.S. milk production was about 169,758,000,000 pounds from about 9,141,000 cows. The number of cows on rbST was about 30% of the national herd, or about 2,742,300 cows. Based on a 305 day lactation at an increase of 8 lbs. per day due to rbST for the 30% of the U.S. herd would mean that 6,691,212,000 pounds extra are produced due to rbST. Bovine growth hormone creates about 8-12 lbs. extra per day. If only a 4 lb. per day increase is considered, there would still be 3,345,106,000 extra pounds being made due to rbST. The original CWT initiative sought to take off about 4,583,466,000 lbs. of milk. When taking off 8 lbs. per day as the average increase (by eliminating the synthetic growth hormone), it is obvious that this goal would have been reached. *(Northeast Dairy Business,* July 2003; letter to the editor, pg. 8)

But the co-ops did not require eliminating rbST, probably due to hidden forces behind the scenes. A simpler, less politically charged solution would be for farmers to simply graze their cows. This would eliminate surplus problems while improving cow health across the United States.

Organic dairy farming hinges upon the following main principle: to produce milk in an environmentally friendly way that also enhances cow health by grazing pastures, while prohibiting antibiotic and hormone use except to save the life of the cow. On an organic dairy farm, a visitor may see calves suckling from their mothers out on pasture, groups of animals being rotated through various well-marked paddocks, streams fenced-off from cows to let vegetation minimize erosion and allow wildlife to flourish, herbal tinctures and homeopathic remedies in the medicine cabinet, and a farmer that seems to truly enjoy farming. Usually, organic farmers enjoy visitors because they are truly happy with their farming style in that they are producing food for society without damaging the environment. They have nothing to hide — it is all there, plain to see.

There are generally two types of organic dairy farms. There are those that have simply never taken up the conventional practices of field herbicides and insecticides nor used much in the way of antibiotics or hormones on their cows. These farms are sometimes called "organic by neglect." Then there are those farms that have been solidly conventional but at some point became dismayed with the ever-increasing reliance on herbicides, insecticides, antibiotics and hormones. These farmers have intentionally shifted their farming practices to become more environmentally friendly and to approach their cows' health in a more holistic manner. These farmers are sometimes called "organic by design."

As a veterinarian, I am called to a wide variety of farms, each one being unique in the way it is managed. I find that those farmers who are "organic by design" generally run a tight ship: they have very good record keeping skills, they often go to meetings to enhance their knowledge of organic production techniques, they have their cows' milk weights and reproductive status checked monthly, they enjoy talking about the latest news in the organic world and they usually have paid extra attention to land stewardship issues (streambank fencing, wetland protection, hedgerow enhancement, etc.). Those farmers that are "organic by neglect" generally do not place much emphasis on the just mentioned topics. The "organic by neglect" group may evoke thoughts of the "Old MacDonald had a farm . . ." scenario and probably would be fairly accurate. Although both kinds of organic farms are definitely in the same camp (in contrast to conventional farms), the farms that are "organic by neglect" create extra challenges for those of us professionals entrenched in an already strictly monitored organic agriculture.

One thing that all organic dairy farmers seem to lack is the time to read all the volumes of rules of the certification agency that certifies them. With the USDA National Organic Program in charge of enforcing the Rule, there is less difference between the various certification agencies (but there still remain some serious gaps). As organic livestock production continues to gain momentum, it is unfortunate to say that there is a general lack of products out there for livestock health care needs that have been reviewed by the appropriate authorities. I am often asked to look at a product that a farmer has just bought from a truck route salesman. And although the salesman claims it is OK for organic farms, there is no statement on the product label that indicates such. In fact, many of the "natural" type products that contain enzymes or *Lactobacillus* (probiotic) products may have been made by using genetically modified organisms (GMO's) in a hidden, intermediate step. It is highly questionable as to whether or not a product like this would be acceptable for a certified farmer. Sadly, I know a farmer who was one day away from shipping Certified Organic fluid milk and then was finally told by his certifier that he had used a prohibited product on his fields a year and a half before. He was therefore denied certification. And yet the salesman of that field product told him it was "OK for organics" a year and a half ago. That set the dairy farmer back nearly $75,000 of lost income — all because some truck route salesman said something was OK for organics *without the actual proof on paper.*

To ask dairy farmers to stay on top of all the rules that pertain to their soils, crops and livestock is a rather monumental task. Ultimately, however, the farmer *is* the one responsible for what he buys and uses on his land and animals. The best way to stay protected is to ask a supplier for the certification number for the product or crop and to keep every single receipt for all purchases made. This is definitely possible. And it *is* an extra burden, but being certified organic is voluntary and has its responsibilities (to protect consumer confidence). The difference between "organic" and Certified Organic is in *the keeping of records* to prove that the production process is in line with the USDA NOP Rule.

Chapter 5

Health Care Dilemmas in Organic Dairying

"The producer of an organic livestock operation must not withhold medical treatment from a sick animal in an effort to preserve its organic status. All appropriate medications must be used to restore an animal to health when methods acceptable to organic production fail. Livestock treated with a prohibited substance must be clearly identified and shall not be sold, labeled, or represented as organically produced" — 7CFR §205.238 (c)(7) USDA National Organic Program, Livestock health care practice standard.

In plain English, the above statement says that a farmer cannot withhold appropriate medical treatment, yet if s/he uses antibiotics, hormones or steroids, etc., the animal must be removed from the herd. In theory, this is a good statement, but it is also a "Catch-22" — you are damned if you do (use antibiotics) and damned if you don't. This chapter will explore this vexing statement.

I have given much time and thought to this as it directly impacts livestock veterinarians out on the front lines. Couple the statement with a woefully inadequate list of allowed substances for livestock health care on the National List (at the time of this writing) and it becomes a chronically infuriating situation — not only for the veterinarian but the farmer as well (not to mention the animal which may be suffering). This particular USDA statement is one of the core reasons for writing this book — in order to give dairy farmers ideas for treatment options before using prohibited materials which implicitly condemn an animal to premature removal from the herd. It also has gotten me very involved with many people in the organic industry on the national level. This chapter will hopefully describe in detail about how this statement, in the end, directly affects USDA Certified Organic livestock any hour of the day or night.

To begin with, it is a nearly monumental task for a certified organic farmer to know which of the hundreds of livestock health treatments that are commercially available are allowed or prohibited. Generally speaking, all *natural* products are allowed. Vitamins, minerals and other nutritional needs can be from any source if given to an animal for health purposes (intravenous, intramuscular and/or orally) if approved by the FDA (Food and Drug Administration) or AAFCO (American Association of Feed Control Officials). They are allowed when an individual animal needs rapid infusion of such things, for example, calcium gluconate (electrolytes) for milk fever, hypertonic saline (electrolytes) for rehydration, dextrose (glucose) for ketosis, etc. Biologics such as vaccines are also allowed. The very popular biologically derived products of colostrum and whey are allowed since they are biologics, but may come under fire if scrutinized, and are required to be only from certified organic sources. But then again, livestock health products will probably always be given more leeway in reviewing them due to the very limited amount of products available to actually treat a certified animal and for the animal to be kept in the herd.

Under the USDA National Organic Program (NOP) Rule, all synthetic substances are prohibited, unless petitioned to, and subsequently allowed by, the USDA National Organic Standards Board (NOSB) and given final approval by the U.S. Secretary of Agriculture. Prohibited synthetic substances include all antibiotics (including but not limited to) — penicillin, ampicillin, any kind of tetracycline, all sulfa drugs, and ceftiofur; steroids; synthetic growth and reproduction hormones such as dexamethasone, isoflupredone, prostaglandins (PG), gonadorelin (GnRH) and recombinant bovine somatotropin (rbST/ rBGH/ bovine growth hormone). Under current rules, if prohibited compounds are used, the animal will be rendered useless for organic milk production. In general, certified farmers can understand this. It would be a relief to farmers and veterinarians if some day a dairy cow that was given an antibiotic or hormone to save its life could still be of (limited) value by employing her as a nurse cow for calves — once the standard regulatory drug withholding time has expired. Hopefully, if such a cow were to be kept in the herd, continuously fed organic feeds and then bred back, her offspring would be able to be certified organic. This kind of addition to the already stringent Rule would be welcome by those of us who aspire to keep animal welfare as the basis for organic livestock husbandry. Unfortunately, I am sure that this will be contested by certain parties who are obsessed with wanting nothing less than absolute purity in the whole organic process. If absolute purity is being demanded, then the use of polluting, toxic diesel fuel should be prohibited and farming with horses required!

The absolute prohibition on using any synthetic growth hormones, reproduction hormones or antibiotics in a milking animal is based on consumer demand as well as dairy processors explicitly stating on the carton that no antibiotics or synthetic hormones have ever been used on the cows that produced that particular product. It is totally created for consumer perception and marketing strategy. It unfortunately brings up some serious welfare issues with which many farmers and veterinarians struggle at any moment.

I remember from being a herdsman on an organic farm that if I knew I could never again be allowed to put milk from a certain cow into the tank when treated with a certain substance, I would try everything possible on the cow to get her well again — short of using the prohibited material. Now as a veterinarian, I see the situation repeat itself when I am tending to a sick cow on a certified farm, and the farmer has to decide how to go ahead treating a particular animal. I can truly empathize with them. It is darn difficult for a farmer to say, right from the start, "OK, treat her with the penicillin and I'll just ship her after the treatment" when she may have the best genetics in the herd, or she may be the farmer's favorite animal for any number of other reasons. Technically, according to the statement at the beginning of the chapter and according to all USDA accredited certification agencies, the farmer is required to use a prohibited substance on the cow if that is *the* substance that will restore the health of the animal. But in the next breath, it says that such a cow treated with a prohibited substance is required to leave the herd. These are two mutually exclusive principles that look great to consumers and regulators, but leave a farmer between a rock and a hard place (and usually at some late hour when there is a crisis). In addition, a farmer can become de-certified if found to be withholding "appropriate" treatment in order to keep the status of the cow as certified organic. This is a very sticky issue and one that I find myself constantly grappling with as a primary care provider while trying to deliver sound health care to my certified organic clients' animals.

I would like to raise a few questions regarding humane treatment and the concept of health care as required for USDA Certified Organic farms.

First, if an effort to restore the animal to health includes using a prohibited substance that will lead to an animal's removal from the herd, where is the *incentive* to actually treat the animal? Unless a farmer is acting entirely from an altruistic basis, s/he will probably forego any prohibited substances and utilize Complementary and Alternative Veterinary Medicine (CAVM) until the last possible moment. This is due to the economic reality of farm animals needing to be useful, productive and profitable creatures. Farming is utilitarian by nature — all inputs need to have functional value towards the end goal of operating a profitable business. If an animal will be removed from production due to a certain treatment, time and energy are spent on more promising parts of the enterprise. It's human nature. I've seen this occur time and time again on all sorts of farms. Secondly, what exactly is meant by "appropriate medication"? Is "appropriate medication" referring to what is generally understood as "conventional" or "standard" medicine and surgery by mainstream veterinarians? The term "appropriate medication" may have very different meanings to herbalists, acupuncturists, homeopaths and chiropractors, depending on what each considers being the right way to restore health. Thirdly, when exactly in the course of disease shall use of an "appropriate medication" be initiated to restore health? Certainly the earlier any kind of treatment is given the better chance it has to work. And exactly who is to say when the "appropriate medication," is to be instituted? Farmer, veterinarian, alternative healthcare provider, regulator? If we consider the above points, one overall question emerges: What exactly is meant by an "appropriate medication" and who will say if and when it will be instituted to restore an animal to health? Then, if appropriate medication is neglected in order to maintain an animal's organic status, who is to initiate de-certification and when (as required by NOP)?

Animals are living, breathing, sentient creatures deserving of humane care at all stages of their life. It is incumbent upon certified organic farmers to optimize their animals' health and welfare. However, doing so cannot involve immediate removal of the animal from the herd as a punitive measure for trying to relieve an animal's pain and suffering. The average consumer of organic products *expects* humane treatment. Veterinarians, who at critical moments are the link between animals' needs and these consumers' expectations, take an oath to relieve pain and suffering when they enter the profession (see Veterinary Oath on page xv). To tie the hands of veterinarians who try to relieve pain and suffering by unreasonable regulations will undoubtedly be met by a loud backlash from many segments of society.

A point to consider is whether or not the condition is life-threatening, acute or chronic. Is the animal breathing with its mouth open? Is the animal down, bloated and kicking at its belly? These are two examples of life-threatening conditions. In other words, if not treated quickly, the animal will die. Acute conditions are those problems that arise quickly and cause an obvious sickness but are not necessarily life-threatening (if tended to properly in a timely fashion). Many types of mastitis, diarrhea, or fever would be examples of an acute condition. Chronic conditions are those problems that the farmer sees develop over a period of time, usually a few weeks time. Various locomotion problems involving the hoof, fetlock, hock, knee and hip would fall into this category, as do reproduction problems, high somatic cell counts (SCC) and Johne's disease. Sometimes an acute condition will not be tended to, and then it either becomes life-threatening or chronic if it doesn't naturally resolve on its own.

It is safe to say that life-threatening emergencies cannot exist for long. This is in contrast to chronic conditions which may be more of a problem for certain types of farm systems (*i.e.,* chronic lameness is obviously worse for cows walking to get to paddocks for intensive grazing). Acute problems, such as mastitis and diarrhea, will be addressed usually within a few milkings due to obvious production decline and body condition decline. These problems all exist regardless of being a conventional or certified organic farm. It is assumed that preventing such problems is any farmer's goal. The difference comes in what methods are used to correct the situation once it occurs.

Unbeknownst to the average consumer of organic foods, a cow may be severely ailing at 10 o'clock at night and need veterinary attention if the farmer's front line defense of probiotics, herbs and homeopathic remedies has failed to restore her to health. Then the veterinarian is called in to help. (It must be assumed that the vast majority of veterinarians know only about synthetic medicinal treatments as taught in veterinary schools.) What then is in the doctor's medicine bag that can help relieve pain and suffering while simultaneously allowing the cow to remain in the herd? The farmer is probably more knowledgeable of possible natural treatments, but then again perhaps the farmer doesn't know of such things. What is the veterinarian to do? Certainly discussing the findings and all possible courses of action is always appropriate. And the farmer should hopefully be able to tell the veterinarian which substances are categorically prohibited.

The veterinarian can simply say that there is nothing that s/he can do because of organic restrictions and proceed to wash up and leave. I've been told personally by a colleague that this happened. This leaves the veterinarian feeling frustrated and awful for not helping the animal, not to speak of the animal's situation not being adequately addressed. This situation, if repeated often enough, may end up with a few possible outcomes. First, the veterinarian lodges a formal complaint with the appropriate authorities regarding perceived systematic neglect of animals needing medical treatment on organic farms, or, the veterinarian begins to "slip-in" substances to relieve pain and suffering that are not routinely tested for in the meat or milk. (I have had a colleague tell me he did this — perhaps it is ethically debatable and legally questionable, but nonetheless it is a real possibility.) A second scenario is that the veterinarian forcefully asserts that s/he knows what's best for the animal, and the farmer finally gives in to the veterinarian's insistence for treatment. This will greatly diminish the animal's productive value to the farmer after treatment if the veterinarian uses a prohibited material (which is likely). In addition, as I have experienced when reviewing files during farmer certification for PCO (Pennsylvania Certified Organic), if there is a substance used that is unknown to the certifiers then it is red-flagged (and rightly so), to be reviewed either internally by members with scientific knowledge or out-sourced to independent professional review organizations such as OMRI (Organic Materials Review Institute). This can become a problem for the farmer when all s/he was doing was allowing the veterinarian to relieve pain and suffering to the best of their professional judgment.

As a veterinarian who actively integrates botanical medicine, homeopathy and some acupuncture, I still have difficulties using CAVM with certain real-life situations and need to use certain synthetics. Again, please keep in mind the legion of my veterinary colleagues who have absolutely no knowledge of alternative medicines yet sincerely desire to do what is most appropriate for the animal *and* the farmer.

On conventional farms, the use of antibiotics, hormones and steroids is allowed, so long as used according to FDA rules. This means that the vet or caregiver can quickly reach for these types of substances in any situation—life-threatening emergencies and acute or chronic conditions. The appropriateness of antibiotics, hormones and steroids varies upon the condition in my view, even when only talking about conventional farms. Certainly, in life-threatening emergencies, such as seen with rapid onset pneumonia, antibiotics and anti-inflammatories can be worth their weight in gold. In acute situations involving mastitis and diarrhea the appropriateness of antibiotics depends more on the personal experience of the farmer or vet (if called in for an opinion). Is the diarrhea due to spoiled feed or due to an infectious cause? Even if infectious, such as with Salmonellosis, the use of antibiotics is debatable among strictly conventional veterinarians due to the potential of setting up a "carrier-state" where the animal recovers but sheds *Salmonella* intermittently (usually when stressed such as around calving time or at peak lactation). An acute case of mastitis, one that develops over the course of 24-36 hours, may respond to intramammary antibiotic therapy or may resolve with extra stripping out of the effected quarter, depending on the type of mastitis. It is up to the individual farmer how they best deal with the situation. Chronic problems, such as lameness and high somatic cell count that come and go in an individual cow probably benefit the least from antibiotic or steroidal treatments. In the case of mastitis, depending on the reaction of the cow's immune system, scarring of various areas of the internal tissue or micro-abscesses within the udder develop around areas of insult, as is the case with *Staph aureus*. Antibiotics, at least the standard commercially available ones, act poorly against these "walled-off" areas. The best hope for such a cow, save possible alternative therapies, is for the resting time of the udder during the dry period. Lameness, due to swollen hocks, periodic knee-cap dislocation and hip problems, of course are not good candidates for antibiotic therapy. Hoof problems that are not obviously strawberry/hairy heel wart should be closely examined for abscesses, ulcers or footrot. Except for a serious case of footrot that quickly can involve the joints of the ankle, antibiotics are unnecessary for the vast majority of hoof problems encountered.

Now we come to the question of exactly when should a USDA Certified Organic farmer actually revert to use of antibiotics and/or hormones as "appropriate medications." It is worth repeating that certification agencies require the removal of an animal treated with a prohibited substance from production *forever*. Yet livestock must be humanely cared for. Most rational farmers agree that a live cow (saved with a prohibited substance) is better than a dead "organic" cow (that was denied a prohibited substance).

The statements regarding removal of an animal due to using prohibited substances, while also stressing the need to treat animals appropriately (due to humane reasons), stem from the fact that the organic industry is strongly consumer driven. Until the USDA National Organic Program became law, there was no government intervention in the organic industry aside from individual state laws regarding sanitation of milking equipment from an inspection standpoint (no difference between organic and conventional). Even with the adoption of the USDA National Organic Program, nearly all rules were initiated due to consumer demand for a certain type of organic product. The problem is that there is still that total "disconnect" and gap between the modern consumer's awareness of livestock agriculture and the everyday realities encountered by family farmers and veterinarians. On the one hand, organic consumers want absolutely no prohibited substances to be used in the animals producing the dairy products that they are consuming. On the other hand, they imagine Bessie the Cow to be happily grazing lush green pastures every minute of her life. The average consumer of organic dairy products (like any other dairy consumer) has very little, if any, knowledge of typical problems that can spring up during a normal 24-hour stretch on a farm. The organic farmers, nearly all of whom were conventional previously, are in most instances aware of what can and cannot be used. However, when exactly should a producer knowingly revert to prohibited substances if alternative therapies have been successfully used for various conditions on the farm in the past? Producers, as with consumers, also come in many types (see Chapter 7). At one end, are the producers that have recently become certified and lament that they can no longer use wormers or fly spray to combat the usual parasitic organisms that follow cattle around. At the other end there are a few producers who it seems would rather actually see an animal die an "organic death" rather than treat it with prohibited substances. The rationale among this kind of producer is that the animal is genetically not fit to be in the herd if it is going to get sick. Should this kind of producer be "de-certified" for withholding an appropriate treatment? If the producer truly believes philosophically that s/he is allowing nature to take its course (Darwinian principle of allowing only the strongest to survive), then, in a sense, that farmer *is* perhaps bettering the gene pool of cattle in the organic industry. Just don't tell that to the animal who is going to be sacrificed (or the consumer). However, as a person involved in animal agriculture and committed to compassionate care of livestock, I maintain that farmers have an ethical imperative to give the best possible care to the animals that provide them with their income. If, however, a producer is withholding treatment to simply stall and keep an economic devaluation of the animal from occurring (organic animals are worth more due to the increased value of their products), then de-certification procedures should take place.

Regardless of farming philosophy among certified organic farmers, the question of when antibiotics and hormones should be reverted to has no simple answer. From experience as a herdsman and now as a vet I can perhaps share some insights. First and foremost as a veterinarian, I have the responsibility to prevent and reduce suffering of animals. I also have the duty to put my professional scientific training to use in sorting out which cases merit emergency attention versus which don't. In addition, as a livestock veterinarian, I need to continuously keep in mind the issues of food quality and food safety. And, although farmers who have voluntarily sought and achieved certified organic status have a severely reduced number of conventional treatment options, as their vet I need to respect the rules they have decided to play by. Hopefully they try to learn alternative types of treatments for self-limiting problems. Granted, some farmers are simply better than others at picking up early signs of a sick cow, knowing which treatments can work well, and also realizing when to call in the vet for a more in-depth work-up or emergency help.

As their veterinarian, I still need to keep in mind my rights and responsibilities as a licensed veterinary practitioner while also playing by the organic rules when asked to tend to sick animals on their farm. As a licensed professional veterinarian, I am entrusted by the consuming public to be a compassionate doctor of livestock in terms of humane treatment during emergency care as well as to ensure a safe food supply. A livestock veterinarian carries equally as much trust from the farmer as responsibility to the consumer. The farm vet is the person who implicitly "bridges" the farmer's need for animal care in the barn with the consuming public's demands in the marketplace. This is true whether it be from the conventional or organic standpoint. My own personal background enhances this "bridging." Having been raised in the suburbs of Philadelphia, then intentionally learning and working both in conventional and organic agriculture gives me clear insights into both consumer demands and producer problems. I know quite well the lifestyle, different philosophies, and environmental concerns the average urban and suburban consumer of organic products possesses. I also know their lack of knowledge regarding daily farming activities and cows. As for farmers, I know exactly how it feels when a cow gets mastitis or needs help to deliver a calf and also when a favorite cow leaves the herd and "goes to market." I also know the farmer's joy of watching a herd of cows that are grazing a new grass paddock in the early evening as the sun is setting. All of these points combine to create a balance that I, as a veterinarian (as well as a consumer of organic products), constantly weigh during my farm calls. I have given this topic considerable thought, discussing it with a wide range of folks from differing perspectives.

So, now, how exactly do we (the consuming public, dairy support personnel, farmers and livestock vets) come up with a way of knowing exactly when to use a prohibited substance in a certified animal? I would say it goes back to the life-threatening situations that I described previously. The two described were an animal that was having difficulty breathing and an animal bloated and kicking at its belly. Fortunately, most farmers know that a bloated animal can be rapidly de-bloated by sharply stabbing a knife in the rumen to allow the gas out. It is an emergency situation needing rapid treatment and that's the treatment if a vet isn't right there to help. (See Bloat in treatment section.) The situation of an animal with respiratory difficulty is entirely different. Not knowing what exactly is causing the labored breathing (heat stroke? bacterial pneumonia? allergic reaction? etc.), we can administer certain conventional medicines if complementary and alternative veterinary treatments afford no immediate relief. For instance, the non-steroidal anti-inflammatory drug, flunixin, can reduce a dangerously high fever (especially if pregnant), improve lung ventilation and relieve pain. Most likely, the animal will feel better quickly, especially if administered intravenously. Clinically, I have seen many animals show relief by starting to eat shortly after its administration when they had not been eating previously. So, we could help relieve the animal's suffering in an emergency situation and not threaten food safety or quality since this medicine is allowed to be used in cattle by veterinarians if deemed appropriate (if the vet has a valid Veterinary-Client-Patient-Relationship with the farm). This may need to be repeated for a couple days along with other alternative treatments including botanical medicines, homeopathic remedies, acupuncture and extra vitamins and minerals. I firmly believe that if we can help a cow simply *feel better* and get her to begin eating while simultaneously enhancing her immune system, she has a much better chance to pull through a crisis *without* resorting to prohibited substances. If, however, her labored breathing in no way improves within 6-12 hours and is diagnosed as bacterial pneumonia by the vet, then I would say that she should be treated with the appropriate antibiotic for pneumonia. If this is compounded by another severely debilitating condition, such as a toxic metritis (bad uterus) or coliform mastitis, the animal's vitality is significantly lowered and antibiotics should be used without delay. To not do so would be to the detriment of the welfare of the animal by allowing it to suffer and also a food safety issue. Although there may be effective alternative treatments to fulminant bacterial pneumonia (see Case Study in Chapter 14), the vast majority of livestock veterinarians unfortunately have no knowledge of them.

The other dilemma that I have experienced is the situation where a known condition, such as hardware disease is diagnosed *(e.g.,* a piece of sharp metal or wire that is lodged internally). The farmer is given various treatment options but is clearly advised that an antibiotic and a magnet would be of best benefit to the cow, but the antibiotic portion of the treatment is declined in order to keep trying alternatives to prohibited substances. Unfortunately, and I have seen this repeatedly, the cow will die if not given the magnet *and antibiotic.* The magnet and antibiotic treatment works well in probably 80-90% of the cases if it is truly hardware disease. Certainly the farmer has the right to decline treatment, but since the farmer is, in effect, dictated by the organic consumer, does s/he have an unfettered "right" to decline what is, in a veterinarian's opinion, the best treatment for the ailing animal? I still believe that the farmer does have that right, but I am also strongly of the opinion that the farmer should always keep in mind what the consumer would say if the consumer were standing right there when such a decision has to be made. Could the farmer defend his/her position? Could the farmer also defend his/her position while allowing the vet to describe the condition and its progression in detail to the consumer with the various treatment options and prognoses?

Does the organic consumer care? In a word — Yes. One of the main reasons consumers will pay a premium for organic dairy products is due not only to forbidding herbicides, insecticides, etc., but also due to *a heightened sense of increased animal welfare.* Urban and suburban consumers generally think of dairy farms as romantically nestled places out in the countryside, not the "factories" which conventional livestock commodity farming is rapidly becoming. Then, to take urban and suburban consumers who are also politically aware and environmentally conscious is a prime recipe for a consumer who demands that certified organic dairy cows are happy and extremely well cared for, not neglected or denied treatment when hurting and in pain. By having addressed and finally endorsed the limited use of certain synthetic veterinary pharmaceuticals, I believe the USDA National Organic Standards Board has studied the issues and taken a balanced approach to the humane care of certified organic livestock.

Chapter 6

NOSB Approved Synthetic Materials for Certified Organic Farms

The USDA National Organic Program is the regulatory agency overseeing Accredited Certifiers mainly from throughout the U.S.A. but also from other countries. They are advised by the National Organic Standards Board (NOSB), which is made up of 15 individuals from various stakeholders groups from within the organic industry and society at large. The NOSB has statutory authority to endorse various materials for use in organic agriculture. Once a material is endorsed, it is to be added to the list of approved materials as set forth by the NOP. However, the NOP is within the USDA and must follow certain procedures while respecting other agencies' enforcement powers, such as the Food and Drug Administration (FDA); thus materials need to be reviewed by lawyers from within the USDA, which can take time. Technically, materials endorsed by the NOSB but not yet appearing on the Federal Register are not yet "approved." The NOSB needs to verify and double check that their annotations for livestock health materials do not supercede the FDA since it is the FDA that has sole enforcement power over any livestock healthcare products that it has licensed. It is a slow and daunting process to get synthetic materials reviewed in the first place, and then they are discussed in excruciating detail by the NOSB, and then voted on. Although there are a few farmers on the Board, they are not necessarily livestock farmers. There are also consumer advocates, environmentalists, scientists, processors, distribution professionals, and folks representing retail organic products. It is a balanced group in general. They are appointed by the Secretary of the USDA after being nominated by various interest groups. However, when technical details arise regarding products — whether they are products for soil, crops, livestock or processing — many hurdles all of a sudden appear. This is mainly due to the same "disconnect" mentioned earlier in regards to the gap between farmers and retailers/consumers. This being said, I have worked through the appropriate channels for a few years, gaining a strong contingent of professional veterinarians from across the U.S. as well as organic processors, to petition for a few synthetic products to be allowed to relieve pain and suffering in emergency situations. My reasons to petition for a few synthetic materials for use with certified organic livestock were based on realities not envisioned by the writers of the National Organic Program Rule. Being a veterinarian who is known to work with many certified organic farmers in my practice (nearly 75 at the time of writing), I am also periodically phoned by other veterinarians and farmers who are seeking to gain an insight into the world of organic dairying. During the Rule writing process, I contacted the USDA regarding the compound oxytocin. I felt that it should be included as "allowed" since it is critical in helping to contract a uterus down to normal size after replacement of a prolapsed uterus (the truest of bovine emergencies). Granted, there are some natural products that work nearly as well (see ergot and caulophyllum in the *Comparative Materia Medica section*). However, not many modern livestock veterinarians other than a small handful (including myself) would ever think of anything other than oxytocin, let alone have on hand the ergot or caulophyllum. Therefore, since it is an international program, veterinarians in other areas need to be taken into account (see previous chapter). In addition, oxytocin is a naturally occurring hormone. Then the Rule became part of the Federal Register, with oxytocin listed as "allowed." (It remains to be seen if it will continue to be allowed, since it is unfortunately technically a hormone. The word "hormone" is fraught with connotations of anabolic steroid abuse, etc. — although oxytocin is not at all, in any way, shape or form, used in such ways. But, alas, it is still unfortunately "a hormone.") After oxytocin became allowed, I also began to think about other instances when a veterinarian may be called out to relieve pain and suffering — one of the main missions of veterinarians who are in practice. So, some other items came to mind, such as analgesics and anesthetics to perform humane surgical procedures or to stabilize very sick cows. Many conditions and odd instances that punctuate a veterinarian's practice life could not have been considered in great detail when the Rule writers in Washington, D.C. sat down at their desks to write the guidelines for organic livestock. So, to summarize, the following list below has been voted on and endorsed by the NOSB — however, at the time of this writing, the NOP has yet to officially add them to the Federal Register. I do believe they will be; it is just a matter of time. But, as always, the certified organic farmer is the one ultimately responsible for what goes into the cows. Readers should also refer to the Organic Materials Review Institute (OMRI) Generic materials list to see what is currently approved.

Compounds that are already in the Federal Register (allowed or prohibited)

§205.603 Synthetic substances allowed for use in organic livestock production:

In accordance with restrictions specified in this section, the following synthetic substances may be used in organic livestock production:

(a) As disinfectants, sanitizer, and medical treatments as applicable.

(1) Alcohols.

(i) Ethanol — disinfectant and sanitizer only, prohibited as a feed additive.

(ii) Isopropanol — disinfectant only.

(2) Aspirin — approved for healthcare use to reduce inflammation.

(3) Chlorine materials — disinfecting and sanitizing facilities' equipment. Residual chlorine levels in the water shall not exceed the maximum residual disinfectant limit under the Safe Drinking Water Act.

(i) Calcium hypochlorite.

(ii) Chlorine dioxide.

(iii) Sodium hypochlorite.

(4) Chlorhexidine — Allowed for surgical procedures conducted by a veterinarian. Allowed for use as a teat dip when alternative germicidal agents and/or physical barriers have lost their effectiveness.

(5) Electrolytes — without antibiotics (author: calciumborogluconate, calcium as boryl esters of gluconic acid, lactated ringers, hypertonic saline, CMPK).

(6) Glucose (author: dextrose and glucose are equivalent).

(7) Glycerin — Allowed as a livestock teat dip, must be produced through the hydrolysis of fats or oils.

(8) Iodine.

(9) Hydrogen peroxide.

(10) Magnesium sulphate (Epsom salts).

(11) Oxytocin — use in post parturition therapeutic applications.

(12) Parasiticides — Ivermectin — prohibited in slaughter stock, allowed in emergency treatment for dairy and breeder stock when organic system plan-approved preventative management does not prevent infestation. Milk or milk products from a treated animal cannot be labeled as provided for in subpart D of this part for 90 days following treatment. In breeder stock, treatment cannot occur during the last third of gestation if the progeny will be sold as organic and must not be used during the lactation period of breeding stock.

(13) Phosphoric acid — allowed as an equipment cleaner, provided that no direct contact with organically managed livestock or land occurs.

(14) Biologics — vaccines (author: colostrum-whey products are biologics; and it does not specifically say "without preservatives"; in addition, all vaccines usually have minute amounts of antibiotics as preservatives, and again it does not specifically say "without preservatives").

(b) As topical treatment, external parasiticide or local anesthetic as applicable.

(1) Iodine.

(2) Lidocaine — as a local anesthetic. Use requires a withdrawal period of 90 days after administering to livestock intended for slaughter and 7 days after administering to dairy animals.

(3) Hydrated lime — (Bordeaux mixtures), not permitted to cauterize physical alterations or to deodorize animal wastes

(4) Mineral oil — for topical use and as a lubricant.

(5) Procaine — as an anesthetic (same annotation as Lidocaine).

(6) Copper sulfate.

(c) As feed supplements — Milk replacers — without antibiotics, as emergency use only, no nonmilk products or products from BST-treated animals.

(d) As feed additives.

(1) Trace minerals, used for enrichment or fortification when FDA approved.

(e) As synthetic inert ingredients as classified by the Environmental Protection Agency (EPA), for use with nonsynthetic substances or a synthetic substance listed in this section and used as an active pesticide ingredient in accordance with any limitations on the use of such substances.

EPA List 4 — Inerts of Minimal Concern
§205.604 Nonsynthetic substances prohibited for use in organic livestock production.
The following nonsynthetic substances may not be used in organic livestock production:

(a) Strychnine

Compounds Voted On By NOSB
But Not Yet in the Federal Register

	NOSB Recommended Category	NOSB Date of Vote	NOSB Recommendation
Activated charcoal	§205.603(a)	9-18-02	Allowed from vegetative sources only
Atropine	§205.603(a)	5-14-03	Allowed
Bismuth subsalicylate	§205.603(a)	9-18-02	Allowed
Butorphanol	§205.603(a)	9-18-02	Allowed
Calcium borogluconate	§205.603(a)	11-00	Allowed for the treatment of milk fever
Calcium propionate	§205.603(a)	5-14-03	Allowed as a mold inhibitor for dried formulated herbal remedies
DL-Methionine	§205.603(d)	10-01	Allowed for poultry until 10/21/05
Epinephrine	§205.604	9-18-02	Prohibited non-synthetic except for emergency treatment of anaphylactic shock
Flunixin	§205.603(a)	10-20-02	Allowed
Ivermectin	§205.603(a)(12)	11-00	Allowed except the slow release formulation
Kaolin pectin	§205.603(a)	9-18-02	Allowed
Magnesium oxide/hydroxide	§205.603(a)	9-18-02	Allowed
Mineral oil	§205.603(a)	9-18-02	Allowed
Peracetic acid	§205.603(a)	11-00	Allowed for facility and processing equipment sanitation
Pheromones	§205.603	10-20-02	Allowed
Poloxalene	§205.603(a)	3-01	Allowed for emergency treatment of bloat
Potassium sorbate	§205.603(a),(b),(d)	3-01	Allowed only for use in aloe vera products for livestock production
Propylene glycol	§205.603(a)	9-19-02	Allowed only for the treatment of acute ketosis
Tolazine	§205.603(a)	9-19-02	Allowed only as antidote to xylazine
Xylazine	§205.603(a)	9-19-02	Allowed for emergency use only

Chapter 7
Real Life Situations, Natural Treatments & Farmers

Working clinically as a veterinary practitioner with natural treatments in dairy cows can be highly rewarding but also quite challenging. The constancy of milking cows twice daily along with the economic goals of dairy farming stimulates the farmer to continually assess whether or not a given treatment is working. In simplest terms, if a cow is resuming milk production or beginning to eat again after a certain treatment is given, then the farmer is usually satisfied that the treatment has been effective. An effective treatment in the mind of a dairy farmer does not necessarily equate with what the scientific method in medical research would call "effective" with a strong bias to studies that are statistically significant. In addition, most farmers are not concerned about the scientific mechanism of action that a given treatment exerts — they just want the treatment to work in their barn. If it doesn't work, they will not buy it any further and go on to another product. On the other hand, farmers will keep buying the products that work on their farms. This is common sense and the economics of reality. In addition, the farmer is definitely happier if milk does not have to be withheld from the milk tank on account of residues from a specific treatment type. After all, a main goal of dairy farming is the production of high quality milk without relying on treatments. Taking this idea to an extreme is the realm of organic dairy farming. As already mentioned (and worthy of continual repeating), if a dairy cow is treated with a prohibited material while on a USDA Certified Organic farm, the cow must be immediately removed from certified organic production forever. There is a negative incentive on organic dairy farms to use conventionally accepted methods of treating common problems such as mastitis, pneumonia, uterine infections and ovarian dysfunction. Prevention becomes critical.

In ecological and organic farming, there is strong emphasis that if everything is biologically healthy in the soil, then the crops harvested will supply the correct nutrition to the cows to maintain a superior immune system. The thinking continues that the cows will remain vigorous and will of course repel attack from infectious organisms (viruses, bacteria, protozoa, fungi, etc.). This philosophy is honorable yet a bit too optimistic and too easily plants false expectations into the minds of gullible people. Granted, if the soil is in bad shape, the crops that go to feed the cows will not be as healthful. However, it does *not* necessarily follow that if the soil is healthy then the cows will automatically be healthy. There are just too many steps between the life in the soil and what ends up going into the cow's mouth as fed by the farmer. The closest philosophy that is defensible is that dairy cows which actively graze are in general healthier than dairy cattle permanently confined in pens indoors and fed silages and concentrates. The study I conducted in Holland (previously cited) confirms this.

Even in grazing herds, which truly seem to need fewer visits by the vet for sick cows, there is always the real possibility that a cow will come down with pasture bloat, lameness, mastitis or some peri-parturient problem. There may be less incidence of illness on organic or grazing farms, *but there will still be illness* — and a variety of it as well (Karreman, "Snapshot"). The potential risk of illness occurring in an animal is a fact of life for those holding livestock. Fortunately for dairy farmers, problems can be recognized early due to the never- ending cycle of milking cows twice daily. Timely intervention is important no matter what mode of treatment is used, but it is perhaps more critical for those relying on natural treatments. This is because many natural treatments do seem to rely on stimulating the animal's vitality in order to overcome problems on their own. Perhaps it is good to point out the basic assumption often implicitly accepted by those preferring natural treatments: there exists an inherent vital force in every living creation. The vital force is not quantifiable but is somewhat observable in the form of each animal's unique interaction and responses to its immediate environment. Under high-density factory farming confinement conditions, such unique character traits are essentially suppressed, and the idea of recognizing an animal's individual "vitality" is nearly obliterated. The vitality of an animal is somewhat akin to its biological constitution, but it also includes unique behaviors in character that sets it apart from all other animals around it.

Some of the following questions may help to recognize different kinds of vitality in animals — but it is all a variation on the theme of all animals having a unique vital force. And it should be noted that nearly all complementary and alternative veterinary medicine (CAVM) modalities take into consideration factors beyond what is normally noted during a standard physical exam. For instance, how does the cow interact with her herd mates, other kinds of animals (dogs, horses, chickens, etc.) or strangers in the barn? Is she peaceful, or is she nervous? Is she curious but shy? Is she mean? How does she react to being examined? How does she react to changes in feed — does she immediately show digestive disturbance; does she prefer certain kinds of feed; does she always have a great appetite regardless of feed quality? How does she do in the heat of summer or with various weather changes? All these types of reactions of a cow to her immediate environment can give an observant caretaker insight into his/her animals' natural tendencies and inherent vitality. (These points are also used in formulating a cow's correct constitutional homeopathic remedy, see Chapter 12). Some animals are hardier than others, given equal surroundings. Hardiness may be due to a cow's breed, genetics, conformation, nutrition or pre-disposed constitution and may or may not fit a particular farm's surroundings. And, even though they are large mammals, some cows are actually quite delicate when it comes to their constitution or vitality. They may be excellent milk animals, but they are easily thrown off balance with slight changes in the daily routine. In the end, each animal has quite a unique character if a person is willing to take the time to learn about the individual animals which make up the entire herd.

Some treatment philosophies emphasize the character of the entire herd rather than the individual animals. Granted, most herbivores are herd-oriented animals, as opposed to more independent animals such as dogs and cats. And indeed there are many times when adjustments or "treatments" are administered to the herd as a whole — feed ration changes, housing changes, vaccination, ventilation, and stall changes, etc. Keep in mind that these treatments are not looking at individual characteristics or even the herd's basic character; they are simply being applied to the whole herd regardless of any kind of unique aspects of individual animals that may actually occur within that particular herd. By combining knowledge of the individual animals and the herd as a whole, experience often guides the best farmers into taking appropriate measures to enhance the well-being of his/her farm animals. This perhaps is how holistic treatments for dairy cows differ from holistic treatments for pets. However, holistic treatments for livestock can be similar to that for pets if taking individualized symptoms into account, bearing in mind, however, that the environment impacts livestock much more since they are directly bound to it. Where holistic treatment differs considerably from pets is the amount of time given for an animal to respond. The economics involved with dairy farming often dictates "quick fixes" for animals in terms of individual treatment. Yet "treatment" may also include perhaps slower, more basic changes in the farm's environment in terms of water and air quality as well as manure management in terms of land health and animal housing. These are factors which cannot always show direct cause and effect but nonetheless affect the farm's overall health. Even when the soil is healthy and everything is humming along, there'll still be times when individual animals become sick, be it due to a weakening of their vital force, a hot challenge posed by factors in the environment, or both. Unforeseen accidents are always a possibility as well. The way the individual farmer handles a situation often determines if an animal regains health quickly enough to remain a productive member of the herd.

Applying CAVM to real life situations

While many different modes of natural treatments exist, only a few modes are in general use among those livestock veterinarians who use CAVM with dairy cows. The description of these few modes as applied to dairy practice is due in part to the author's own experiences as well as being the easiest to administer. By presenting only a few and explaining their basis, my professional veterinary peers will hopefully better understand how some alternatives can fit into dairy practice. In addition, individuals learning from scratch may find it easier to start with only a couple alternative modes rather than the many that actually do exist.

It should be pointed out that most modes of treatment come with a basic set of principles which guide the use and understanding of the specific treatment regime. Most also come with very intellectual philosophies that may also spill over into the everyday living of a purist practitioner dedicated to a specific type of thinking. Though the strictest adherents of each modality usually cringe when therapeutic recipes are given in a "cookbook" format, it is exactly the cookbook approach that can help others to try out new therapies. Then, once gaining some confidence through small victories in the real world, most students of natural medicine will be stimulated to inquire further and delve into particular philosophies and fine tune new avenues of therapy to their liking. It is exactly from a cookbook approach that I was initially exposed to homeopathic natural treatments. But due to careful examination of available literature on the rationale for homeopathy and plant medicines, medicinal plants have now become my focus for ongoing study. It also seems that medicinal plants lend themselves well to recipe approaches. Actually, when using medicinal plants in the context of the historical use of them by veterinarians, there are no eccentric or "odd" intellectual philosophies to absorb and learn other than relying on known physiologic responses exerted by the various active constituents in the plants. Each practitioner of complementary and alternative veterinary medicine tends to find a realm of therapies that intuitively resonate with that person's approach to cases seen clinically. For example, due to my private study of homeopathy — how its remedies are made from various source materials (plants, minerals, insects, animal poisons and disease discharges) and their indications for use — I really enjoyed studying immunology during veterinary school as it pertains to animals responding to low doses of antigenic and therapeutic agents. Some modes of treatment may have more scientific basis than others, but all are only as good as the practitioner.

Most of the demand for a full spectrum of CAVM therapeutic approaches is within the small animal and equine realm where cost of treatment is not as constrained as with the tied-to-economic-performance world of dairy cows. Either a cow "pays her way," or she is replaced. There is simply not as much time allowed for various treatments to work (including conventional therapies). However, there are usually some select cows that receive extraordinary treatment, but that often depends on previous performance, her potential and also her general disposition.

Another reason that I only use a few modalities is that in a full service dairy practice there is also the need to allocate time to conventional interventions — particularly for rapidly correcting metabolic imbalances with fluids as well as time for surgeries, obstetrics, life-threatening emergencies (where conventional medications may be a necessity) and hoof work/ lameness. The nicest aspect of dairy practice, in my opinion, is that medical problems are much more common than surgery and, as such, anyone interested in alternative medicines will have lots of opportunities with dairy cows. Also in support of using natural medicines in cows is the reality that farmers despise dumping milk down the drain due to the withholding times of antibiotics.

When giving consideration to treatment approaches, we need to keep in mind *homeostasis* — the tendency for an organism to regain normal equilibrium, which reflects the animal's inherent vitality. Animals will of course always try to regain functional equilibrium as an outcome to any insult against their system. Their inherent constitution will dictate whether or not they succumb to a given insult in the first place. And likewise, their ability for resilience will dictate whether or not they overcome the insult and survive the likely course of disease *if left unchecked*. This being said, it is necessary to know the normal course of a disease. Then, by clinical experience, it is possible to sort out an animal which would normally have self-cured from one which was cured due to intervention. This is a cardinal point to remember, especially before ascribing outright success with any particular therapy. (Of course innate resilience aided by intervention is often the main reason for an animal regaining homeostasis quickly.) In dairy cattle, the question of an animal merely surviving versus an animal which resumes milk production is paramount. Thus the reason was given earlier that perhaps the effects of interventions can be gauged by resumption of milk production or achieving conception if problematic infertility previously existed. It is, of course, always uplifting to the farmer or the veterinarian when an intervention appears to have rapidly improved a specific animal. Whether or not a given intervention was the prime cause for improvement is in the realm of statistics, so central to agricultural veterinary research. Although practitioners are taught statistics during veterinary training, the art of practice is such that an accumulation of clinical responses by the private practitioner often ends up guiding decision making "out in the trenches." Certainly one of the outcomes of being medically trained is to formulate a differential diagnosis (a list of possible "rule-outs") by appropriate physical examination and history taking, then implement a treatment strategy for the most likely pathologic condition. This must include the most appropriate and potentially effective options, which must also take into account the farmer's abilities for any follow-up treatments as well as economic ramifications. The essence of being a good livestock veterinarian is to explain the problem clearly and to give the owner options from which to choose, thereby fully including him/her in the decision making process. To do so, the doctor should be willing to teach the farmer about a certain disease's progression and explain pertinent reasoning for various therapeutic options. It is not uncommon, however, for a farmer to have jumped to a conclusion about the case prior to the veterinarian's participation in the case. This is when a medically trained professional can point out other possibilities regarding the case, while giving due credit to the farmer for having thought about the situation and trying to rectify it. This builds rapport and trust between the farmer and the veterinarian and instills in the farmer a sense of confidence as to the analytical decision-making capabilities of the veterinarian. It is extremely important that the veterinarian *not* give the farmer false hope or expectations but state the likely outcomes of a disease under the various treatment options. The best way to guard against giving false hope is for the vet to examine the animal in order to pick the right treatment for the case at hand and be available for direct follow-up. This establishes a valid "Veterinary Client Patient Relationship" (VCPR), which gives the veterinarian certain legal rights, as well as responsibilities, in regards to the case at hand. As an example of how a VCPR works, a veterinarian can prescribe a medication for a cow that is technically only labeled for horses (or humans) if the veterinarian has personally examined the cow and there are no medications specifically labeled for cows with that condition. However, the veterinarian must also be available for direct follow-up on the case, to check for progress or regress of the condition for which the drug was prescribed. For a practitioner new to CAVM, direct follow-up visits will also help to see exactly which alternative treatments work best in given cases — and which do not. Successes are wonderful to talk about, but perhaps our failures teach us the most in the long run. This does not imply that we try one specific medicine at a time and see which works and which doesn't. Actually, using a multi-prong, eclectic approach is probably the best way to approach problems in dairy cows. This will probably increase the chance of success, especially considering that there may only be one chance given for the vet to intervene on a given case (due to economics). In the living laboratory (the barn), a keen observational mind combined with a medicine box filled with an array of natural and conventional treatments can lead to a rewarding and fulfilling life in healing dairy cows. There will be many successes while relying on natural remedies if one rationally takes a holistic view of commonly encountered problems.

The kinds of farmers who use CAVM

Let us consider some of the kinds of farmers who use natural treatments. Farmers range from automatic acceptance of natural medicines to being extremely skeptical of them. This may or may not depend on education level. The acceptance or interest into scientific explanations of medicines is of interest to many people, whereas the rejection of them can be due to innocent ignorance or deliberate defiance. If simple testimonials are all that are needed to sway an individual farmer, then that farmer is usually open to try out all kinds of natural treatments. Having even more bearing than testimonials is the farmer's own upbringing and whether or not natural medicines were part of growing up. On the other hand, conscious rejection of conventional medicines is usually due to bad personal experiences with them, rebelling against the powerful medical establishment or rejection of science-based knowledge out of hand.

Looking at the most rebellious kind of farmer (against all conventional medicines) would be a farmer that has basically an empty medicine cabinet and relies on "survival of the fittest." Granted, agricultural livestock production always seeks to progress through genetic improvement, but this kind of farmer does not seem to mind too terribly if animals die, for their genes will be eliminated from the gene pool. This kind of farmer tends to be highly suspicious of most consultants that espouse mainstream agricultural views, including those of a long-term, trusted herd veterinarian. (Remember that few large animal vets are involved with alternative philosophies and treatments.) Usually it takes a severe disease outbreak before a rational approach to correct the problem is carried out. This kind of farmer usually doesn't even have many natural treatments on hand since treatments of any kind just don't fit into the farmer's management framework. If anything, there may be one or two trusted products that the farmer thinks are effective for any situation. Of course no farmer relishes the idea of doing extra labor, but this particular kind of farmer almost abhors doing anything that means taking the extra little bit of time to medicate individual cows. Among all kinds of farmers, these farmers are in the distinct minority.

Another kind of farmer, most likely certified organic, is one who likes intervention and treats animals by any and all available natural treatments. This kind of farmer really believes in his/her heart that it is just better to use natural treatments. They fully realize that antibiotics and other conventional treatments are available and can be useful, but they strongly prefer treating with anything but conventional medicines ("Whatever it takes, Doc . . . but please no antibiotics or hormones.") It is not a rebellion against conventional medications, it is just a deeply held belief that alternative remedies can do as well as conventional ones. When standing in a barn with this kind of farmer, it is an eye-opener to see such an array of natural remedies in the medicine cabinet or to listen to what the latest, greatest remedy is. Advertisements from traveling sales people will be presented in hopeful anticipation of a glowing comment by me as their "open-minded vet." If regularly visiting a farmer like this, more often than not there will be half used containers of various remedies gathering dust since s/he gave up on it and switched to a newer remedy that a sales person foisted upon them more recently. Sadly, purveyors of truly worthless junk prey upon this kind of farmer. However, over time, one will notice a renewed supply of certain remedies which the farmer has found to be consistently useful on his/her particular farm. It is these particular remedies which can pique the curiosity of a herd veterinarian, especially when direct follow-up observations are carried out and consistent results are evident.

A third kind of farmer is one who somewhat begrudgingly uses alternative veterinary medicine. This is usually only seen on farms that recently have shifted to certified organic methods for more purely economic reasons than deeply held philosophical beliefs. This kind of farmer really likes conventional treatments (*i.e.,* antibiotics and hormones) but simply is not allowed to use them since they are prohibited for use on organic farms. Use of natural treatments is generally taken on very slowly, with much more confidence placed on material amounts of natural substances such as plant compounds and biological derivatives (colostrum-whey products) than energy medicines such as homeopathic remedies or acupuncture. Often times there will be admission on the part of the farmer that antibiotics and hormones don't always give the expected results and therefore, in the process of learning, it shouldn't be held against natural treatments if they may not work also. When this kind of farmer first uses a natural treatment on a case that I prescribe for, and then sees it work, it is especially gratifying. For then the farmer has a small victory from which to continue building. These farmers may remain skeptical about natural treatments for quite a long time. However, as they see more and more positive results, they become more at ease with them. Usually they will not seek out literature on their own to read but will rely on discussions with me as their veterinarian, since veterinarians are generally respected for discussing problems from a neutral point of view and in confidence. But it is essentially only a string of successes that will change this kind of farmer's attitude.

A fourth type of farmer is one that likes to use both conventional and natural treatments, and chooses interchangeably among them as each case demands. As a veterinarian, these farmers are the easiest with which to work. They tend to have great confidence in the veterinarian to make the best choice for their cow. When a natural treatment is discussed for a particular case, this farmer will on occasion recall when s/he saw a parent or other respected person use that same kind of treatment. Often, after all options for treatment will have been honestly presented (i.e., discussing positives and negatives of each), the farmer will rely on intuition and decide which route to go, at least for the time being — especially knowing that I am always available for further direct follow-up and possible antibiotic usage, if needed. These farmers are usually the best communicators in a "give and take" sense. They are open to hear about a variety of options and are grateful for the options presented. They tend not to become certified organic, at least with the animals, since they want to be free to use all possible options at all times without the certified organic Rule hanging over their heads. However, these farmers seem to always be interested in how my organic herds make out in terms of health care. Occasionally, one will surprise me and say that the herd is going to "go organic."

Of course there are more kinds of farmers than the four just described. There are certainly different shades of the four presented as well. These examples are simply illustrations of the kinds of farmers that I work with in my daily practice. Regardless of the point of view of the farmer, any natural treatment success is in large part based on accurate and timely observations of specific symptoms and knowing how to use certain modes of treatment. For a veterinarian, this means conducting a thorough physical exam — using conventional with perhaps non-conventional diagnostics. Just as important is to obtain a good history of the cow's problem and what the farmer has done before calling in the vet. Rushing through an examination will lead to less success, no matter what mode of treatment but especially when using natural treatments. My average time on a farm is about an hour, when looking at all kinds of farm calls with which I am involved — therefore I don't do more than 6-7 farm calls a day on average which also includes driving between farms. It cannot be stressed enough that there is absolutely no substitute for personally examining an animal. In addition, the physical exam will indicate whether or not a cow needs immediate help in the form of intravenous fluids for rehydration or perhaps pumping her stomach with a nutritional slurry mix to stimulate her rumen. Intravenous fluids and pumping stomachs are mainstream supportive therapies which farmers readily allow and should be carried out anytime needed. This also allows for any natural treatment to work better since the cow will be returning to adequate hydration and therefore improved general circulation. Tending to a cow's immediate circulatory needs by stabilizing her metabolically will allow natural treatments to be better utilized by the animal and give her a better potential outcome without resorting to antibiotics, steroids or hormones. Of course knowing when to call in for professional help before it is too late goes without saying.

PART 2
STRATEGIES

Chapter 8

Clinical Practice Data & DHIA Data

I'll begin the "Strategies" section of this book with a chapter focused on numbers — objective data that reflect the production and reproduction of herds with which I work. The presented data will hopefully lend credence to the idea that the natural treatments I use on these farms do exert effects. Some folks, especially skeptics of any paradigm that is "outside the box," are always proclaiming, "show us the data" — well here it is. I have been fortunate that many of my USDA Certified Organic dairy farmers have allowed me to access their Dairy Herd Improvement Association (DHIA) data. DHIA is objective and cannot be disputed. Folks may or may not like what they see. The comparison study which was carried out in conjunction with Penn State University Extension is especially revealing of the strengths and weaknesses of organic and conventional farms in Lancaster County, Pennsylvania. I will present that DHIA data after I give you a glimpse of my own practice data which shows the everyday work I do on certified organic farms. The data from my practice reveal the frequency of procedures carried out and medicines used with my certified organic farmers. Since I am a solo practitioner, the numbers reflect my own work and not of any other veterinarian. Some folks think that there will come a point in time on an organic dairy when the farmers can "get beyond medicine, vaccines, and healthcare problems." This is usually stated by the same folks that proclaim that if the soil is healthy then the animals will automatically be healthy. Sorry to be the wet blanket on that philosophy, but the data does not bear out such thoughts. Perhaps the number of visits to an organic farm will be fewer than to conventional farms, but they will still occur.

In the study, "Snapshot of a Holistic Dairy Practice" (Karreman), I show my total procedures and medicine use as they separated out between conventional, transitional and organic farms for a two-year period, 12/99-12/01. I have been incorporating natural treatments into practice since graduation from veterinary school in 1995, but it is only since 1999 that I began my own practice. I have included part of that study's comparison here (see Table 1 and Table 2) to show the frequency and proportion of work done on different farm types (organic, transitional, conventional) to give the reader data from which to draw their own conclusions. The total of 102 farms, as seen in Table 1, is from many conventional farms in that time period. The transition farms seen in Table 1 were in first or second year transition, and as can be seen in Table 3, the farms previously shown as transition have since become certified organic. There are yet more in transition at the time of this writing. From January 2002 to the present, I have only been keeping track of work on organic farms, which combined with the previous study's findings, will give a time span to better see consistent patterns of work emerging on such farms. I plan to update this set of data annually for potential research purposes over the long run.

Procedures that either implicitly or explicitly include complementary and alternative veterinary medicine (CAVM) in my practice are treatments for reproductive problems, mastitis, pneumonia, fever and lameness, to name just a few items. Standard procedures reviewed (procedures common to all dairy veterinarians) include physical exams, pregnancy confirmation, infertility diagnosis, obstetric manipulations, metabolic corrections, lameness diagnosis and dispensing medicines. The obvious majority of surgeries are for correction of displaced abomasum with occasional suturing of teat lacerations, milk vein lacerations or other hemorrhages and occasionally a toe amputation. Other surgical needs are rare in my all dairy practice. Surgical procedures obviously need sedation and analgesia, and the sedatives and analgesics are now endorsed by the USDA NOSB. Conventional medications potentially administered include fluid electrolytes to treat metabolic disturbances and dehydration, anti-inflammatory/pain relief products, antidotes, anticoagulants, hormones, antibiotics and steroids.

It should be noted that in tabulating my practice data, certain areas were not monitored prior to January 2002. This is reflected in the table by dashed lines in those columns. Also, many times when I am administering therapy to a cow, especially IV, there will be a series of different fluids and medicines being given. In the Medications table, the column labeled Administered "A" would mean that every medicine administered during the IV was allowable for certified organic use. Whereas the Administered "P" signifies that at least one of the medicines administered was prohibited for certified organic use (even though all the rest given at the time may have been "allowed" for certified organic use). For both the Administered and Dispensed columns (whether "A" or "P"), the exact quantities of medicine are not shown *i.e.,* not the total volume of fluids or medicines — in other words, simply the act of administering medicine was tallied. The same holds for the Dispensed columns. However, every time a homeopathic remedy or botanical compound was dispensed, it is shown, since this is a unique part of my practice. In other words, there were 2418 one ounce size bottles of homeopathic remedies dispensed to organic farmers in the time period shown. The botanicals are more varied (tinctures of different amounts, ointments, tablets, etc.) so that each and every botanical, no matter what exact product, tallied a score when tabulating information from my farmers' bills. This should give readers an insight as to what is needed to be done on organic farms and how often.

I provide complete herd health care (reproduction, medicine, emergencies, obstetrics, surgery, diagnostics, etc.) for about 95 dairy farms. The numbers shown in Table 3 are only for my organic farms. The reason I have elected to do this is because original data pertaining to procedures carried out on organic farms has not been shown previously in the literature, to the best of my knowledge. The time covered is almost 4 years, a fair amount of time from which the reader can draw reasonable conclusions about veterinary work on organic farms.

From looking at the data, it can be asked how is it that any prohibited materials (antibiotics, hormones, steroids, etc.) are shown to be used on organic farms? Keep in mind that the numbers presented are from a snapshot in time. Some of the farms that are now certified organic may not have been previously (they were in transition). Thus, certain treatments that may be prohibited on certified organic farms may have been recorded in my file for a farmer prior to becoming certified. In addition, the rules regarding the conventional treatment of young stock (using prohibited materials in an emergency) are still being developing by the National Organic Program. At this time, a transitioning farmer's file might reveal that young calves suffering from severe enzootic calf pneumonia may have been treated with a common antibiotic (*i.e.,* tilmicosin) in order to save them. It should be pointed out, however, that most transitional and organic farmers will use botanical, biologic and/or homeopathic remedies at the first signs of respiratory involvement in order to get used to working with CAVM treatments. (See case study in Chapter 14 regarding natural treatment of bacterial pneumonia.)

It can be seen that organic farms certainly do have surgeries performed. In no cases did I follow with an antibiotic. However, the animals were of course given sedation and analgesia, as well as a reversal and occasionally flunixin (pharmacologically related to aspirin) or butorphanol (synthetic morphine) immediately post-op as a pain reliever.

In the first two years shown for lameness, there was a relatively high amount of prohibited treatment that was used (usually powdered oxytetracycline applied topically). This is in part due to the high prevalence of the very painful condition called interdigital dermatitis (aka "hairy heel," "hairy wart," "strawberry heel"), which is highly responsive to such therapy. However, over time, there have been some new botanical methods of treating this condition which I have incorporated (see Hoof Problems in the treatment section). Thus, closer analysis of a certified organic farmer's individual files would probably show my use of topical oxytetracycline rapidly declined to zero.

As for herd data, Table 4 and Table 5 are perhaps the most revealing, for they compare traditional farms to certified organic farms in Lancaster County. Although the power of the study may not be as strong as some statisticians would argue (Student's T-test was used and p<.05 was considered significant), it is perhaps one of the first comparison studies carried out between traditional and certified organic farms using purely objective data. By only comparing farms within Lancaster county, differences in soil type, climate, animal husbandry practices and agronomy practices were minimized. Although the "treatment group" (the organic farms) was small (n=16), at least there were valid statistical measures applied to the data. It should be noted that 15 of the 16 (nearly 94%) certified organic farms used artificial insemination. It is highly doubtful that 94% of the 1059 conventional farms used artifical insemination.

It is noteworthy to point out that there were only three significant differences between the two types of farms. They were 1) rolling herd average, 2) SCC score for 41-100 days in milk and 3) the cull rate. This means that *all other parameters were statistically equivalent.* That is actually an amazing finding since (as the reader should be well aware by now) no prohibited materials can ever be used on dairy cows to remain in the herd. Therefore, all DHIA data is only from cows that have had either no treatments ever, or, have had only natural or allowed treatments. To have equivalent results in all reproductive parameters is especially interesting when considering that the majority, if not nearly all, of the traditional farms most likely rely on routine use of hormonal injections as integral parts of their reproductive programs while the organic farms use none. The equivalency between these farm types will perhaps pique the curiosity of readers into carefully reading and using the treatments described later which help to account for these reproduction and somatic cell count figures. On a point closely leading into the treatment modalities and compounds — having data like this supports the idea that we need to look at objective numbers in order to make the case that natural treatments are working out in the barns, rather than by basing conclusions on wishful thinking and conjecture. Having this data may hopefully make some skeptics of natural treatments pause and realize that these results can't all be due to placebo effects. Although indirect, these data argue in favor of natural treatments working in general. Although it is not possible to say from the data which exact treatments directly affected individual cows, it is possible to see that an association is potentially present.

The data in Table 6 were downloaded from the DRMS website (www.drms. org). These are the 25 farms which do monthly herd testing, their files are uploaded on test day by the technician when finished at the farm and are immediately available to consultants for evaluation. As can be seen, there are many more parameters available with these farms to compare within the group than parameters to compare within the group of 31 in Table 7. The 16 farms in the comparison study are within this set of 25 farms, and, had the comparison study been done at the time of this writing, it would have included 25 farms instead of just 16.

The data in Table 7 show a combination of certified organic farms — 25 of which are enrolled in standard DHIA testing, processed through DRMS in Raleigh, NC, and 6 more which are on the Lancaster DHIA "Basic" testing, which do not have their data processed through DRMS. However, all 31 farms have their milk analyzed in the laboratory at Lancaster DHIA in Manheim, PA. This gives 31 farms within Pennsylvania from which to glean objective information. The remainder of the organic farms are not on any kind of monthly testing program. The "Basic" test farms are only issued a DHIA "TDU 143" for average herd parameters, but only for the specific testing month. Farms using the "Basic" testing service do not automatically have their data accumulated through the months as do farms going through the DRMS in Raleigh. The finding from the 31 farms that the average age of cows is just over 4 years old (meaning cows, on average, live only into their third lactation) — is surprisingly low. However, the average age for cows in their third lactation or higher is 69 months (about 5 ¾ years old). Within the 3+ lactation age data, there is a range between 82 months (nearly 7 year old average) as the highest average age on one farm within this class to just 61 months (just over 5 years old average) as a low herd average on another farm for this class. This is also still somewhat surprisingly lower than I would have predicted for the older cow class.

It is hoped that other regions develop benchmarks using DHIA data so that the organic dairy sector becomes well described and people can speak from an objective perspective when discussing certified organic dairy farming with other agricultural professions.

TABLE 1: Practice Activity Among Different Farm Management Types (12/99-12/01)

PROCEDURES

Compilation:						Repro Tx				Lameness			
Client Type	Reg.	ES	PE	Repro	A	P	OB	AP	A	P	DA	HP
Organic	35	615	147	322	3071	260	7	51	41	78	26	24	1154
Trans.	19	322	54	99	2019	43	23	18	23	33	8	15	630
Conv.	49	1000	174	681	4696	170	128	70	39	79	53	84	427
Total	103	1937	375	1202	9786	473	158	139	103	190	87	123	2211

Compilation:						Repro Tx				Lameness			
Client Type	Reg.	ES	PE	Repro	A	P	OB	AP	A	P	DA	HP
Organic	35	31.8	39.2	26.8	31.4	55.0	4.4	36.7	39.8	41.0	29.9	19.5	52.2
Trans.	18	16.6	14.4	16.6	20.6	9.1	14.6	13.0	22.3	17.4	9.2	12.2	22.3
Conv.	47	51.6	46.4	56.5	48.0	35.9	81.0	50.3	37.9	41.6	60.9	68.3	19.3
Total	100%	100	100	100	100	100	100	100	100	100	100	100	100

TABLE 2: Practice Activity on USDA Certified Organic Farms (2002-2005)

Year	Farms	Reg.	ES	OB	Repro	A	P	Physical Exam	Meds Admin A	Meds Admin P
2002	36	290	70	24	1447	96	0	134	92	20
2003	53	421	103	57	2366	139	0	246	176	7
2004	63	513	62	40	2759	137	0	266	177	24
2005	76	647	59	44	3686	109	1	389	141	61
Average	57	467.75	73.5	41.25	2564.5	120.25	0.25	258.75	146.5	28

Year	Lame A	Lame P	Disp. A	HP	Botanical	Biologic	Disp. P	DA	Sx other	AP	Lab work
2002	44	7	170	780	113		29	9	6	4	19
2003	87	0	217	605	150		11	13	18	14	34
2004	52	0	212	840	344	390	38	13	5	8	680
2005	63	1	406	481	327	446	72	26	25	8	164
Average	61.5	2	251.25	676.5	233.5	418	37.5	15.25	13.5	8.5	224.25

> **Key (Table 1 & 2): Reg.** = regular farm call; **ES** = emergency or late call; **PE** = physical exam; **Repro** = pregnancy exam; **Repro A** = Allowed treatment for Certified Organic; **Repro P** = oxytetracycline pills or other antibiotic; **OB** = obstetrics/dystocia, direct vaginal examination; **AP** = acupuncture; **Lameness A** = Allowed treatment for Certified Organic; **Lameness P** = Prohibited (antibiotic); **Surgery DA** = displaced abomasum (left or right); **Surgery other** = Caslick's, C-section, lacerations, teat, etc. **HP** = homeopathic remedy dispensed.

TABLE 3: Medicines Used on USDA Certified Organic Farms (12/99-9/03)

MEDICATIONS

Total medications used:

Year	Farms	Administered A	P	Dispensed A	P	Dispensed Homeopathics	Dispensed Botanicals	Laboratory Diagnostics
12/99-12/01	35	274	56	--	--	1154	--	--
1/02-12/02	36	92	20	170	29	780	113	19
1/03-9/03	53	141	7	174	9	484	120	127
12/99-9/03	124	507	83	344	38	2418	233	146

> **Key (Table 3): Administered A** = Allowed medicines for C/O given IV, IM, SQ or PO. **Administered P** = Prohibited medicines for C/O given IV, IM, SQ or PO; **Dispensed A** = Allowed medicines for C/O dispensed for follow-up treatment; **Dispensed P** = Prohibited medicines for C/O dispensed for follow-up treatment; **Homeopathic remedies** dispensed as 1 oz. bottles with #35 pellet; **Botanicals** dispensed as 6 or 8 oz. tincture bottles to be given orally or botanical ointments or botanical nutraceuticals; **Laboratory diagnostics:** fecal flotation for parasites, blood serology, abortion work-ups, necropsy.

DHIA STUDY
A Comparison of Conventional and Certified Organic Herds in Lancaster County, PA

1059 Conventional Farms vs. 16 Certified Organic Farms

TABLE 4: Production
(September 2002)

Parameter	Conventional	Organic
Total Cows	49	44
Rolling Herd Average **	**21,520**	**17,876**
Fat %	3.68	3.68
Protein%	2.97	2.94
Actual SCC	330,000	391,000
SCC <40 Days in milk	3.18	3.12
SCC 41-100 Days in milk **	**2.89**	**3.69**
SCC > 300 Days in milk	3.79	4.26
SCC 1st lactation	2.71	2.76
SCC 2nd lactation	3.03	3.12
SCC 3+ lactations	3.67	3.94
% Herd SCC < 3	57	53
% Herd >6	9	7

** $p<.05$ (using T-test), showing statistically significant differences between conventional and organic farms

TABLE 5: Reproduction
(September 2002)

Parameter	Conventional	Organic
Average age at 1st calving	25.5	25.0
Calving Interval (months)	14.2	13.9
Days to 1st service	95	92
Days open	153	144
Services per pregnancy	2.7	2.7
Cull rate**	**31**	**24**

** $p<.05$ (using T-test), showing statistically significant differences between conventional and organic farms

(Data was downloaded for individual cows from DRMS in Raleigh, NC and statistical analyses were run by Penn State University Extension Agents Beth Grove and Dr. Gregory Martin, Lancaster County)

TABLE 6: Dairymetrics Data for 25 Certified Organic Dairy Herds

Dairy Records Management Systems, Raleigh, NC (June 2003)

ITEM	AVG. OF HERDS	STD. DEV.	LOWEST HERD	HIGHEST HERD
PRODUCTION				
Number of cows	46	14	28	84
Number 1st lactation cows	14	5	5	30
Number of 2nd lactation cows	10	3	3	20
Number of 3rd lactation herds	21	8	10	49
Days in milk	186	35	77	241
Age of 1st lactation cows	24	1	23	27
Rolling Herd Average	16461	2425	12320	21586
Rolling Protein	490	66	387	626
Daily % Protein	2.9	0.1	2.7	3.1
Rolling Fat	616	81	482	777
Daily % Fat	3.6	0.2	3.1	4.3
Summit milk 1st lactation	55	9	39	73
Summit milk 2nd lactation	70	10	54	92
Summit milk 3rd lactation	78	10	58	97
Projected 305 day ME milk	18216	2728	14372	23536
Standard 150 day milk	54.3	8.1	40.6	72.6
REPRODUCTION				
Projected calving interval	14	0	12	14
Days open-projected min–all	146	21	85	173
Days open-projected min–1st lact.	142	31	77	205
Days open-projected min–2nd lact.	144	38	71	247
Days open-proj. min–3rd+ lact.	150	25	92	196
Days to 1st service (%herd<VWP)	20	13	4	54
Days to 1st serv. (%herd to 100D)	45	12	19	69
Days to 1st service (%herd>100D)	33	15	2	63
Conception Rate–1st service	49	13	26	87
Conception Rate–2nd service	47	19	22	100
Conception Rate–3rd+ service	52	20	22	100
Number of calvings in past year	40	17	6	90
UDDER HEALTH				
SCC Actual	420	216	158	1040
SCC Score	3.5	0.5	2.6	4.7
%cows (SCCS of 0-3)	49	8	27	64
%cows (<41D with SCCS>4)	36	36	0	100
% 1st lact. (SCCS of 0-3)	59	21	0	83
% 2nd lact. (SCCS of 0-3)	50	14	20	77
% 3rd lact. (SCCS of 0-3)	40	14	14	67
SCC score for 1st lact.	3	1	1.5	5.9
SCC score for 2nd lact.	3.3	0.6	2.2	4.7
SCC score for 3rd lact.	3.9	0.7	2.9	5.4
SCC score for cows 41-100 DIM	2.8	1.1	1.2	5.9
SCC score for 101-199 DIM	3.2	0.8	1.8	4.8
SCC score for 200-305 DIM	3.9	0.8	2.5	5.5
SCC score for 306+ DIM	4.1	1	1	9

TABLE 7: DHIA Test Data for 31 individual Certified Organic Dairy Farms
June 2003 PCDart General Herd Summary (TDU 143)[a]

Farm	Cows	Milk/ cow (lbs.)	RHA	Actual SCC x 1000	Age (all)	Age (3+ lact)	Peak milk (all)	Days to 1st service	Services per preg.	Calving Interval	Min. Days Open
1	44	43.9	13,304	316	48	65	57	Bull			
2[b]	41	57.3	14,618	201	52	66	73	Bull			
3	35	66.8	21,199	216	53	66	92	93	1.4	13.0	114
4	38	55.9	15,502	712	54	74	78	100	1.8	14.2	151
5	45	47.8	15,878	262	43	63	70	Bull			
6	31	52.0	16,954	158	53	80	77	92	1.8	13.7	135
7	30	41.3	15,042	411	50	71	64	Bull			
8	43	57.0	18,622	183	45	63	80	Bull			
9	68	48.9	16,878	359	45	64	70	Bull			
10	65	55.3	18,644	540	56	79	82	121	2.0	14.6	164
11	80	53.4	17,995	325	42	62	80	77	2.3	13.7	136
12	21	55.4	16,129	119	53	77	81	Bull			
13	84	52.4	12,319	309	51	67	62	Bull			
14[b]	29	54.7	(c)	784	68	78	70	Bull			
15	37	48.3	15,231	221	45	63	68	Bull			
16	44	45.3	15,976	211	51	76	72	Bull			
17	42	54.2	17,655	297	53	65	77	Bull			
18	49	50.8	17,732	480	47	65	76	111	2.0	14.4	157
19	50	50.9	16,900	668	57	82	70	Bull			
20	58	38.8	15,410	765	49	72	65	77	1.8	13.1	117
21	35	39.9	14,341	297	46	63	65	105	2.0	14.9	173
22	27	42.6	12,713	804	46	63	56	100	1.7	14.8	171
23[b]	49	53.3	13,593	344	56	72	39	Bull			
24	51	66.9	20,630	442	44	66	87	91	1.8	14.3	155
25	31	62.2	21,586	195	46	77	95	112	1.8	14.0	146
26	38	49.9	15,947	301	45	61	73	91	2.0	14.0	145
27	36	56.4	18,396	395	40	67	80	100	2.1	14.4	159
28	44	51.9	13,961	311	48	67	61	Bull			
29	45	60.4	15,994	505	51	71	76	Bull			
30	40	57.8	15,848	234	50	68	72	92	2.6	14.3	154
31	47	46.1	14,093	589	54	77	60	Bull			
Avg.	**44**	**52.2**	**16,303**	**386**	**50**	**69**	**72**	**97**	**1.9**	**14.1**	**148**
Holstein Only	**45**	**51.9**	**17,069**	**379**	**49**	**69**	**73**	**97**	**1.9**	**14.1**	**148**

(a) Includes herds processed through DRMS in Raleigh, NC (n=25) and herds on the Lancaster DHIA "Basic" program (n=6)

(b) Majority of herd is anything but Holstein (c) Not on test yet for full year to give RHA projection

Chapter 9
Holistic Pharmacology

We will now begin to examine the types of treatments being used that may help to account for the numbers which were shown in the data section. Let us first remember that prior to the advent of synthetic antibiotics, steroids and reproductive hormones, animals were treated by veterinarians and farmers with other substances. There was widespread use of medicinals derived from plants and minerals, with heavy reliance on antibacterials, antiseptics and germicides such as thymol (essential oil of thyme), iodine and hydrogen peroxide. Materials were used based on their physiologic effects as observed by clinical practitioners. This is called biochemical pharmacy — a compound is known to give certain physiologic effects, and thus it is used clinically. For instance, although we now have very high tech anesthetics available, ether and chloroform were the only agents used clinically for general anesthesia. They were effective. After surgery a patient's temperature was closely monitored for fever which would indicate probable infection (or lack of fever and no infection). This is still standard practice. However, various plant or mineral compounds were used to combat fever while dressings were changed frequently, often relying on iodine-soaked gauze to keep the incisional wound clean. Due to my work with organic cattle, I have had a strong incentive to read the old veterinary and medical textbooks regarding the treatments for infectious diseases, reproductive problems and gastro-intestinal conditions. It seems rational to look back upon what was employed in the medical world prior to all the synthetics today. Granted, there were deaths due to lack of modern antibiotics, but there were also a great many successes using the methods of the time. Many case studies presented by respected doctors of the time can be found in the annual volumes of the various medical societies. In "The Transactions of the National Eclectic Medical Association" from the Eclectic school of physicians, many detailed accounts of physicians showed how they used plant tinctures, fluidextracts and inorganic minerals in their daily practices. The Regular school doctors and surgeons (as the conventional/allopathic school graduates were called by the Eclectics) would always try to use purified active ingredients in as strong a manner as possible whenever they could. This is in contrast to the Eclectic practitioners who always wanted to use as balanced an approach as possible by utilizing all the constituents of a plant instead of only the isolated active ingredient. In essence, the Eclectic school of medicine was the foundation from which modern day pharmacognosy is derived. Pharmacognosy is the study of medicines derived from nature. The Eclectics studied what is now known by pharmacognocists as "classical pharmacognosy" — the botanical identification, geographical location of habitat, rudimentary chemistry, and application for clinical use. Modern pharmacognosy is very high tech and uses the latest analytical methods to screen for substances and their activities. The Eclectic and Regular physicians only had plant medicinals and mineral derivatives as their primary anti-infective means for treating systemic infections, traumatic wounds and surgical incisions. I am convinced that by reading technical information from both the Eclectic and Regular medical textbooks — by taking what was known and tempering it by modern veterinary science — present-day practitioners will be better able to treat organic dairy cows. Reading such medical literature has helped me to re-define my own understanding of those remedies which I used as a herdsman in the late 1980's before my veterinary training.

Veterinary medicine is a great comparative science — many species are studied, not just the human species as is the case in medical school. Veterinary school forces its students to study many animal species' anatomy, physiology and pathology; the many comparisons and contrasts enables students to see that there are many angles of approach in framing solutions to problems. Thus, my previous background of using homeopathic remedies as a herdsman coupled with an intense interest in learning why they may work (possibly in the physical realm as reflected through immunological changes) led me back to basic courses in organic chemistry and biochemistry — courses most veterinary students sooner wish to forget than to further investigate. I prefer studying and concentrating on ways to further develop natural medicines from plants for the great variety of conditions that are medically treatable in dairy cows. I am fascinated and intrigued by the myriad of organic chemical constituents already known and still being discovered in plants. It therefore seems wise to describe what I believe are both rational and reasonable natural treatments, based on practitioners from the past, current alternative therapy literature and ideas based on experience.

As just mentioned, my first experiences with natural treatments were by using homeopathic remedies as a dairy herdsman. But to keep the record straight I should state that while I prescribe and dispense standard homeopathic remedies, I would never consider myself a "homeopathic veterinarian." As a herdsman I began using homeopathic remedies in a cookbook fashion (a "this-for-that" approach) and still do as a clinical veterinarian. It's a practical and applied approach that seems to be helpful. By memorization of a limited homeopathic materia medica, I have come to know which remedies are needed when I see a few clinical symptoms. There is a major difference between using homeopathic remedies in a cookbook fashion as I do, versus fully ascribing to the whole Hahnemannian philosophy of repertorizing, which guides its medical doctrine. That doctrine, according to many of its adherents, has its greatest strength in chronic disease and entails accumulating excruciating detail of subjective symptoms relayed to the doctor from the patient, giving a single potentized remedy, and then waiting for a period of time to see how it worked (often a few weeks between each administration of a remedy). As a herdsman, I saw a physical problem in a milk producing cow and it needed to be addressed immediately; in other words, acute problems that needed attention quickly and sometimes drastically in order to get the cow quickly back into production. It may have involved the udder, liver, rumen, lungs, eyes, blood or reproductive tract. Usually I treated for one problem at a time. Starting out this way and seeing a problem clear due to a specific remedy can build confidence. It also allows one to start wondering just how remedies containing such diluted materials may actually work.

Whenever I peruse a homeopathic materia medica I am still amazed at the extent to which it details symptoms, both mundane and quite peculiar. It makes for fascinating reading. However, using highly diluted homeopathic remedies does take a leap of faith, which I will not impose upon you, the reader. The only reason that I was initiated into their use was because my former boss and dear friend, David Griffiths, told me that I *had* to use them on his Biodynamic farm. I had no choice in the matter. But you do, and that is why I want to specifically look into the issue of concentrations and potencies — the major hurdle to the general acceptance of homeopathic remedies. My general aim in this book is to educate you so that you will make rational and logical choices in treating your cows. I don't want people to blindly use remedies without thinking about the concentrations (potencies) of the dose given. This I say because when most farmers find out that a homeopathic remedy in high potency has no actual material remaining from the mother tincture, they are somewhat shocked and yet amazed at the same time. Homeopathic strength is not what it seems to be to the casual observer — a remedy which has a higher potency is stronger only in a homeopathic sense, not in material reality. In homeopathy, the term "potency" takes on a fundamentally different meaning from what it is normally understood by people not specifically trained in its philosophy. It is critical to understand the various potencies of remedies in order to know how to use them with cattle in acute situations needing immediate attention. Potency refers to the degree to which a crude substance undergoes progressive dilution and succussion. Dilution and succussion are terms for the routine process undertaken during the manufacture of homeopathic remedies. In homeopathic pharmacy, manufacturing is standardized by the Homeopathic Pharmacopeia of the United States (HPUS) as enacted by Congress in the first half of the 20th century and still (technically) regulated by the Food and Drug Administration (FDA). The more highly diluted and succussed a remedy is, the higher the homeopathic potency designation (*i.e.,* a 200C is a higher potency than a 30C). This is also considered more powerful, but only from the homeopathic viewpoint.

The process of manufacturing a homeopathic remedy begins with any crude substance, usually derived from plant and mineral origin, but sometimes animal sources such as venoms. For instance, certain parts of plants will be collected, mashed in alcohol and allowed to sit in a container for about two weeks (as prescribed by the HPUS). With plants, there are very specific times of the season and plant parts which are to be used as starting material. For example, leaves and flowers are usually gathered in the summer while root parts are gathered in the autumn. The final product of this slurry-mash is called the "mother tincture." This is then filtered. One part of the mother tincture is then diluted in nine parts of pure water. This mixture is then succussed (vigorously shaken), and the resulting mixture has a 1X potency. This means a 1 in 10 dilution. (Roman numerals are used — "X" for 10 and "C" for 100.) Alternatively, if one part of the mother tincture were to be placed in 99 drops of water and then succussed, the resultant mixture is 1C potency. This would be a 1 in 100 dilution. Succussion is essential to homeopathic pharmacy in that it purportedly releases the crude material's inherent energy into the waiting water. Generally speaking, 1C potency is more energetic in terms of homeopathic medicinal value than is 1X. Obviously, there is less crude material in a 1C than in a 1X, but the 1C has more energy of the original mother tincture, as it is explained in homeopathy.

Now, what happens with a 1X or 1C dilution? One part of a 1X will be added to nine more parts of water to create a 2X. After succussion, a 2X dilution is then obtained. Then taking one part of a 2X and adding nine more parts of water we will end up with a 3X product, and so forth. The dilution and succussion process can be repeated for as long as the manufacturer wants. Basically, the number in front of the letter indicates how many times the process has occurred. For example, a 30C would mean that the process of taking one part of the material and diluting it in 99 parts of water has been repeated sequentially for 30 times. One may rightfully ask: at what point is there no more detectable crude substance left? The potency of 12C or 24X is the limit at which a chemist could detect any molecules of the original starting substance (Avogadro's number: 6.02×10^{23}). One more dilution will technically render the mixture void of original starting substance. And yet in homeopathy this is considered one step more powerful. So, even though a 30X remedy would mean there is only one part of original starting material in a solution having 30 zeros (1×10^{30}), or a 30C would have one part of original starting material in a solution having 60 zeros (1×10^{60}); these are actually considered more energetic because the succession of each step releases the original material's inherent energy. That is why homeopathy is considered "energy medicine." But this concept is why homeopathy is a tough pill to swallow for those inquiring about it.

In classical homeopathic provings, those experiments taken on voluntarily by persons to ascertain the potential medicinal value of a remedy, the usual starting potency used is generally low, such as a 3C or 6X or something similar. "Provings" are carefully recorded experiences of a person (the prover) after administration of the remedy to be assessed. These are extremely descriptive and qualitative by nature. Effects are noted by occurrence at time of day, body side affected, emotions, thoughts, behavior, reactions to atmospheric changes, as well as direct physical effects. The descriptions are more influenced by our higher cerebral activities than would be an animal. I would strongly assert that observation of strictly physical and objectively measurable responses of domestic animals would be of much greater importance in assessing the value of a remedy for the non-human animal kingdom than simple extrapolation from established homeopathic materia medica for humans. This would also be more satisfying to the question of efficacy so important for clinical research. In any event, it is interesting that fairly low potency is used to prove remedies (*i.e.,* material amounts are still present), at least in the initial proving stages.

In homeopathy as taught in North America, common potencies prescribed are usually 30C, 200C and 1M. However, in Europe, where homeopathy occupies a much bigger place in the medical world, French doctors (*i.e.,* Demarque, Joly) emphasize single remedy use but in fairly low potencies, such as 5C, 8C and 12C (with 30C considered to be very high potency). In the theory and practice of homotoxicology, as delineated by Reckeweg, use is made of one remedy in various potencies (*i.e.,* Belladonna 2X, 6X, 12X, 30X) but often it is multi-combination remedies in multi-potencies (shot gun approach).

Before prescribing a homeopathic remedy in people, great care is taken to repertorize the case correctly in order to get the patient's whole state of being when the case is taken by a person trained in classical homeopathy. The classical homeopath strives to accumulate unique, perhaps even peculiar traits of the patient. Emotional and mental symptoms of the patient may be more important than the actual physical complaint to a homeopath. The more unique a set of symptoms, the more precisely a prescription can be arrived at to fit the totality of symptoms that matches a "remedy picture" (as recorded during a remedy's proving). Ideally, veterinarians do this as well. It would seem likely that a person would know their pet dog or cat's special characteristics much more than a person would know highly individual traits of the cows in a herd. We always do strive to spot the unique character of our individual cows if we are going to treat them with homeopathy or other holistic modalities. However, it is not easy to do this within a herd of animals as their individual characters blend together to make up more of a "herd character." When a "herd character" is observed (and realized), it is likely that a few remedies will always seem to work specific to that farm. I have seen this repeatedly. On the farm where I first learned about homeopathic remedies, a review of my records made me realize that *Apis mellifica* (honey bee venom) worked consistently well. I know of other farmers that echo this idea and find a certain remedy works consistently well for them. Certain cows of course do stand out with certain peculiarities, and these features should definitely be noted if observed when trying to arrive at a treatment remedy for that particular cow. Yet, with animals, we are often left with looking only at the obvious physical symptoms. Thus livestock tend to be treated in a more shallow way, homeopathically speaking, giving a certain remedy for a certain malady every time (cookbook approach) without delving deeply into a cow's entire character to truly repertorize the case (so crucial for classical homeopathy). If a person cannot delve into a case in depth as classical homeopathy requires, it must be realized that other modes of treatment are probably needed for the situation. If electing to use homeopathy in cattle, the lower potencies, especially the decimal or "X" potencies, which seem to have more effect on crude physical symptoms, would make sense for cows since they are very rooted to the physical earth and its material elements.

Having originally been taught that homeopathic pellets should be put on or under the tongue got me interested as to the importance of contact location of a remedy with the patient. The mucus membranes of the oral cavity allow for easy melting and absorption. It seemed to me that the lymph nodes in our oral cavity could in some way be involved in the processing of the remedy. Our mouth is one of the main entry ways for outside materials to come into our system. Another prime entry way into an animal are the nostrils, which would be suited to scents and smells used in aromatherapy as well as other mists or inhaled stimuli. Lymph nodes surrounding the area react and swell when there is an antigenic challenge (usually viral or bacterial). A challenge can be very small but it still causes an effect — swelling of the lymph nodes indicates reactivity (whether visually observable and detectable by touch or only by microscopic detection). The mouth and accompanying digestive tract has lymph nodes so our body can screen incoming foods and nutrients for potential danger. Lymph nodes are basically the frontline "gate-keepers" of our body's immune system — as far as our environmentally interactive digestive and respiratory tract is concerned. When giving a remedy in the mouth, I think this may trigger local lymph nodes and activate the immune system, due to the reasons just mentioned. It is also possible to give remedies on other mucous membranes, such as the vaginal mucous membranes. This makes for easier labor for the dairy farmer. The contact here is not as near lymph nodes; however, all mucous membranes in the body have the immunoglobulin IgA integrated into them. IgA, like IgG, IgM and IgE, is a type of circulating antibody which will communicate and activate the immune system to trigger its cascade of events involving interleukins, interferon, B-cells, T-cells and all other needed messenger cytokines. Therefore, placing a remedy upon the vaginal mucous membranes would also make sense.

This activation of the immune system is different for each individual remedy, just as it is for different bacteria and viruses coming into contact with our organism. All are foreign in some way to the organism. Depending on the remedy, the animal's body gears up and virtually says, "Hey, what's *this*? I better react to this." Thus the immune system is triggered, stirring the animal's vitality into action. If the animal's inherent vitality is intact, the animal becomes stimulated and "energized," and therefore strengthened to attempt restoration of normal balance and being well. The immune system is probably the closest physical reflection of an individual's "vital force" that can reasonably be offered for this term (often scoffed at by Western medicine). The immune system separates "self" from "non-self" (the rest of the world). Aside from genetic differences when born, an animal's heart, lungs, liver, kidneys, pancreas, bones and muscles all pretty much function materially the same from one animal to the next. Mathematical equations can describe the electrical activity of the heart. Biochemical pathways can describe the function of the liver and pancreas. Electrolyte chemistry can describe the kidney process. Biomechanics can describe the bones, muscle and locomotion. But try to describe the immune system and you would need to describe a cascade of events continually fine-tuning and re-shaping itself as demanded by outside antigenic challenges. The closest system, which is itself linked to the immune system, is the endocrine (hormone) system. The endocrine system is highly dynamic, and its activity often stems from the pituitary gland and hypothalamus, both located in the brain. Being centered and generated from the brain certainly is reflective of a system unique to a specific person or animal. Indeed, all of our systems are dynamic and interdependent; however, I believe the immune system is the most self-defining, unique biologic characteristic each organism has. It says, in essence, "halt, I need to review your access code." Organs seem to be passively reflexive compared to this. The immune system is the physical manifestation of our inner strength — our vitality. And this is what the energy medicine known as homeopathy is supposed to trigger — our vitality.

An animal's individual immune system is based on its genetic blueprint which, even for its species as a whole, has developed over time. Animals and humans have been (and will be) forever tied to their immediate landscape and environment. We live from the landscape — its waters, its soil, its plants, insects and other creatures (and pollution). All "higher" forms of life depend on "lower" forms of life to meet their nutritional demands. The biologically active soil, those first 6-12 inches of earth, rich in rock-derived minerals as well as organic matter, supports nearly all land-bound life. In the soil are bacteria and fungi that can "fix" nitrogen from the atmosphere and thereby make it available in the ground for plant uptake. Bacteria and fungi also breakdown dead and decaying material rendering it useful as humus — that stable fraction of soil organic matter that helps to continually nourish plant life. Plants, whose basic physiology is to create primary metabolites such as carbohydrates (energy) from sunlight and carbon dioxide, as well as make various celluloses and lignins for structural support, also make secondary metabolites for self protection which are unique to their respective species. Herbivorous and omnivorous animals, man included, have partaken of the bounty of Earth's plant constituents forever. Over incredibly lengthy periods of time, the genetic programming of animal life has been affected by the environment. Enzymatic processes within grazing animals could certainly have shaped and refined animals' adaptive responses. Likewise, adaptations could have created enzymatic processes to more easily degrade ingested materials. It is not inconceivable that these adaptive processes became genetically encoded and became a "species' memory- response" to material substances tried and tested by a particular species, both beneficial and toxic. Perhaps this is why animals are sometimes seen seeking out seemingly odd substances, due to their species' memory-response guiding them to those materials that can somehow benefit their health. A primitive part of the brain, the hippocampus — that part of the brain shared commonly to some degree by all land oriented animals (amphibians, reptiles, and mammals) and tied directly to olfaction (sense of smell) — is perhaps triggered by deficits when in ill-health. Not unreasonably, this primitive sense may be genetically guided to pick up certain scents to what is beneficial (and also harmful). This area of interest and research is the nascent, fascinating field of zoopharmacognosy. It is often said that odors and aromas can trigger memories from the distant past. Why then couldn't an animal's sense of smell, via its species' memory-response, imperceptibly guide it to plants that will help restore its internal equilibrium? And why not specifically to those substances which its species has "known" and encountered over countless generations? After all, the generation that gives birth to the next passes along all genetic information from its ancestors. Even when the genetic codes are all discovered, the specific interaction of chromosomal genes with other chromosomal genes is nearly endless. It will unfortunately be near impossible to determine exactly if and how animals do indeed genetically "remember" various beneficial or harmful plants. Questions regarding specific doses needed to elicit a response from the actual plants will make for a lifetime of research. When considering all this, I think it is wise to consider low potency homeopathy or, better yet, somewhat dilute, low concentration plant extracts from which to create helpful medicines for animals — especially grazing animals which have "known" plants forever. I further believe that any medicine will be more easily assimilated if in its natural form than in a synthetic form. This may be due to the ability of the body's enzyme systems to more easily interact with naturally occurring configurations of molecules presented rather than synthetically manufactured configurations to which the animal's enzyme system must adjust. It is well known that any synthetic version of a natural molecule is a racemic mixture, 50% levorotary and 50% dextrorotary, indicating which way light is bent by the molecular configuration in the presence of plane polarized light. The body's enzyme system is an incredibly complex three-dimensional system and only locks onto specific molecular configurations. By introducing 50/50 synthetic mixtures, maximal enzyme efficiency is hindered and the foreign molecules can occupy enzyme binding sites but not allow the enzymatic process to occur at normal rates, if at all. In the realm of therapeutic medicines, let's also not forget that most, if not all, synthetic and semi-synthetic modern pharmaceutical products are definitely "foreign" (or potentially toxic) to the internal workings of mammals and are toxic, especially at increasing concentrations. Conventional treatments, when within their therapeutic range, can possess a range of side-effects (nausea, vertigo, etc.). This is why many laboratory animals are used (and suffer) during the testing of new pharmaceutical drugs. Animal testing as done by pharmaceutical companies derive an LD50 — the lethal dose at which 50% of the animals will die when administered the substance at a certain concentration in question.

From quite the opposite tack, everyday feeds and nutritional supplements containing herbs and plant nutraceutacals are a part of the recipe for maintaining and/or re-gaining health. In agriculture, we need to provide biologically active substances since the animals under our watch are usually fed monotonous diets devoid of fresh, lively components. They do not forage as their counterparts in the wild would do. I would say that even intensively grazed cows do not get to take in a variety of healthful fresh greens — they indeed consume fresh plants, but only a certain few.

Perhaps the mechanisms behind classical homeopathy as an energy medicine will someday be explained; we just may not, as of yet, have the appropriate laboratory instrumentation. As technology grows by leaps and bounds, I wouldn't be at all surprised if we someday have the appropriate laboratory techniques. Still, many of homeopathy's greatest supporters will steadfastly say it can never be proven because it addresses the whole patient and is extremely specific to the individual being treated. This is theoretically true. And therefore, laboratory work, which so often takes things apart one by one to see how each piece of the puzzle acts, will probably never get at the heart of any holistic mechanism of healing. Yet, I have faith that we will find the needed means to show how homeopathic remedies generally work. Certainly, when there is a cure, a physical response has happened. When there is a consistent, temporally recurring association between administration of treatment and subsequent physical change, investigators will more likely elucidate the mechanism of action. While the effects of homeopathic remedies may not be understood or readily apparent at times, at least there are no harmful side effects (as opposed to conventional medicine where hundreds of adverse reactions are recorded yearly).

The idea of low exposure environmental stimulus is actually the basis upon how I now justify using homeopathic remedies. I figure that if there is a genetically programmed species' memory-response programmed into animals' immune systems, then introducing compounds in *low concentration* will trigger the individual's system to react and thus attempt to restore equilibrium on the energetic level — as well as to oppose the morbid process by physical quantity. The concentration would depend upon the actual known pharmacologic-toxicity spectrum. Substances more strongly toxic in crude form need to be given in smaller doses than those substances less toxic to normal functioning. Plants that contain potent alkaloids (potentially toxic) probably have their most beneficial effects upon the entire system when used in low concentration. For instance, using only 2-5cc of Aconite tincture (1:10 or 10% concentration) every few hours should suffice while using safer compounds such as garlic or echinacea would require 20cc 3 times daily for pronounced effect.

But why use any potentially toxic (especially alkaloidal) plant substances to begin with? Mainly, it would be to quickly trigger a response in the animal. Hopefully we can pick exactly what material will fit what situation (the art of "specific diagnosis" in Eclectic medicine). If a substance is not toxic or potentially so, then it could be classified by default as a foodstuff or a neutraceutical. By using naturally derived substances of low concentration, I believe it is possible to alert the animal's system to a potential poison (not all substances will be toxic, of course) without actually harming the animal. Instead, with a low concentration medicinal, the animal's biochemistry and protective immune systems are triggered and a cascade of events follows. A response in the animal occurs — hopefully stimulating the immune system and physiologically curative. This process may reflect the encoded species memory-response of the organism due to its own species interaction with that substance (or similar substances) over time, especially if it is a plant derivative, and the animal is an herbivore. Again, I think it's not unreasonable to consider that the bioactive aspect of a naturally derived substance is more easily processed and recognized by an organism's cellular receptors than a synthetic substance. Using a natural substance in low concentration can enable the animal's immune system to handle the administered remedy because it gently reminds the animal's inherent "memory" to come to action, rather than assaulting the current state of physical affairs in concentrated, truly toxic amounts.

Remedies are traditionally made from mineral, plant or animal origin. Some remedies are also now made from modern pharmaceutical compounds and radioactive compounds. Others, most notably homeopathic nosodes, are made from the discharges of specific disease processes. (Beware of using nosodes for prevention — are they made from the actual disease discharge or simply derived from an already manufactured vaccine? If made from a vaccine, potentization of the original vaccine will also potentize the preservative and any other carriers in it.) In my experience, those conditions which arise quickly have responded well to animal and plant based remedies, while longer standing conditions seem to respond more favorably to mineral based remedies. Perhaps this is due to the age-duration in nature of the base remedy and its affinity to parallel structures and disease course in animals and man.

For instance, if we consider a mineral-based homeopathic remedy such as silica (derived from the geologic development of quartz and other minerals), we see its main actions on long-standing cases of scarring, chronic drainage and expelling foreign bodies. This is a slow working, deep acting remedy and is generally given once or twice daily over a few weeks time. In contrast, the remedy Apis mellifica derived from the very active honey bee, acts on acute cases of swelling or inflammation and given frequently enough (4-5 times daily), I've seen Apis help bring down soft tissue swellings in a day or two. Plant derived compounds work quickly or more gradually, I believe, due to their inherent toxicity level. Aconitum napellus and Belladonna (having the potent alkaloids aconitine and atropine, respectively) tend to act very quickly on conditions of rapid onset, fevers and inflammation. Contrasting this are Bryonia and Chelidonium which act somewhat more gradually on respiratory and digestive conditions, respectively.

The type of body part affected also should also be considered in choosing a remedy. The "vegetative" soft-tissue structures which rhythmically continue without conscious effort (heart, lungs, smooth muscle of digestion, pancreas, liver and kidney) seem to respond to plant derived remedies of low concentration and low potency in my experience. Perhaps this is due to the inherent parallel cyclic form of plant-life mirroring the "vegetative" cyclic function of visceral organs. Skeletal involvement of bones and conditions affecting them seem to respond to mineral substances *i.e.,* calc phos, calc carb, calc fluor, etc. Unique functions of the organism such as sexual reproduction tend to respond best to animal derived remedies (apis, lachesis, sepia, folliculinum, ovarian). There are of course overlapping conditions and the respective remedies based on integrating the full range of a remedy's characteristics and known provings.

The most objective way to clinically investigate natural remedies would be to observe the effects in animals — to avoid any placebo effect. I don't think animals consciously feel better simply because the doctor has examined them, as people certainly can do. If anything, in my experience with dairy cows, animals can give the doctor a tough time and are seemingly thankless for our efforts. Humans, on the other hand, know they just went to the doctor and perhaps feel relieved simply because they were seen by a doctor. So that anything the doctor did, give a suggestion or give a water injection or sugar pills, may evoke a positive response — the placebo effect. Not so with animals! By and large, I don't think animals enjoy a visit to the doctor's office and like veterinary examinations even less. Therefore, as the mainstream of medical science opens up to other possibilities, we should keep in mind that the best patients in which to evaluate whether or not remedies work are probably with our own pets and livestock. This is true also for acupuncture and other forms of medicine which are not yet scientifically proven through clinical studies using the p<.05 statistic. By the way, the term "p<.05" simply means that there is less than a 1 in 20 chance that the observed event would occur in nature spontaneously *i.e.,* if p<.05, then there is something occurring due to the treatment. There is no really good reason why the medical establishment has crowned p<.05 as the gold standard for reviewing clinical research data. Nonetheless, it is now convention that p<.05 in order for the results of a study to carry any weight among medical professionals.

As a herdsman, I could have cared less about what potencies I was using, so long as the remedies worked. I also was not interested in scientific documentation or statistics; I just wanted a cow to get well. It is easy to get into the swing of homeopathy on a daily basis as a herdsman and use remedies on a variety of situations. Trial and error is actually the main way one should go about learning homeopathy. Learning all the various homeopathic materia medica are important, *but not as important as first-hand experience with the remedies in real-life situations.* Perhaps being generally inquisitive and curious, I also wanted to know *why* they work. It wasn't enough for me to blindly accept "of course the remedies work." I needed to know *why,* especially to help me form a stronger basis in using them. I wanted to make sense of the system called homeopathy — if that can be done. I knew that the remedies worked — well enough in fact that I wanted to go on to learn standard veterinary medicine, so that I could mix and match whatever treatment types would be needed when out in the barns after graduation. This is exactly what I do now as a practitioner. But, all along the way, I always wondered how these remedies actually exerted their effects — *their mechanism of action*. This has always held my attention. I want to know the nitty-gritty details! So during veterinary school I always kept my ears open for possible openings by professors of the basic sciences during their lectures as to how I could blend the reality of homeopathic successes that I already had with dairy cows with all the theories that we students had to learn to become licensed, professional veterinarians. During veterinary school I privately kept studying the topic "on the side" (while also studying for a multitude of exams!), reading both scientific and lay books on homeopathic remedies and plant medicines. (My intrigue with plant medicine pre-dates any use of homeopathy and is due to my college training as a soil scientist, which of course includes identification of plants and their positions in the landscape.)

Interestingly, what I did not hear from my own veterinary professors I may well have heard from medically trained professionals around the turn of the 20th century. During the period between roughly 1880 and 1940, there were extensive materia medicas in the three recognized professional medical schools — the Regular school, the Eclectic school and the Homeopathic schools. The Homeopaths had a strong following in the late 1800's, but the leap of faith needed to accept it in its totally kept many well intentioned doctors away. The leap needed, as mentioned previously, was to believe that the more dilute and succussed a remedy, the better it would be at treating symptoms the remedy would have caused had it been given in the actual crude form. This is one of homeopathy's major tenets. The Eclectics (briefly introduced in the opening of this chapter) believed that plant tinctures and fluidextracts, those compounds with real amounts of constituents but given in small doses yet frequently, were the key to treatment. Their belief in using actual amounts of medicinal plant derivatives to oppose disease was more easily accepted by the mainstream than the basic premise of the Homeopaths. The Eclectics were highly attuned to an exhaustive physical examination (including different types of pulses, coatings of the tongue and eye examination in humans).The homeopaths honed in on highly individual patterns of symptoms noting physical complaints but, as mentioned previously, paying closer attention to behavioral and mental symptoms, as well as symptoms peculiar to homeopathic case taking — the sides of the body effected, time of day when better or worse, weather condition making better or worse, etc. However different in their case taking protocols, both schools strived to diagnose a *specific remedy* for the *specific symptoms* presented. John Scudder, M.D., one of the founding fathers of the Eclectic School put it this way:

> We propose studying the expressions or symptoms of disease with reference to the administration of remedies. It is a matter of interest to know the exact character of a lesion, but it is much more important to know the exact relationship of drug action to disease expression, and how the one will oppose the other, and restore health. If I can point out an expression of disease which will be almost invariably met by one drug, and health restored, I have made one step in a rational practice of medicine. (Scudder, *Specific Diagnosis*, p. 14-15)

The thrust of Eclectic treatment was to directly antagonize and subdue a given condition with low concentrations of the correctly diagnosed (chosen) remedy. "Antagonize" and "subdue" are terms not associated with homeopathy. They are, however, terms which aptly describe using material amounts of substances. Again, Scudder:

> The first lesson in therapeutics is, that all remedies are uniform in their action; the conditions being the same, the action is always the same . . . Then we study the action of drugs upon the sick, and when we find them exerting an influence opposed to disease and in favor of health, we want to know the relation between the drug and the disease — between disease expression and drug action.
>
> I do not say that we should not study a drug action in health — indeed I think it is a very important study. You may, on your own person, study a wholly unknown drug, and determine its proximate medicinal action. How? Easy enough. You will feel where it acts; that points out the local action of the drug, and as a matter of common sense, you would use it in disease of that part, and not of a part on which it had no action. You will feel how it acts — stimulant, depressing, altering the innervation, circulation, nutrition and function. If now you want to use it in disease, use it to do the very things it did in health, and not as our Homeopathic brethren would say, to do the very opposite things. (Scudder, *Specific Diagnosis*, p. 16)

Although many modern homeopaths, especially the classical sect, prescribe individual remedies in high potency (30C, 200C, 1M, etc.), their colleagues at the turn of the 20th century often prescribed in low potency (1X, 2X, 3X, etc.)

where there were actual small amounts of succussed crude material remaining in the remedy. In Boericke's famous homeopathic, *Materia Medica and Repertory,* an overwhelming number of entries explicitly state to use remedies in a concentration ranging anywhere from the mother tincture to the homeopathic 30th potency, with an obvious majority being in low homeopathic potency (*i.e.,* 3X, 6X, etc.).

As an example, for active congestion of the breast (mastitis), the French homeopath Jousset writes:

> The principal remedies are Belladonna, Phytolacca, Bryonia and Chamomilla. Belladonna — shooting pain in the breast, erysipelatous redness. Clinically we know that this is the best remedy. Dose: The three lower dilutions. Phytolacca is highly recommended. Its symptoms are very characteristic: inflammation of the breast, which are stony hard; mammary glands full of hard and painful nodes; nipple sore and fissured. Dose: 1st dilution (1X) and the mother tincture (Ø). Bryonia — mammae hard, with lancinating pain, great heat, and redness. It may render service should the preceding remedies fail. Dose: 1st dilution (1X). Chamomilla — mammae hard and tender to touch, with intense pain. In this case it could be given in alternation with Belladonna. Dose: 3rd dilution (3X). (Jousset, p.1002)

Dr. F. Humphreys, a medical doctor who later turned his attention to veterinary medicine, used the term "homeopathic veterinary specifics," which was an abbreviated form of homeopathy combining low potency homeopathic preparations (mainly of plants). Usually 5-10 drops would be placed on the back of the tongue. Although his long books were essentially advertisements (due to never divulging the ingredients of his "AA" or "BB" veterinary specifics), once the mixtures were revealed to me by pharmacist Jack Bornemann, it was easy to understood why Humphreys had advised his various combinations of "veterinary specifics" for the different conditions mentioned.

Humphreys' low potency homeopathic remedies had true pharmacologic activity since there was actual material still present. And they were mildly "energized" due to one or a few succession steps. These succussed, low dilutions were not strong enough to wreak havoc with the animal's system as large volumes of the same remedy most likely would (especially if repeatedly administered at the dosage prescribed by the Regular/conventional school). Whether or not these individuals (Jousset, Boericke, Humphreys) are considered true homeopaths by modern day homeopathic practitioners matters little. What matters is that there is historical evidence of when and how to use low concentrations of homeopathic remedies in human and veterinary medicine. What is even more intriguing is that the Eclectic practitioners concluded from their own materia medica *the same exact remedies* for many of the same physical conditions as this set of homeopaths did. (See Chapter 10 Comparative Materia Medica). But the Eclectics used the same remedies in material doses (albeit usually small ones). The more one reads about the Eclectics' use of medicinal plants, the more one may unfortunately become increasingly confused about the same homeopathically prepared remedies being used for the same conditions. For isn't a main guiding principle of homeopathic remedies that they are to be used for the opposite of what the crude plant would do to a person in health? Complicating the subject further is that many of the symptoms seen on the physical level with dairy cows are treated in high potency by modern homeopaths. Hopefully it makes sense to the reader to use at least some homeopathic remedies in low potency, especially if all the physical symptoms in a case guide us to the exact same specific remedy in both the Eclectic and Homeopathic materia medicas. Why should we limit ourselves or prescribe to use a homeopathic remedy of Nux vomica, which has only 1 part of Nux vomica in 400 zeroes of solution (Nux 200C) when we could use an Eclectic remedy with 1 part in 3 zeroes of solution (Nux 3X = 1 part in 1000 = 0.1%) with known physiologic effects for nearly the exact same physical symptoms? It seems to make little sense to use an extremely diluted form of a remedy (with absolutely no starting material left in it) when the same exact remedy in low concentration is known to exert the needed physiologic effects for the case presented. It just makes logical sense to use the actual material — *if it is specific to the condition, and it is used wisely in low concentration.* In addition, if the prescriber or farmer is not certain enough in his/her choice of remedies, then by using a low potency remedy there may be some usefulness even if the remedy was not perfectly diagnosed (specific to the symptoms). In the world of dairy cows, when there are many chores to get done in a day between milkings, there is not a whole lot of time to do in depth case studies of cows to ascertain their exact remedy needs as true classical homeopathy requires. It is just a practical reality that most dairy farmers usually can identify about 3-4 physical symptoms from which to prescribe, but to go on into further homeopathic repertorizing is not efficient use of the farmer's time or patience in most cases.

TABLE 8: HUMPHREYS HOMEOPATHIC VETERINARY SPECIFICS*

AA = Acon 2X, Verat vir 3X, Bell 3X
Fever, Inflammation, Congestion

BB = Rhus tox 2X, Ruta 2X, Arnica 2X, Calc F 3X
Strains, Injuries, Lameness (Tendons, Ligaments, Joints)

CC = Phytolacca 2X, Merc iod F 13X, Kali bich 13X
Influenza, Quinsy, Nasal Gleet, Catarrh

DD = Acon 2X, Ars alb 4X, Ferr sulph 12X
Worms, Bots, Grubs

EE = Ars iod 12X, Bell 2X, Phos 2X
Coughs, Bronchitis, Inflamed Lungs

FF = Colocynth 2X, Colch 2X, Bell 2X, Nux 2X, Dioscorea 3X
Colic, Belly-Ache, Wind-Blown (bloat), Diarrhea

GG** = Cimicfuga 2X, Caulophyllum 2X, Secale 4X
Miscarriage, Imperfect Cleansing or Hemorrhage

HH = Apis 2X, Chim umb 2X, Canth 2X
Urinary & Kidney Diseases, and Dropsy (Increase urine output)

II = Rhustox 2X, Ars iod 4X, Hepar 4X
Eruptions, Ulcers, Mange, Grease, Farcy, Abscesses, Fistulas, Unhealthy Skin, Eczema

JJ = Nux 2X, Ant crud 12X, Sulph 5X, Lycopod 4X
Indigestion, Over-Feed, Bad Condition, Paralysis, Stomach Staggers, Loss of appetite

*Thanks to Jack Bornemann, R.Ph. for deciphering these patent medicine formulae.
**Ingredients postulated by author (but not verified), however, the indications were as stated.

Although I strongly believe that low potency has the most merit in remedies derived from plants, high potency homeopathy (30C and higher) certainly does have a place with dairy cows. Its place, in my experience, is mainly in the realm of reproduction. Why? The reproductive process, conception — the generation of new life — is as close to "pure energy" in a biological system as can be imagined. Thus true energy medicine would seem to be highly appropriate. Another time when high potency is appropriate is to "finish off" a problem. By this I mean that if a result is being seen with a low potency remedy, then a more lasting cure often may be obtained by going to high potency, but giving it less frequently. Finally, starting a case with high potency makes sense when considering a remedy like *Pyrogenium* (made from rotten meat). This is commonly prescribed for a cow with retained placenta and fever. *Pyrogenium* is as nearly homeopathic to a decomposing retained placenta as can be imagined. The problem in beginning a case with high potency, especially for those just learning about homeopathy, is that if the remedy chosen is incorrect, it will have absolutely no effect and valuable time will be lost in helping heal the cow and getting her back into milk. Think of choosing the right remedy as a dart game. The higher the potency you wish to use, the better you need to be in hitting the bulls-eye in order to get the desired effect. But by using low potency, a hit near the bulls-eye can still help, but it may not finish off a problem totally. However, if you do see some improvement you will then know to go to high potency with that particular remedy to finish off the problem. (There are some remedies which have different effects if given in low or high potency, such as hepar sulph and folliculinum and they need to be kept in mind — see Chapter 12.)

For comparison and to keep things in perspective, let us briefly touch on some conventional compounds used in low concentrations whose biological effects are certainly well known. The following examples are all drawn from real life. The popular procedure of injecting botulinum toxin (Bo-Tox) into wrinkles to make them disappear definitely uses extremely small quantities of that deadly substance to give its desired effects. The relatively new scientific field of hormesis is based on the supposition that very low levels of background radiation in the environment cause effects upon biological systems. The psychedelic compound, LSD, exerts its profound hallucinogenic effects at dose concentrations of only 25 micrograms (25×10^{-6}), which would be the same as roughly a 6X homeopathic remedy. Another example would be the ability of humans to pick up very faint scents in the air — molecules very diluted by the air currents they are floating on. Internally, hormones circulate around the body in low concentrations, roughly $10^{-6} — 10^{-8}$ / ml blood (equivalent to 1cc of a 6X-8X or 3C-4C remedy). Even a pharmaceutical compound like butorphanol (synthetic morphine) is administered to a cow at about 10 mg per dose, usually just one time for minor surgery. That is 10×10^{-3} of a drug into a 1200 pound animal. Not much when you think about it and its anesthetic effects are very noticeable when you put scalpel to skin. Biological systems, whether they be a 1200 pound cow or a 2 pound chihuahua are in delicate homeostatic balance and anything foreign introduced can have dramatic effects, even in extremely low concentration.

So I think it is very reasonable and rational to use remedies in low concentrations, especially the potent plant alkaloids. We need not consider, confuse or misname these medicinal compounds as "homeopathic remedies." They can simply be called "remedies" to avoid confusion (or carrying the potential stigma that skeptics assign to homeopathy). And in any event, not every medicine used needs to be "homeopathic" nor does every situation call for "energy medicine" alone. For instance, if we do not administer fluids to a dehydrated cow to correct her circulation, our attempts at using homeopathy (or other treatments) will be less successful. Remedies with potency less than 12C or 24X (such as a 3C or 6X) simply are not "energy medicine" as is popularly talked about by holistic consumers. Low potency remedies fortunately *do* retain the biochemical characteristics of their starting material — and that is good! For clinically ill cows we should want the whole, combined and orchestrated effects of all the constituents of the *Aconitum napellus* plant to decrease a fever by reducing the heart rate, not only the "essence" of Aconite (as with Aconite 200C). Why would we only want to administer 10 pellets saturated with the "essence" of aconite (Aconite 30C) to a 1200 pound cow when the exact same symptoms in the Eclectic school call for it also, but in a small material quantity so we insure some of the aconitine alkaloid is present to reduce the pounding heart? Cows are large creatures, and I firmly believe they could benefit from material quantities of whole plant mixtures, sometimes in small amounts as with the alkaloidal plants and sometimes in larger amounts as with ginger, gentian, aloe or peppermint.

Perhaps the best of both worlds is using multi-potency *homeochords.* These are individual remedies that come in multi-potency strength. For instance, an arnica homeochord may contain 1X, 2X, 4X, 12C, 30C and 200C. Using this method, you can let the cow's own system select which potency it will respond to. The correct remedy still needs to be chosen, but then you are covered as far as if it is the exact right remedy, for the high potency is there to act. Yet if it is not the exact right remedy, the low potency will hopefully still be somewhat helpful. The 1X, 2X, 4X, 12C, 30C and 200C combination is the homeochord mixture that I like to use with certain remedies — especially those plant medicines that have essentially the same physical indications in both the Eclectic and Homeopathic Schools. Truly, the best combination for the rapid cure of the animal would be to use medicinals that correct the animal's material and energetic level. This would mean using the correct homeopathic remedy in potency (with a homeochord) to stimulate the animal's vitality and administer a measurable amount of a correct plant tincture to directly oppose and subdue specific symptoms. When this method is used judiciously with other supportive therapy (IV or oral fluids, if needed), I believe a practitioner can truly enhance the well-being of the cow by stimulating their vitality as well as help oppose dangerous physical changes. In short, the practice of truly holistic dairy medicine can be accomplished.

Hopefully I have sketched out why I believe low potency remedies or multipotency homeochords should be used for dairy cows when electing to use remedies and why they would be a logical choice for a beginner. I would like to also caution against using the full dose regimens as written about by veterinarians in textbooks of the time; however, many used by the Regular school veterinarians were USP plant fluidextracts which are no longer available anyway since the advent of purified synthetics in the 1930's and 1940's. By careful study and following some of the reference material presented in this work, veterinarians or farmers may become adept at use of available plant tinctures, which have all the constituents still locked into the mixture. If a veterinary practitioner is intrigued about this whole field of medicine, especially as practiced by the Regular school, please refer to chart by Pierre Fish (see chapter 11) which may be used as a guide to dosing with plant medicine tinctures (crude drugs). The chart shows both fluidextract and tincture doses for various plants and intended species. Dr. Fish created a table which, as it stands now, is a critical link to understand how veterinarians of yesteryear used plants medicinally. However, strength or power always seemed to be foremost on the minds of practitioners trained in the Regular school (be it veterinarians or physicians). The Eclectic practitioners, on the other hand, were always hesitant about outright strength of plant medicines and always kept in mind the whole array of constituents of a plant as critical to wise usage of such materials (as opposed to focusing on the one dominant active principle). John Uri Lloyd, the great Eclectic pharmacognosy expert, delineated the issue of *quality versus strength* quite well. While endeavoring to use plant tinctures or their derivatives, we would do well to keep his words in mind. Lloyd:

John Uri Lloyd

An error common to a superficial, as well as to a one-sided or fragmentary conception of pharmacy, is that of considering *strength* and *quality* as synonymous terms. As we have said, it is a common error, but it is established by very high authority. The truth is that, although more or less related, the constituent that gives the factor *strength* is often less important than are the attributes that go to make up *quality,* which, perhaps more than does strength, leads to high excellence. Let us define *strength* as a dominating something that stands out boldly, and which, in toxic drugs, produces a violent or energetic action, as does the poisonous something that produces death when an overdose of a toxic drug is administered. Let us define *quality* as a balanced combination of other something, with just enough of the toxic agent to make a complex product that, as a whole, has wider functions than are possible if the single death-dealing substance dominates. But we need not confine ourselves to toxic drugs, for, from all time, in many familiar directions, such as tea, coffee, spices, tobacco, etc., standards of strength have been differentiated from those of quality.

The dominating, poisoning agent in nux vomica is a strychnine compound, and on this substance rests the official (U.S.P.) strength of the drug. But nux vomica contains other alkaloidal structures and essential oils, as well as other organic complexities, which, balanced in Eclectic pharmacy and then used in Eclectic therapy, are necessary to the quality of the Eclectic nux vomica. In the standardizing of nux vomica, the U.S. Pharmacopoeia recognizes strychnine only, whilst the Eclectic physician considers

strychnine, in undue proportion, objectionable in that it dangerously overbalances *quality.* (Lloyd, *The Eclectic)*

Unfortunately there weren't many veterinarians back then that recorded
their work from which to draw information regarding low potency remedies
or Eclectic tincture dosing for livestock. Dr. Humphreys left us his *Manual of*
Specific Veterinary Homeopathy, and we have the conventional veterinary textbooks of the time. The veterinary texts are a wealth of information regarding
plant derived medicinals (especially the books of Dun, Winslow, Fish, Udall
and Milks) — and these are all specific to the veterinary field. However, to fully
grasp and utilize the idea of low concentration remedies, we need to look to
the Eclectic school's use of plant tinctures and fluidextracts in the realm of
human health. We need to extrapolate (as is commonly done by veterinarians)
and study their materia medica. Among the best known Eclectic practitioners
were Finley Ellingwood, Harvey Felter and Eli G. Jones. These physicians used
whatever the case required, regardless of medical doctrine. They were truly
eclectic in their approaches to taking care of their patients and enjoyed a high
rate of success.

The low potency homeopathic doses are nearly the same as Eclectic doses
to treat the exact same conditions, but homeopathic mother tinctures are by
definition 1:10 dilution whereas Eclectic tinctures varied anywhere from 1:2
— 1:10 (with fluidextracts always being 1:1). However, it was recognized
among the Eclectics that a Homeopathic mother tincture was of high quality
due to the HPUS standards. If a homeopath would prescribe a 2X potency of
a remedy, this would be roughly equivalent to using one part of an Eclectic
plant tincture placed into 10 parts water to be given orally (each medicine
being a 1 in 100 concentration). The difference being that a homeopathic remedy is diluted and succussed (to release the starting material's energy) while
the Eclectic tincture, if put into water, is simply diluted but retains its original
constituents and biological vitality. The Eclectics often prescribed a combination of tinctures or fluidextracts and would recommend 40 drops placed into
4 ounces of water and administer a tablespoon of the mix every hour or two
"until it operated." That would be for humans. By looking at the doses of Humphreys (10-20 drops per dose for a horse or cow, in no water or further dilution), we can extrapolate that tinctures as Eclectics would use them and low
potency homeopathic remedies would be similar doses for large animals like
cows. The frequency of administration of low concentration medicines would
be more often, however, about 3-5 times daily as needed for a day or two, or 3
large doses a day as indicated by the Regular school for their use of plant
medicines in full strength (fluidextracts). Using low concentration plant
medicines also keeps us well out of the danger areas delineated for various
plants by Osweiler (see Chapter 11). Therefore, I would again submit that it is

safe to use low concentration plant medicinals for use in dairy cows and also logical to do so. In summary, by using low concentration medicinals we get 1) true pharmacologic activity owing to the plant constituents being present albeit in small quantities, 2) we get the holistic benefit of all the constituents by using tinctures, and 3) by using them according to historical veterinary precedent we can feel confident in using them for their indicated conditions.

The one time predominance of plant medicine in the therapeutic arena, however, is unfortunately ignored by mainstream veterinary medicine today. Plants should be respected for they contain a vast source of therapeutic compounds and should not only to be looked at with doubt due to their potential toxicities as mainstream veterinary medicine insinuates. By consulting the veterinary texts between 1900 and 1940, we can get a very clear description of what, when, how and why plant tinctures and fluidextracts were used. Therapeutic use and toxicity are partitioned from each other simply by the dose administered. It should be emphasized that historical conventional veterinary use of plant medicines and the Eclectic use differ fundamentally in that when conventional medicine was able to obtain the purified, synthetic active principles of plants, they immediately began to use them. (The cheaper mass produced synthetic versions that became available were, it must be recognized, based on the natural substances which were shown to have worked clinically.) The Eclectics never aimed to only use isolated active ingredients but rather use the whole juice of the plant, with all its biologically complex constituents still bound together in order to obtain the entire orchestrated effect of the original plant material. Thus, an Eclectic approach would appear to be an ideal basis for the holistic treatment of acute medical conditions in certified organic dairy cows.

Chapter 10
Comparative
Materia Medica

The following list of comparisons between remedies is necessary to learn when each should be used. It is important to study compounds in various materia medica due to the possible confusion of somewhat similar indications set forth by the three schools, with each school prescribing them in radically different doses. I also believe people should try to have a good idea as to what the remedies are as found in nature and why they are to be considered for use. Experience will guide the user in how much to use and how often to use a compound. Since I began my path into alternative medicines from the homeopathic "energy medicine" standpoint as a herdsman, I will confine the comparative list to those plant based remedies used in homeopathic medicine which have also been used in the other two schools that use actual material amounts of the same compounds (Eclectic and Regular/Allopathic). For those interested in a more in-depth knowledge of remedies (from whichever school's point of view), please refer to the bibliography. With the need to conserve space, I will limit the indications for each remedy up to the most commonly known *physical* indications as promoted by each school. As stated earlier when compared to the Regular school, both the Eclectic and Homeopathic schools of thought take into consideration the more subtle symptoms. If one of these more subtle symptoms is a hallmark of when to use a remedy, it will then be included. Otherwise, only the most prominent physical symptoms will be described. This is in keeping with my general concept that livestock are observed much more easily from the physical standpoint than the more in-depth unique symptoms that are able to be perceived with companion animals. Moreover, since all animals lack the ability to speak and give exact detailed descriptions of their problems, the physical realm would seem the most logical to stay within. This may also counter claims of a placebo effect taking place.

The remedies shown within the Regular school are based on physiological experiments as observed by veterinary scientists of that time. This means that the medicines advocated by the main stream veterinary community at the turn of the last century were based solely on the resulting physiologic effects. This often is still the case in current clinical practice. Claims that all clinical studies must be based on statistically sound double blind randomized controlled trials were unheard of prior to the pharmaceutical revolution which began in earnest only after World War II. There were some studies with a handful of animals that compared animals' reaction to a given medicine to those that were given none, but in general, treatment effects were observed and clinicians recorded these and built confidence in remedies by repeatedly seeing the same effects. Thus clinicians gained insight and wisdom about the medicines that were available from their accumulated experience "out on the front lines." This is still the *modus operandi* of most clinicians in private practice.

Where the reader finds that the indications as given by the schools are nearly equivalent, care should be taken to consider the general safety or toxicity of the remedy before deciding which school of thought to follow in figuring out dosing schedule. A good rule of thumb would be to use the Regular school's dose for very safe medicines, such as for peppermint, ginger and licorice and go with the Eclectic or Homeopathic schools' dosing for strongly energetic compounds (such as alkaloids) like aconite, belladonna, gelsemium, veratrum, etc. It must be remembered that dairy cows are large creatures, usually weighing between 1000 — 1400 pounds. Some conditions simply need more quantity of a naturally derived plant medicine on a physical level to deliver what a creature of that size weighs. However, highly active alkaloidal compounds generally are not needed in large quantities and should be given in probably smaller physical quantity (yet more frequently) than what the Regular school prescribed. It probably is best to keep in mind the difference between a fluidextract and a tincture. Fluidextracts could be too strong when used at the doses of the Regular (conventional) school. By researching the literature on tinctures, I have found that extrapolating the human dose (as used by the Eclectics) to dairy cows, the corresponding tincture doses actually match well with the tinctures listed in Fish's table. A fluidextract contains 1000 gm of the desired part of the medicinal plant in 1000cc of an alcoholic solution. The part of the plant used was dictated by the USP, of which certain tests for identity were required before pharmaceutical preparation of the fluidextract. Of importance to note is that the starting plant material was to have been dried, macerated and able to pass through a certain size sieve. Drying all plants before fluidextract manufacture may harm constituents, if the constituents are more available and active when fresh. So, although fluidextracts would yield as exact a standard product as was possible, the process of drying each and every original plant material may have negatively altered the biological vitality, if not the medicinal chemistry, of the starting plant material. Since fluidextracts are no longer manufactured, this should suffice to give the reader an idea of what they were. However, many tinctures are currently made. Hence, I will quote Ivon Rachon, owner of Herbal Vitality, Inc. (Sedona, Arizona):

> The starting materials for our fresh plant tinctures are sustainably wild crafted or organically grown plants shipped to us by next day courier (most leafy and flowering plants) or by second day air (most roots). We use professional wild crafters who are committed to ethical harvesting techniques, and to safe guard wild plant populations. These individuals are expert botanists who use additional academic resources when needed to ensure exact identification of the plant species sent to us.
>
> Once the plants are received at our facility, we confirm plant species identification, and then begin the process of meticulously cleaning and processing the herb. For leafy plants, the cleaning process includes: a) removing foreign plant species — which may be present in significant amounts with plants such as cleavers herb (Gallium asparine), squaw vine herb (Mitchella repens), or dandelion leaf (Taraxacum officinale), b) discarding necrotic leaves and older non-photosynthetic stem, c) washing the leaves in filtered water if necessary, and d) spinning excess water off before extraction.
>
> For roots, the cleaning process includes: a) washing and air drying the roots, and b) removing all moldy or necrotic areas — these areas can amount up to 20% of the root weight for plants such as black cohosh (Cimicifuga racemosa) or stoneseed (Lithospermum ruderale).
>
> The resulting cleaned herb — free of foreign materials, and concentrated to contain only the plant parts with the highest proportion of known active ingredients — is then extracted in species-specific concentrations of pure grain alcohol (organic) and pure distilled water. We use species-specific concentrations of ethanol during extraction to insure the highest possible con-

centrations of the known active constituents in our liquid extracts. The carrying capacity of dissolved molecules within a solution is limited, and over-saturation of the solution can lead to precipitation of specific components of the mixture, or dilution of the active constituents. For example, it has been reported that a 30% ethanol extract of cramp bark (Viburnum opulus) is 5 times more anti-spasmodic than a 60% ethanol extract. We are pursuing research of our own aimed at developing extraction protocols resulting in products with the highest levels of vital active constituents.

It is certainly worthwhile to treat the animal both on its physical and energy levels, which may mean using more than one particular remedy or mode of treatment. When using products made as described above, the resulting tinctures will have biological vitality and pharmacologic effects — in short, a potent holistic medicinal.

Cost is always a consideration in dairy farming and without doubt, homeopathic remedies are the cheapest (and totally non-toxic due to their extreme dilutions). Homeopathic remedies as I dispense them cost $10 per 1 ounce bottle of size #35 wetted pellets — each bottle containing about 250 pellets. Cows usually are given 10 pellets per dose, so there are 25 doses per $10 bottle. Doses of fluidextracts or tinctures used according to the Regular school are "real amounts" of compounds; an ounce (~30cc) of tincture usually will cost the farmer about $6 (as supplied by Herbal Vitality, Inc.). With the Eclectic theory teaching to use quantifiable but small amounts of whole plant tinctures (all the plant's original constituents present), the Eclectic approach is truly holistic while offering safe medicinal doses of known effective bioactive compounds. However, the total quantity in a bottle, if accidentally ingested in its entirety may be potentially toxic — thus it is wise to use child resistant screw cap lids. Although a farmer may have a six- or eight-ounce bottle of tincture on hand, the amount administered orally to a cow is very small (10-20cc) and not toxic in the Eclectic approach. Another way to embark on the Eclectic approach for those who have experience in homeopathy is to use low potency dilutions as previously discussed. These may be bought pre-made from a homeopathic pharmacy in order to ensure the proper quantity of a medicinal is being administered. This makes very much sense to me and is quite defendable. Alternatively, giving a liquid dose of a homeochord remedy (3-5cc) may again be the best of both worlds (material and energetic).

I admit that it is unfair to each of the schools' *materia medica* to give such a brief synopsis of plant medicines. However, this section is definitely intended to show some of the great similarities in the main *physical indications* of a compound among the three different schools. In addition, perhaps the reader will appreciate why the author started to think about how compounds were being used for similar symptoms in different schools, and why the question of concentration or potency started to blur the lines between the Schools. There is abundant information on botanical medicines and other natural medicines (pharmacognosy) which unfortunately has been long forgotten or simply tossed aside to make way for purified, synthetic pharmaceuticals. It is my sincere hope that some of the readers of this work will be stimulated to delve into the fascinating world of pharmacognosy and how it can apply to herbivorous animals such as dairy cows.

The following botanical medicines include the common name, natural order, part used and primary active constituents. The average range of concentrations of various constituents are shown, when known (USDA, ARS, National Genetic Resources Program. *Phytochemical and Ethno-botanical Databases* [online database]. National Germplasm Resources Laboratory, Beltsville, Maryland, accessed 8/15/03 and 11/8/03.) Medical indications will be given by each school. The main reference work for the Eclectic descriptions is *King's American Dispensatory* by Felter and Lloyd, 1898. The homeopathic indications are for remedies made according to the official HPUS — Homeopathic Pharmacopeia of the United States. The Regular school description denotes how and what to use the remedy for (found in the works of Dun, Winslow, and Milks), while the Eclectic and Homeopathic description denotes the appearance of the condition (indications) calling for the remedy. This is a subtle difference but necessary to point out. Indications oriented towards conditions of special interest for dairy cows will be highlighted if the particular compound has specific action upon, for instance, the mammary or reproductive systems. It should be noted, however, that the information from the Eclectic literature needs to be extrapolated to dairy cows, but that is not too difficult since the Eclectics did rely and refer to physical symptoms to a great extent. The information from the Regular school of previous years is specific to veterinary medicine, and the Homeopathic school is specific to dairy cattle.

The USP annotation is included if the substance was found to be USP listed. The term "USP" stands for United States Pharmacopeia; USP crude drugs (medicinal plants) were subjected to standardized processing, conforming to certain descriptions and tests *i.e.,* standards of identity. If no USP annotation is shown, the remedy was not part of the USP. Also note that "NF" stands for "National Formulary," a commonly accepted listing of substances used by pharmacists and doctors of sufficient importance to render standardization desirable. They were recognized as official by the National Pure Drug Act of 1906. All these were the medicines of the time, relied upon heavily by all physicians and veterinarians. Mineral medicines such as mercury, arsenic and lead were also used widely, but will be omitted in keeping with the Eclectics main thrust of botanically derived medicinals as well as my own bias towards plants as crude drugs. Emphasis on botanical medicines is also done to be totally compatible for certified organic use, whereas not all the mineral medicines would be. Perhaps I will take up the myriad of mineral medicines in the future. For those interested in mineral medicines, quick reference is made to them (as used in homeopathy) and can be found in The Homeopathic Remedy Guide for Dairy Farms (see chapter 12).

Fluidextracts are no longer obtainable, while tinctures are easily obtainable. If, by chance, a tincture is labeled as (1:1), as can be found for ginger, this is still a tincture since fluidextracts are "extinct," and the ginger tincture would have most likely been made with fresh ginger root. Also, make sure that you know the concentration of the tincture you are buying. If the processor will not tell you, you have no idea how strong or weak the tincture is and you therefore also don't know how much you are paying for. All legitimate herbal companies state clearly in their brochures what the concentrations of their products are. Therefore, if you are not told what the concentration of constituents are in a mixture or as a single remedy, find a different supplier! See the list of resource contacts in the appendix.

If the reader understands and learns the following 21 remedies, this will be a great foundation in learning to use natural medicines in dairy cows. Although there are some other plant medicines and homeopathic remedies that are definitely needed for use in the barn, the main ones will have been addressed here. See the Recommended Remedies in the Appendix for the entire listing of remedies which are essential to keep on hand.

Note: ppm = mg/l

Aconite (USP) — *Aconitum napellus.* Monkshood, Wolfsbane. Nat. Ord. — Ranunculaceae. Contains alkaloids in leaves (1200-9600ppm) and root (2000-15,000ppm) of which aconitine is the best known; also calcium aconitate, mannite, cane sugar, glucose, resin, fats. Leaves: aconitine, gum, albumen, sugar, tannin, aconitic acid, nappeline (possible decomposition product of aconitine).

Regular: To reduce fever; reduce restlessness and pain; employ at first stages of febrile (feverish) respiratory disease (Winslow).

Eclectic: Asthenic (weak, non-dynamic) febrile state with small and frequent pulse, skin hot and dry with chilly sensations; with or without restlessness; irritation of mucous membranes with vascular excitation and redness.

Homeopathic: Sudden physical or mental shock, sudden or early fever, usually with anxiety; sudden bleeds; shuddering or shivering; usually see red, inflamed watery eyes; watery nasal discharge; tense, rapid pulse; urine retention and thirst (Day); Sudden fever after stress, shipping, or a chill accompanied by fear, restlessness and want for water (Sheaffer).

Arnica — *Arnica montana*. Arnica, Leopard's bane, Mountain tobacco. Nat. Ord. — Compositae. Contains inulin (rhizome: 90,000-120,000ppm), catechin-tannins (rhizome: 23,000ppm), essential oils (rhizome: 5000-63,100ppm; root: 17,700-37,400ppm; flower: 400-1400ppm). Other components include volatile oil, acrid resin, extractive, gum and woody fiber in the root, the flowers contain resin, a bitter substance resembling cytisin, gallic acid, yellow coloring matter, albumen, gum potassium chloride and phosphate, traces of sulphates, calcium carbonate, and silica.

Regular: Used externally for strains, bruises and wounds, especially broken knees and sore shoulders; corneal ulcers (ulcers of the outer layer of the eye) affecting weakly dogs recovering from distemper (Dun).

Eclectic: muscular soreness and pains from strains or over-exertion, especially when limbs are moved; lack of control over urine and feces, cystitis (inflammation of the urinary bladder) with bruised feeling in bladder; headache with bruised feeling; hematuria (urine contains blood or red blood cells) with dull, aching lumbar pain.

Homeopathic: All conditions involving injury; bruising, fear of being touched; blood in milk; hematoma (localized mass of blood) and laminitis in heifers (painful inflammation of lamina to which hoof is attached) (Day); Injury, trauma, bruising to soft tissues, body and mind; concussions, hematomas and mild nose bleeds (Sheaffer).

EMEA: All food producing species; at concentrations in the products not exceeding one part per ten only (1X)

Belladonna (USP) — *Atropa belladonna*. Deadly nightshade, Dwale. Nat. Ord. — Solanaceae. Contains alkaloids (leaves: 1000-12,000ppm; root: 4000-8800ppm). The root has hyoscyamine (2000-8712ppm), L-hyoscyamine (3080-7656ppm), scopolamine (40-88ppm), hellaradine (20ppm) and tannin (120,000ppm). Other components include succinic acid, malates, oxalates, salts of sodium, potassium and magnesium, gum, wax, asparagin, chlorophyll (leaves), starch and albuminous bodies.

Regular: Used internally to stimulate respiration and circulation; to diminish secretion; to relieve spasm and pain; to dilate the pupil; used externally for mastitis to relax spasm, contract blood vessels, and lessens inflammation and congestion; it paralyzes the secretory nerves and thus diminishes the amount of milk, vascular tension, pain and glandular activity.

Eclectic: Dull, expressionless face, dilated pupils; congestion with dilated capillaries; deep redness of skin, effaced by finger leaves white streak with blood slowly returning to part; circulation sluggish with soft, full pulse; slow, labored and imperfect breathing; sleeping with eyes partially open.

Homeopathic: Febrile and inflammatory states with much heat, redness, fullness and pain; pulse full and bounding; minor symptoms of thirst, dry mouth and dilated pupil (Day); Full bounding pulse in any feverish condition which may or may not accompany excitable states; acute mastitis.

EMEA: All food producing species; at concentrations not exceeding one part per hundred only (2X or 1C).

Chenopodium (USP) — *Chenopodium ambrosioides.* Wormseed, American wormseed. Nat. Ord. — Chenopodiaceae. The plant, leaves and root contain the most quantified information, but only the seeds (fruit) have been official (USP): alpha-pinene (plant: 440-4800ppm), aritasone (plant: 9-1400ppm), ascaridole (leaf: 185-18,000ppm), ascorbic acid (leaf: 110-1020ppm), beta-carotene (leaf: 35-165ppm), essential oils (fruit: 1830-25,000pm; leaf: 2000-3000ppm), L-pinocarvone (plant: 1040-11,400ppm), P-cymene (leaf: 365-4400ppm), P-cymol (plant: 730-8000ppm), saponins (root: 25,000ppm), terpinyl acetate (plant: 75ppm), terpinyl salicylate (plant: 75ppm).

Regular: To expel round worms and hook worms, but fast for 12-24 hours first and use a purgative (produces increased discharges from the bowels) immediately before or after dosing; almost 100% efficient against ascarids, except the horse, and for strongyles and pin worms of horses; ranks next to tetrachlorethylene and carbon tetrachloride against hookworms (Milks).

Eclectic: To expel lumbricoid worms; sometimes mixed with oil of tansy, spirits of turpentine and added to castor oil.

Homeopathic: The oil for hookworm and roundworm (Boericke). Helpful in cases of strongyles and hookworms; use as a vermifuge (agent that expels worms), often in combination with homeopathic abrotanum and santoninum (Sheaffer).

Cinchona (USP) — *Cinchona officinalis.* Peruvian bark, Cinchona bar, Quinine. Nat. Ord. — Rubiaceae. The Cinchona spp. contain many alkaloids (bark: 60,000-160,000ppm; leaf: 10,000) chief among them are: quinine, quinidine, cinchonine, cinchonidine, quinamine, homoquinine, cupreine, hydroquinine, hydroquinidine and hydrocinchonine.

Regular: Internal — sharpens the appetite; chiefly a tonic by increasing the number of red corpuscles and stimulating the nervous system generally; anorexia (diminished appetite) and atonic dyspepsia (impaired gastric function) secondary to exhaustion, overwork, anemia, or following acute diseases; acute diseases such as influenza, bronchitis and pneumonia (used in full doses at the outset of colds or inflammatory diseases of the respiratory tract); puerperal fever (relating to the period after birth); muscular rheumatism. Externally (as quinine and urea hydrochloride) — local anesthetic, 1% solution forms a satisfactory substitute for cocaine. It has three advantages over cocaine: it is non-toxic, it can be boiled in solution and its anesthetic effect is often prolonged for hours or days, lessening pain and spasm after operation and aiding dressing of wounds. Anesthesia comes on within 5 to 30 minutes after injection into tissues. In 25% solution it is used to anesthetize mucous membrane (but is not as potent as cocaine). It can be used for intra-abdominal operations, anesthetizing the skin pre-operatively and the parietal peritoneum (belly lining) intra-operatively.

Eclectic: periodicity, when pulse is soft, tongue moist, skin soft and moist, and nervous system free from irritation; gastric debility, anemia, debility from chronic suppuration (formation of pus); afternoon febrile conditions; weakness with pale surface, loss of appetite, feeble digestion and deficient recuperative powers.

Homeopathic (also called *China* in homeopathy): Intermittent fevers; exhaustion and debility with drained constitution, especially after long illness; dehydration, flatulence, colic, tympanic abdomen and painless yellow diarrhea (Day); Loss of vital fluids (diarrhea, hemorrhage) resulting in a thin, dehydrated state; weakness after fluid loss over a period of time; symptoms come and go in a periodic fashion (Sheaffer).

Ergot (USP) — *Claviceps purpurea.* Ergot of rye (Secale cereale), spur, spurred rye, smut rye. Nat. Ord. Fungi — Ascomycetes. It contains ergotinine/ecboline (alkaloid), ergotoxine, tyramine, histamine, ergamine, isoamylamine, lysergic acid, cornutine (alkaloid), sclerotic acid, scleromucin, chrysotoxin, secaletoxine and sphacelotoxin.

Regular: uterine inertia (inactivity or lack of force) where there is no malpresentation or mechanical obstruction (pelvic deformity or rigid os uteri); very small doses to intensify the force of the contractions without inducing spasm of the uterus; prevent or arrest post-partum hemorrhage in cows and ewes; if given before delivery of placenta, ergot may give rise to tonic contraction of the womb and retention of afterbirth; to aid the expulsion of uterine cysts, to contract the uterus and its blood vessels in hypertrophy, subinvolution, chronic metritis; internal hemorrhage when surgical measures are impossible (bleeding from nose, mouth, stomach, intestines, lungs, uterus and kidneys); predominantly useful for uterine hemorrhage.

Eclectic: Uterine inertia during labor, when conditions are otherwise favorable for a safe delivery; hemorrhage due to atony, with weak pulse and cold surface; post-partum hemorrhage (large doses); to expel loosened foreign particles from the womb; congestion or hyperemia (increased amount of blood) of any part; venous fullness; hyperemic or congestive eye and ear disorders; congestive headache.

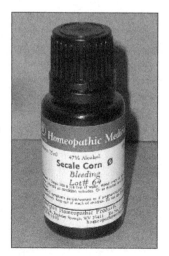

Homeopathic: (called *Secale cornutum* in homeopathy). Damage to circulation to feet, tails, ears and teats; ischemic necrosis (very poor circulation and lack of oxygen); help involution of womb when there is discharge of dark, stinking blood (Day); Dark bloody, putrid uterine discharges; skin becomes dry and shriveled with tendency to gangrene; stools are dark green, alternating with dysentery (Macleod); Animal is thin, scrawny and emaciated; vessels flabby and all tissues lack healthy elasticity; dark, watery post-partum hemorrhages; antidotes some of the toxic effects of endophyte infected fescue pasture and hay (Sheaffer).

Gelsemium (USP) — *Gelsemium sempervirens.* Yellow jasmine, Yellow jessamine, Carolina jessamine or jasmine. Nat. Ord. — Loganiaceae. The root holds most of the known constituents: essential oil (5000ppm), gelsedine, gelsemicine, gelsemidine, gelsemine, gelsevirine, scopoletin, and tannin. Gelsemine is a known analgesic, cardio depressant, CNS-depressant, CNS-stimulant, hypotensive and mydriatic. It also is a known anti-inflammatory, antiaflatoxin, anti-leukotrienogenic, antiseptic, antiprostaglandin, phytohormone and uterosedative.

Regular: Similar action to conium; spinal meningitis and tetanus; cardiac depressant and antipyretic (reduce body temperature) in acute cases (pneumonia and pleurisy), but inferior to aconite and veratrum viride; in spasmodic diseases, irritable cough, vesical irritation, tetanus, chorea — but less satisfactory than opium and belladonna. In tetanus, give gelsemine (.08 gm) every half hour under the skin till muscular relaxation occurs (lobeline is, however, in favor for this purpose); mydriatic, but produces some pain as compared to atropine.

Eclectic: Bright eyes, contracted pupils, great heat, restlessness; mental irritability; insomnia, with excitation; irritation of the urinary tract; pinched, contracted tissues; thin, dry, unyielding os uteri (uterine cervix), with dry vaginal walls; arterial throbbing; hyperemia.

Homeopathic: Strongly indicated when anxiety is present; muscular weakness in viremic conditions, droopy eyelids, general weakness of limbs and incoordination, muscle tremors (Day); Supportive measure in grass tetany, aiding restoration of normal movement; conditions usually show weakness and muscle tremors (Macleod); Dullness, drowsiness, muscular weakness, lack of thirst; in viral diseases; prevention of milk fever and grass tetany (Sheaffer).

Hydrastis (USP) — *Hydrastis candensis.* Goldenseal. Nat. Ord. — Ranunculaceae. Contains the alkaloids berberine (root: 5,000-60,000ppm), hydrastine (plant 15,000-40,000ppm) and canadine (root: 2500ppm); as well as starch, albuminous matter, resin, sugar, fatty matter, and inorganic salts.

Regular: Act as bitters (stimulation of the gastro-intestinal lining) and stomachics (stimulates functional activity of the stomach), in small doses, by improving the appetite and stimulating the secretion, motion and vascularity of the stomach. Uterine hemorrhage, especially conjoined with ergot — but may induce abortion if pregnant. Atonic and inflammatory conditions of the gastro-intestinal organs with great benefit. Externally, the fluidextracts of hydrastis (1:8-1:2) are serviceable as local stimulants in the treatment of sub-acute stages of inflammatory diseases of mucous membranes, and in the relaxed or atonic conditions of these tissues. The solutions are applied as injections, or lotions, in leucorrhea (white or yellowish vaginal discharge), endometritis (inflammation of the uterine lining), balanitis (inflammation of the glans penis or glans clitoris), otorrhea (discharge from the ear), stomatitis (inflammation of the mucous membranes of the mouth), etc., and indolent ulcers.

Eclectic: Catarrhal states (simple inflammation) of the mucous membranes, when unaccompanied by acute inflammation (except in acute purulent otitis media [pus in the middle ear]); gastric irritability; irritation of parts with feeble circulation; muscular soreness and tenderness; worse under pressure or motion; passive hemorrhages from uterus and other pelvic tissues; skin diseases depending on a gastric abnormality.

Homeopathic: Catarrhal discharges of the womb having yellowish color; liver sensitivity with light colored, pale manure (Day); Secretions are thick and yellow; any catarrhal condition resulting in muco-purulent discharge (mild forms of metritis [inflammation of the uterus]or sinusitis [inflammation of the mucous membranes lining the sinuses]) (Macleod); Thick yellow, odorous uterine discharge; ovaries small, hard and painful; chronic digestive problems; cows that look older than their age; lean and worn out appearance; looks like the cow could use a good rest and recuperation (Sheaffer).

Nux vomica (USP) — *Strychnos nux vomica.* Nux vomica, Nux vomica seed, Quaker buttons, Poison nut. Nat. Ord. — Loganiaceae. This seems to be a very well studied plant in terms of quantified inherent compounds. It is rich in alkaloids (bark: 99,000ppm; root: 9900ppm; seed: 2500-50,000ppm). Quantified constituents include: arachidic acid (seed: 2800-3500ppm), behenic acid (seed: 680-850ppm), beta-colubrine (fruit: 40ppm), brucine (bark: 48,000ppm; fruit: 300-650ppm; leaf: 8000ppm; root: 2770ppm; seed: 16,000ppm), C-mava-curine (rootbark: 1000ppm), hydroxystrychnine (fruit: 7-20ppm), icajine (fruit: 20-60ppm); linoleic acid (seed: 360-4600ppm); myristic acid (seed: 360-450ppm), oleic acid (seed: 2480-31,000ppm), palmitic acid (seed: 5040-6300ppm), strychnine (bark: 15,800ppm; leaf: 8000ppm; root: 7030ppm; seed: 400-12,000ppm), vomicine (fruit: 130-910ppm).

Regular: Acts as bitter stomachic in increasing appetite, gastric secretion and motion; improves local tone of alimentary canal, probably by exciting the various spinal centers; stimulates the intestinal muscular tunic and thereby increases peristaltic (wave) action; stimulates the respiratory and vasomotor centers and nervous system; stimulates the spinal cord, specifically the motor cells of the inferior cornua.

Eclectic: Atonic states; yellowness of the conjunctiva; yellow or sallow countenance and yellowish or sallow line around the mouth; fullness and dull pain in right hypochondrium; pain in shoulder; colicky pains pointing to the umbilicus; menstrual colic; constipation; diarrhea of atony; functional forms of paralysis.

Homeopathic: Any condition due to rich feeds; sluggish rumen, mild bloat, liver over-burdened; cow is highly reactive and bad tempered; manure hard and dry and cow strains or manure can be very loose and fermentative; ketosis (Day); Rumenal stasis (non-functioning), hard stools; preliminary treatment in plant poisoning (Macleod); Digestive upset from excessive grains or silage; constipated and strains; colic and bloat; toxemia from conventional drugs, poisons, vaccines and spoiled food; disorders of liver or abomasum (Sheaffer).

Podophyllum (USP) — *Podophyllum peltatum.* May apple, Wild lemon, Raccoon-berry, Wild mandrake. Nat. Ord. — Berbericeae. The root and rhizome account for most of its properties. They contain — alpha-peltatin (rhizome: 3500-6000ppm), beta-peltatin (rhizome: 1750-3000ppm), desoxypodophyllotoxin (root: 1000ppm), podophyllotoxin (rhizome: 7000-12,000ppm), quercetin (resin, exudates, sap: 50,000ppm), resin (plant: 30,000-60,000ppm).

Regular: Powerful though slow acting cathartic (quickens or increases evacuations from the intestines), exerts its effects after absorption during excretion from the bowel, and therefore about ten hours are required to produce purgation; chronic constipation associated with jaundice and hepatic disorders; the result of its action more favorable when the fecal discharges are dark colored; owes its activity partly to the presence of bile, which seems to be a solvent for it. The fecal movements, after medicinal doses of podophyllin, are liquid, often stained with bile and may be accompanied by some nausea and griping.

Eclectic: Fullness of tissues, and particularly by fullness of superficial veins; oppressed full pulse; acts as a certain, but slow cathartic. With a cathartic dose, the intestinal and hepatic secretions are augmented and after a considerable time copious alvine evacuations result. Considerable pain and griping may attend its action, which, however, may be modified by such agents as leptandra, hyoscyamus and belladonna.

Homeopathic: Profuse, painless diarrhea with colicky pains; manure is offensive; rectal prolapse (Day); Diarrhea immediately after eating; "pipestream" diarrhea with little straining; stools contain large volumes of water; the coat is soft and flabby with moisture; morning diarrhea with normal manure in evening (Sheaffer).

Veratrum viride (USP) — *Veratrum viride.* American hellebore, swamp hellebore, Indian poke, itch weed, Green hellebore, Green Veratrum. Nat. Ord. — Liliaceae. The rhizome contains veratrine/cevadine, jervine, pseudojervine, rubijervine, and veratralbine (all alkaloids), gum, resin, starch, red coloring matter, wax, sugar, a bitter principle analogous to veratrine and gallic acid.

Regular: Therapeutic value depends on its effect on the circulation — it first lowers the force (jervine), and, if continued, the frequency (veratrine and jervine) of the pulse, and reduces vascular tension (jervine); applicable as a circulatory sedative at the outset of sthenic (strong, active) diseases afflicting strong animals; causes vascular dilatation. Considerably more depressant to the circulation than aconite, which does not lessen arterial pressure (but aconite does possess anodyne [diminishes sensibility to pain], diaphoretic [producing sweat] and diuretic [increases the secretion of urine]properties).

Eclectic: Pulse frequent and full; tissues full, not shrunken; surface flushed with blood; pulse rapid, bounding; marked arterial throbbing; increased arterial tension with bloodshot eyes; sthenic fevers and inflammations; irritation of nerve centers, due to an excited circulation.

Homeopathic: Pulse slow, soft, weak, irregular, intermittent; rapid pulse, low tension; beating of pulses throughout body, especially in right thigh; hot sweating; effects of sunstroke; best in full-blooded patients (Boericke).

The previous eleven remedies represented the ways in which all three schools used the same remedy, with the implicit assumption that the dosages as used among the three differed greatly. The following list of ten remedies will contain comparisons mainly between the Eclectic and Homeopathic schools because only these two schools used such botanical medicines extensively. Where there is information from the Regular school, it will also be included. Why the Regular School chose to use certain plants and omit others is not easily understood since many of the plants omitted were of the USP or NF (officially recognized as medicinal plants). Perhaps it had to do more with the dosage that they used, since many plant medicines used routinely as fluidextracts at full strength could indeed be toxic. At the other end of the dosage/ strength spectrum (*i.e.,* low concentration Eclectic and extreme dilution Homeopathic remedies), the reader can see that an astonishing number of indications are nearly identical on the physical level when reading of the Eclectic and Homeopathic entries. It is up to the reader to decide which School's remedy would best fit their ideas regarding medicinal compounds in general and which would likely be more efficacious in the barn.

Baptisia — *Baptisia tinctoria*. Wild Indigo, Indigo weed. Nat. Ord. — Leguminoseae.

Eclectic: Fullness of tissue, with dusky, leaden, purplish or livid discoloration; tendency to ulceration and decay; sepsis (various pus forming and other pathogenic organisms present in the blood); typhoid conditions (salmonella); enfeebled capillary circulation; color effaced by pressure and returns slowly; patient's face swollen and bluish; fetid discharges, with agony, and gangrene.

Homeopathic: Septic conditions of the blood; lethargy, low-grade fevers and offensive secretions; salmonella septicemia (blood poisoning); manure is offensive and dark; animal appears delirious and moves with stiffness and difficulty (Day); Useful in poisoning from food or what and where dysenteric stools have a putrid odor (Sheaffer).

Bryonia (USP, NF) — *Bryonia dioica* and *Bryonia alba*. Bryony, Snakeweed, Devil's turnip, Bastard turnip, Parsnip turnip. Nat. Ord. — Cucurbitaceae. The root contains bryonin (glycoside) and also an amorphous alkaloid, bryoresin (root: 20,000-68,000ppm), a volatile oil and an alcohol.

Eclectic: Acute and chronic serous maladies (the fluid lining the joints), in glandular enlargements, chronic orchitis (inflammation of the testicles), chronic pleuritic and pulmonic disorders, fevers and to overcome constipation. Sharp, cutting, pain with harsh cough; equally valuable when the pain is tensive and tearing, especially when aggravated by motion; the parts feel stiff, sore, bruised and there is a large quantity of mucous in the bronchioles, as evidenced loud mucous rales; pulse hard, frequent, and vibratile, and the temperature elevated.

Homeopathic: Dry, hard cough; dry rubbing sounds when lungs are listened to with stethoscope; thirst, dry mouth; painful swollen joints; mastitis cows will stand still for examination and also lie upon side of the mastitis; pneumonia calves will stand still while other calves run around; animal will be seen lying down most of day in effort to relieve pain (Day); Indicated in cows when there is painful udder swelling that also don't want to move; dryness of throat and all membranes with a thirst for a lot of water; hard, dry coughs (Sheaffer).

Regular: Drastic purgative; in overdose act as irritants, causing symptoms of acute enteritis (Milks).

Caulophyllum (USP) — *Caulophyllum thalictroides*. Blue cohosh, Squawroot, Pappoose-root. Nat. Ord. — Berberidaceae. The rhizome and roots contain caulophyllin (resinoid), leontin (saponin), albumin, starch, gum, coloring matters, calcium phosphate and sulphates, salts of iron, potassium and magnesium, phosphoric acid and silica.

Eclectic: Uterine pain with fullness, weight, and pain in the legs; emmenagogue (a drug to stimulate menstruation) promoting delivery, menstruation and dropsical (edema) discharges; relieves false pains during labor and coordinates muscular contractions, at the same time increasing their action; stimulates normal contraction instead of inducing spasmodic (continual) uterine contractions (as with ergot); valuable where parturition is delayed by debility, fatigue or lack of uterine nervous energy and for deficient contractions; afterpains, especially if spasmodic in character; decoction (a tea) given for 2-3 weeks previous to labor to facilitate birth; articular pain.

Homeopathic: Very useful in pregnancy and calving; helps regulate pregnancy and help prepare cow for calving; corrects opening of cervix and sets flagging contractions back into motion; good to help expel retained fetal membranes and help involute uterus (Day); Revives labor pains, possible alternative to oxytocin once cervix is open; useful in ringwomb (cervix not fully dilated) and in cases of uterine twist or displacement (author's note — get veterinary assistance in cases of uterine twist or displacement!); in animals having history of abortion, it can help establish normal pregnancy (Macleod); Useful to prepare for labor and during delivery; labors that do not progress; cervix does not dilate; help expel placenta; rheumatism of small joints with erratic, shifting lameness (Sheaffer).

Echinacea — *Echinacea angustifolia.* Purple cone-flower, Cone-flower, Narrow-leaved cone-flower, Black sampson. Nat. Ord. — Compositae. This is another well quantified plant: alkylamides (root: 40-1510ppm), ascorbic acid (leaf: 2140ppm; root: 843ppm), cichoric acid (flower: 12,000-31,000ppm; root: 6000-21,000ppm), cichoric acid methyl-ester (root: 6000-21,000ppm), echinacein (root: 10-100ppm), echinacoside (root: 3000-17,000ppm), essential oils (flower: 600-6000ppm; leaf: 100-6000ppm; root: 50-40,000ppm); flavanoids (leaf: 3800-4800ppm); heteroxylan (root: 800ppm), inulin (root: 59,000-200,000ppm), L-pentadecene (root: 400ppm), penta-(1,8Z)-diene (root: 400ppm), pentadeca-(8Z)-en-2-one (root: 4000ppm).

Eclectic: Tendency to sepsis (presence of various pus forming and other pathogenic organisms, or their toxins, in the blood) and malignancy, as in gangrene, sloughing, carbuncles (deep seated pyogenic infection of the skin and subcutaneous tissues, usually arising in several contiguous hair follicles, with the formation of connecting sinuses; often preceded by fever or malaise), abscesses, boils and various forms of septicemia; foul discharges, with weakness and emaciation; deepened, bluish or purplish discoloration of skin or mucous membranes, with a low form of inflammation; as a cleansing wash in surgical operations (purified).

Homeopathic: Toxemia or septicemia due to mastitis, metritis, cellulitis (inflammation of cellular or connective tissue); weakness, fever, chilliness, prostration; discharges are foul; possibly infused into udder (Day); Acute toxemias with septic involvement; post-partum puerperal (relating to the period after birth) conditions with sepsis; generalized septic states originating from bites or stings; best in low decimal potencies (Macleod); Acute sepsis, abscesses/boils, puerperal fever (bad uterus); often combined with other remedies — alternate with lachesis for gangrene; alternate with pyrogen in mastitis or metritis; use with vitamin C and probiotics in infections (Sheaffer).

Hypericum — *Hypericum perforatum*. St. John's wort. Nat. Ord. — Clusiaceae. The flowering tops and leaves contain: ascorbic acid (plant: 1300ppm; seed: 395ppm), carotene (seed: 165ppm), essential oils (plant: 600-3500; seed: 3300ppm), flavanoids (flower: 117,100), hypericins (plant: 95-4660ppm), mannitol (plant: 11,000-20,000ppm); pro-vitamin A (plant: 130ppm), tannins (flower: 162,000; leaf: 124,000; plant: 51,400-92,700ppm; seed: 121,000ppm; stem: 38,000ppm) as well as volatile oil, resins, and coloring matter.

Eclectic: Spinal injuries, shocks or concussions; throbbing of the whole body without fever; spinal irritation, eliciting tenderness and burning pain upon slight pressure; spinal injuries, and lacerated and punctured wounds of the extremities, with excruciating pain.

Homeopathic: Cases involving injury to areas rich with nerve endings; spinal injury; soothing effect on nerve endings; possible anti-tetanus properties; lacerations, scrapes and laminitis; photosensitization, underlying liver damage is ameliorated (improved) and tissue damage minimized; the "homeopathic morphine" (Day); Spinal injuries, especially the coccygeal area; its action on nerves suggests its use in tetanus, if given early, helps prevent the spread of toxin; externally applied on lacerated wounds along with calendula (Macleod); Bruised nerves of lower legs and lower back, painful puncture wounds, injuries to tails and toes, tramped teats; weakness and paralysis; sunburn of white areas of coat (Sheaffer).

Lycopodium (USP) — *Lycopodium clavatum*. Club moss, antler herb, Wolf's claws, lamb's tail. Nat. Ord. — Lycopodiaceae. The pollen and spores are the constituents most studied: 9-hexadecanoic-acid (pollen or spore: 120,000-175,000ppm), alkaloids (pollen or spore: 1000-2000ppm), Lycopodine (plant: 870ppm), oleic acid (pollen or spore: 165,000-300,000ppm), sporonin (pollen or spore: 200,000-450,000ppm).

Eclectic: Intractable forms of fever, not an active type, showing obscure periodicity, with afternoon exacerbation, and the voiding of a high-colored red urine; dyspepsia (gastric indigestion; impaired gastric function) and indigestion with the same urinary symptoms, or with red, sandy deposits in the urine;

constipation, borborygmy (rumbling noises produced by movement of gas in the gut and audible at a distance); cystic catarrh (bladder inflammation) in adults, with painful micturition (passing urine); urine loaded with mucous or blood, or both, or deposits of red sand or phosphates.

Homeopathic: Digestive disturbance, there is a tendency to flatus, flatulent distension of the stomach, ketosis, seemingly good appetite but rejected after first contact, constipation, bloating, ascites (accumulation of serous fluid in the peritoneal [belly] cavity); rapid breathing and movement of nostrils at every breath, with possible coughing and dyspnea; thin and anxious but mild in temperament, tend to look older than their age, with early graying; skin tends to be dry and looks weathered (Day); General lack of gastric function and very little food satisfies; abdomen bloated with tenderness over liver area; glyco-genic (sugar building) function of liver is impaired; useful in various digestive, urinary and respiratory conditions (Macleod); Poor function of liver and kid-neys; bloat and gas after eating; cow is full or satisfied from small amounts of food and is hungry again soon; indigestion or ketosis soon after calving (Sheaf-fer).

Regular: Used as a protective in irritations of the skin and used in phar-macy as a coating for pills so that they will not stick together (Milks).

Phytolacca — *Phytolacca decandra*. Poke, Poke-weed, Poke-root, Virginia poke, Garget-weed. Nat. Ord. — Phytolaccaceae. The plant and berries contain most of the quantified constituents: alkaloids (fruit: 22,000ppm), anthocyanin (fruit: 93,000ppm), ascorbic acid (shoot: 1360-16,184ppm). Phytolaccin and poke weed anti-viral protein are also found in the plant.

Eclectic: Physiologically, phytolacca acts upon the skin, the glandular structures, especially those of the buccal (cheek) cavity, throat, sexual system, and very markedly upon the mammary glands; no other remedy equals phyto-lacca in acute mastitis, if employed early it prevents suppuration (formation of pus); it is best given internally, alternating with aconite; locally, phytolacca and glycerin may be applied when suppuration has not begun (or the powdered root moistened with water); sore nipples and mammary tenderness; morbid sensitiveness of the mammary during the menstrual period; ovaritis; sub-involution of the uterus, uterine and vaginal leucorrhea (white or yellowish viscid fluid from vagina containing mucus and pus cells).

Homeopathic: A glandular remedy, when lymph nodes become swollen and painful; mammary gland becomes hot, swollen, painful, inflamed and hard; secretion is stringy, thickened and yellowy; cow is restless and weak (Day); Hard, painful udder with moderate fever; cow kicks when attempting to milk her (Sheaffer).

Pulsatilla (USP) — *Pulsatilla nigricans.* Pasque flower, Wind flower, Meadow anemone, Nat. Ord. — Ranunculaceae. The fresh herb contains anemonin and isoanemonic acid as well as iron-greening tannin. The anemonin may be related to cantharidin.

Eclectic: Nervousness and despondency, unnatural fear, neuralgia in anemic, debilitated patients; stomach disorders from indulgence in fats; thick, bland, inoffensive discharges from mucous surfaces; alternating diarrhea and constipation, with venous congestion; amenorrhea (absence of cycle) and dysmenorrhea (abnormal cycle) with gloomy mentality; styes.

Homeopathic: Cows that are shy but friendly and gentle; prone to bland, catarrhal inflammation of all mucous membranes, which show cloudy mucous, greenish-yellow, mucopurulent (mucus and pus combined) material of the respiratory system, eyes, womb or udder; drinks less than herdmates; infertility due to ovarian inactivity, and calving may be delayed due to venous congestion of the uterus; cow is sensitive and can be contrary, but is always likable (Day); Thick muco-purulent discharges; proved useful in treating ovarian hypofunction (inactivity) and in retained palcenta (MacLeod); All body discharges are thick, creamy and yellowish green; mastitis at calving; post-partum discharge with creamy discharge; infertility and no heats shown; likely history of calf not in normal position at time of birth; sinusitis (inflammation of the mucous membranes lining the sinuses) and bland eye discharges (Sheaffer).

Sabina (USP) — *Juniperus sabina* (Sabina officinalis, U.S.P.). Savin-tops. Nat. Ord. — Coniferae. Tops and leaves contain oleum sabinae (volatile oil), fixed oil, gum, resin, gallic acid, chlorophyll, lignin and calcareous salts and salts of potassium.

Eclectic: Emmenogogue (stimulates menstruation), diuretic (increases secretion of urine), diaphoretic (induces sweating), and anthelmintic (remedy for destroying or expelling worms); not to be given if general or local inflammation or if there is an excited circulation; menorrhagia (excessive bleeding with cycle), atonic leucorrhea (on going white or yellowish viscid fluid discharging from the vagina), amenorrhea (absence or abnormal stopping of the menstrual cycle), irritative urethral diseases; suppresses menses with colicky pains; do not use in pregnancy on account of its tendency to abort, but it is reputed efficient in checking the tendency to abort in small doses; with pink and senna it can be given to remove worms; can induce abortion by combining with oil of tansy, pennyroyal or hemlock.

Homeopathic: Lack of tone of womb with bloody discharges; not indicated if sepsis is present; womb loses tone, involution is incomplete and the result is retained placenta; cystitis (inflammation of the urinary bladder) with blood and straining; genital warts (Day); Uterine conditions including retained placenta; persistent post-partum bleeding (MacLeod); Abortion with hemorrhage of bright red blood that remains fluid; lack of uterine tone and inflammation of ovaries; skin warts (Sheaffer).

Regular: Volatile oil used as an ecbolic to stimulate the gravid (pregnant) uterus to the expulsion of the fetus; abortifacient (an agent causing premature birth of the young) (Milks).

Ustilago — *Ustilago maydis.* Corn-smut, corn-ergot, corn-brand. Nat. Ord. — Fungi — Ustilagineae. It contains ustilagine (alkaloid), sclerotic acid, wax, pectin, and salts.

Eclectic: Enfeebled spinal and sympathetic innervation (poor nerve function); feeble capillary and venous circulation; uterine derangements, with excessive sanguinous (bloody) or other discharges, lax or flabby uterine, vaginal and perineal tissues, with uterine pain; uterine inertia (inactivity or lack of force) during labor; post-partum and passive hemorrhages. Its effects have been compared with those of ergot and nux vomica combined. Preferred to ergot in obstetrical cases because it produces clonic (alternate contraction and relaxation) instead of tonic (continuous, unremitting action) uterine contractions.

Homeopathic: Womb loses dark blood; atony of the post-partum womb (Day); Flabby, fluid-filled uterus after calving; dark, brownish-black stringy vaginal mucus with poor uterine involution; painful left ovary accompanied by light-colored diarrhea; loss of hair and weak hooves (Sheaffer).

Regular: Ecbolic (producing abortion) with ergot-like activity; abortifacient (causing premature birth of young) (Milks).

Table 9: Herb Analysis Chart

	Alfalfa	Dandelion	Lamb's Qtr	Chicory	Comfrey	Plaintain	Leaf Nettle	Burdock	Cleavers
Protein	20.97%	25.00%	31.70%	19.5	23.7	19.6	25.7	29.0	11.7
Digestable Protein				14.7	18.5	14.7	20.4	23.5	7.3
Soluble Protein				4.7	2.7	2.9	4.3	3.9	1.2
Protein Solubility	50.07%	24.40%	18.10%	24.2	11.4	15.0	16.8	13.4	9.9
Nitrogen/Sulfur Ratio	11:1	10:1	12:1	8:1	14:1	6:1	4:1	5:1	7:1
Acid Detergent Fiber	32.10%	19.20%	15.00%	32.8	29.8	34.1	22.6	25.1	40.6
Neutral Detergent Fiber	43.61%	30.00%	21.90%	46.8	42.2	45.8	34.4	36.5	49.1
Relative Feed Value	136.20%	229.00%	329.00%	126	145	127	193	177	108
TDN* (est)	63.89%	80.90%	85.60%	63.5	66.8	64.4	74.5	71.8	57.1
ME (mcal/b)		1.33	1.41	1.04	1.10	1.06	1.22	1.18	0.94
Est.Net Energy (themstc wt)		69.9	74.3	54.0	57.0	54.7	64	61.6	48
NE Lact (mcal/b)	0.65	0.085	0.9	0.65	0.69	0.66	0.77	0.075	0.58
NE Maint (mcalb)		0.895	0.959	0.648	0.697	0.661	0.806	0.0768	0.551
NE/Gain (mcalb)		0.6	0.655	0.383	0.426	0.394	0.523	0.490	0.295
Calcium	1.58%	1.04%	1.10%	0.89	273	1.84	4.38	2.10	1.3
Phosphorous	0.37%	0.33%	0.39%	0.31	020	0.26	0.41	0.34	0.39
Potassium	2.05%	4.46%	7.66%	3.59	394	2.97	3.01	3.28	2.46
Magnesium	0.46%	0.26%	0.55%	0.26	0.39	0.17	0.39	0.43	0.25
Sodium	759ppm				0.04	0.04	0.04	0.011	0.005	0.014
Sulfur-total	0.31%	0.41%	0.43%	0.37	0.27	0.53	0.94	0.90	0.26
ppm Iron	171	657	91	195	176	83	349	149	70
ppm Copper	15	15	8	14	29	12	11	26	13
ppm Zinc	30	34	46	43	46	44	40	32	127
ppm Manganese	23	35	138	36	192	30	36	47	66
ppm Boron	50	30	44	28	42	29	67	32	15

	Alfalfa	DayLily Leaf	DayLily Blossom	Echinacea Leaf	Wild Grape Leaf	Wild Rasp Leaf	Willow Leaf	Hazlenut Leaf	Mulberry Leaf	Chinese Chestn Lf
Protein	20.97%	20.6	23.4	15.7	22.1	15.2	19.8	14.1	26.2	21.8
Digestable Protein		15.7	18.3	11.1	17.1	10.6	14.9	9.6	20.9	16.7
Soluble Protein		5.4	14.8	1.8	1.2	0.4	1.5	0.7	3.6	14.7
Protein Solubility	50.07%	26.4	63.0	11.4	5.6	2.8	7.5	4.9	13.7	67.7
Nitrogen/Sulfur Ratio	11:1	19:1	20:1	12:1	14:1	16:1	7:1	14:1	17:1	11:1
Acid Detergent Fiber 41.2		32.10%	28.2	17.0	20.0	19.5	22.6	24.9	20.2	21.5
Neutral Detergent Fiber	43.01%	35.7	23.5	29.3	34.6	43.1	37.6	42.3	43.2	70.9
Relative Feed Value	136.20%	175	299	233	198	154	172	161	197	75
TDN (est)	63.89%	70.9	83.4	77.3	77.8	74.5	72.0	77.1	75.7	54.6
ME (mcal/b)		1.16	1.37	1.27	1.28	1.22	1.18	1.27	1.24	0.9
Est.Net Energy (themstc wt)		60.7	72.2	66.6	67.1	64.0	61.8	66.4	65.1	45.7
NE Lact (mcal/b)	0.65	0.74	0.87	0.81	0.81	0.77	0.75	0.8	0.79	0.55
NE Maint (mcalb)		0.756	0.929	0.845	0.853	0.806	0.771	0.842	0.823	0.513
NE/Gain (mcalb)		0.479	0.629	0.557	0.564	0.523	0.493	0.555	0.538	0.259
Calcium	1.58%	0.81	0.39	2.57	1.91	0.85	1.45	1.44	3.09	1.37
Phosphorous	0.37%	0.25	0.43	0.25	0.32	0.16	0.23	0.12	0.26	0.2
Potassium	2.05%	2.24	2.17	2.22	0.95	1.6	1.71	0.75	1.85	0.84
Magnesium	0.46%	0.20	0.17	0.88	0.25	0.29	0.27	0.31	0.34	0.37
Sodium	759ppm	0.025	0.05	0.02	0.02	0.01	0.011	0.04	0.016	0.015
Sulfur-total	0.31%	0.17	0.19	0.21	0.25	0.15	0.44	0.16	0.24	0.31
ppm Iron	171	203	86	131	502	100	117	118	154	120
ppm Copper	15	10	22	21	16	18	13	19	12	15
ppm Zinc	30	25	66	32	32	35	105	27	36	61
ppm Manganese	23	54	40	132	89	210	101	373	63	160
ppm Boron	50	49	16	66	31	23	34	28	36	72

Source: Jerry Brunetti, Agri-Dynamics, used by permission.

Chapter 11

Documented Historical Use of Medicinal Plants in Veterinary Medicine

The following *materia medica* is to give the reader a glimpse at the indications of many of the remedies that were used by mainstream veterinarians (Regular school) as well as Eclectic practitioners. The list is in reference to the table "Veterinary Doses" compiled by Pierre Fish from Cornell University in 1930. Dr. Pierre Fish's table is a goldmine of information when used in conjunction with the known physiologic effects of the crude drugs listed. It wouldn't be fair to readers who are interested in using plant medicines to give such a listing without also presenting the basic indications of when the remedies would be used. The reader should look to the doses in that table for amounts to be administered. However, bear in mind that the table would rep-

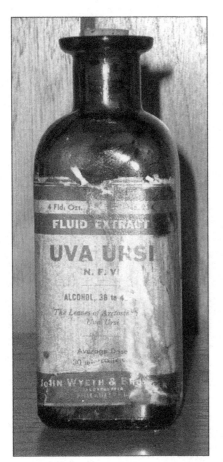

resent the dosage used by the Regular (conventional) school. The materia medica presented here is, in a sense, my effort to merge medicinal plants that were recognized by both the conventional veterinarians and the Eclectic human practitioners. When Dr. Fish's table was compiled in 1930, some of these medicinal plants were no longer widely used by veterinarians and thus *King's American Dispensatory* is just about the only place to retrieve in-depth information on their uses, albeit from a human perspective. There were three recognized dispensatories 1) King's American 2) National and 3) United States. Dispensatories contain all that the USP states regarding official drugs and much added information on lesser known compounds. Please note that the Table 10 and *materia medica* only show plant medicines and not the many mineral medicines (with the addition of hydrogen peroxide and iodine as germicides). Descriptions are for the remedy's Eclectic Specific Indications and Uses — which is the complementary aspect of the Eclectics' Specific Diagnosis. When Specific Indications are not presented in *King's Dispensatory,* information is then used from the section labeled Action, Medical Uses and Dosages. Once again, mainly physical symptoms will be listed in order to keep with standard physical examinations undertaken by veterinarians.

The two main reference works are *King's American Dispensatory* and H.J. Milks' *Veterinary Pharmacology, Materia Medica and Therapeutics* (1940). The publication date of 1940 is important in that it rapidly approaches the time when medicinal plants fell into disuse among mainstream veterinarians. This was due to the progress and advancements made in synthetic pharmaceuticals. Even the 1940 publication by Milks makes some relatively negative remarks about various medicinal plants (as contrasted by the new synthetics at the time), plants which were listed with dosages in Fish's table just 10 years before. Therefore, some of the following list will be referenced to the physiologic effects attributed to the plants as found in Udall's *Veterinarian's Handbook of Materia Medica and Therapeutics,* published in 1922. Some readers may wonder what the need is to dig into such old information. The answer is simple: medicinal plants were part of the regular armamentarium of medical professionals, and older texts help to understand their pharmacology and applied uses. What is really interesting is that the uses for many of these plants are quite similar between the Regular and the Eclectic schools. Part of the reason that medicinal plants may have fallen into disuse in the Regular school is that, in addition to the advent of synthetics, the pharmaceutically precise fluidextracts could have caused toxicity more quickly with conventional dosages whereas the Eclectics used smaller doses and did not have such toxicity problems or concerns. Unfortunately there were not many Eclectic veterinary practitioners and so *King's*

Dispensatory again gives the reader the best approximation of Eclectic use of medicinal plants for humans while Milks' standard veterinary text will give real veterinary uses for the same compounds. Bear in mind that the references from Milks' work pertain to many species seen by veterinarians, not only bovine. However, I feel it is important to show the indications and uses of as many of these natural crude drugs as possible so that as many options as possible open up for treatment of Certified Organic livestock.

In addition to the medicinal properties of plants, some have been analyzed for their nutritional content. The analysis shown in Table 9 was carried out by Jerry Brunetti of Agri-Dynamics (used by permission). Alfalfa is shown as the standard reference to which the others can be compared. All the plants shown can be found in *King's Dispensatory,* if they are not in the materia medica shown in this book. In the final analysis, it is evident that non-agricultural plants can be both nutritional and medicinal.

On the other hand, the potential toxicity of some plants known to be used by the Regular, Eclectic or Homeopathic schools is from Osweiler's book, *Toxicology.* The vast majority of entries in this book are not of plant origin. The plant listing mainly includes plants of which no medicinal uses are known by any of the three schools being discussed. The listing in Table 11 is to give the reader an idea about the amounts of plant needed to induce toxicity. As can be seen, the vast majority of the plants in the Eclectic *Materia Medica* shown in the materia medica that follows here are not considered toxic. Some of the entries will not coincide with any of the plants found in the materia medicas included in this book; however, some can be found when researching either the Eclectic literature or Homeopathic literature. The body weight (BW) of most dairy cows range from 1000-1400 lbs. As can be seen, even with toxicities beginning at 0.5% BW, a 1000 lb. animal would need to ingest 5 lbs. of the plant material. This is far in excess of what is considered a *medicinal dose* in the Regular school (the school most likely to have used "full" dosing regimes which potentially bordered on toxicity).

Obviously, a working knowledge of concentrations of plant medicinals administered is important. My original basis in alternative treatments was with homeopathy, which is without the safest due to the dilutions. However, by balancing the realities of physiologic effects with low concentrations, we can be more certain of treatment without lapsing into the realm of potential toxicity as the Regular school of veterinary medicine may have encountered in earlier years.

BOTANICAL *MATERIA MEDICA*

TO ACCOMPANY "VETERINARY DOSES"

(The doses of most of the following are found in Table 10)

Reminder: the Electic (Felter) descriptions show what indications would trigger the use of the remedy. The Regular (Milks & Winslants) descriptions show what the remedies do.

Aconite (USP) — See Chapter 10

Aloe — Tonic, purgative (produces increased discharges from the bowels), emmenagogue (stimulates menstruation) and anthelmintic (destroys or expels worms or prevents their development) (Felter).

Slightly stomachic (stimulates functional activity of the stomach) in small doses; increase secretion of bile; increased peristalsis (wave action of gut); cathartic (quickens or increases intestinal evacuations); colic, hidebound, overloaded stomach or bowels, to expel worms after a vermicide (agent to destroy parasitic worms), to promote the excretion of waste products from the bowels or blood: in fact, in those conditions demanding strong purgation; bitter tonic; as a stimulant to wounds, especially the compound tincture of aloe and myrrh; internally, ruminants are not so susceptible to aloe as horses. Contra-indicated in inflammations of the digestive tract, high fever, weak and debilitated animals, cystitis (inflammation of the urinary bladder), inflammation of the pelvic organs, pregnancy, and in milking animals the milk will have a bitter taste (Milks).

Areca nut (Betel nut) — Astringent (causing contraction or constriction of tissues) and teniacide (agent to expel tape worms) (Felter).

Anthelmintic for all animals, but especially dogs; tapeworms in horses (Milks).

Arnica —See Chapter 10

Asafetida (USP) — Nervous irritation; functional wrongs of the stomach, gastro-intestinal irritation, with flatulence (gas) and palpitation of the heart; dry, deep, choking, bronchial cough (Felter).

Feeble carminative (allays pain by causing the expulsion of gas from the alimentary canal), circulatory and nervous stimulant, and antispasmodic; useful in colics and convulsions of the young; can be administered as an enema; used externally to stop feather pulling of birds and to prevent bandage chewing with dogs (Milks).

The extract/essential oil is GRAS.

Balsam of Copaiba (USP) — Vesical pressure and tenesmus, frequent desire to urinate, the urine coming in drops; urethral mucoid discharges; cough with thick expectoration, accompanied by loud mucous rales (Felter).

In small doses it is stimulant to the mucous membranes generally; diuretic (increases urinary secretion) and expectorant acts upon the pulmonary mucous membrane to increase or alter its secretions); antiseptic (antagonizes sepsis or putrefaction), rendering the urine and bronchial secretions antiseptic; very disagreeable, should be administered in capsule or emulsion (Milks).

Balsam of Peru (USP) — Occasions some cutaneous heat (relating to the skin), increases the rapidity of circulation, and augments the renal (kidney) secretion, with irritation of the kidneys (Felter).

Expectorant (medicine that increases or alters the secretions of the pulmonary mucous membranes); parasiticide (a remedy for the destruction of parasites) and skin stimulant; useful externally as a stimulant to indolent wounds, and in mange; for mange, use in equal parts of alcohol, castor oil or in combination with sulfur (Milks).

The extract/essential oil is GRAS.

EMEA: All food producing species; for topical use only.

Balsam of Tolu (USP) — (Like Balsam of Peru); stimulant, tonic, and expectorant, and may be used as a substitute for it in chronic catarrhs, and other pulmonary affections not actively inflammatory in their character (Felter).

Belladonna (USP) — See Chapter 10

Bryonia (USP, NF) — See Chapter 10

Buchu (USP) — Abnormally acid urine, with constant desire to urinate, with but little relief from micturition; vesico-renal irritation; copious mucous, or muco-purulent (combination of pus and mucus) discharges (Felter).

Useful in the less severe cases of subacute and chronic cystitis (inflammation of the urinary bladder) and is particularly valuable in chronic irritability of the bladder shown by a frequent desire to urinate (Milks).

Calendula (Marigold, Wild Marigold) (USP) — Locally, to wounds and injuries to prevent suppuration (formation of pus) and promote rapid healing; internally, to aid local action, and in chronic suppuration, capillary engorgement, varicose veins, old ulcers, splenic and hepatic congestion (Felter).

GRAS

EMEA: All food producing species; in the products not exceeding one part in ten only (1X).

Calumba (Columbo) (USP) — Enfeebled stomach, with indigestion, or feeble digestion; anorexia, and general debility. Acts upon the stomach much like hydrastis; dyspepsia, chronic diarrhea and dysentery (Felter).

Simple bitters; stomachic (Milks).

Cannabis indica (Indian hemp) (USP) — Irritation of the genitor-urinary tract; painful micturition, with tenesmus; scalding, burning, frequent micturition (Felter).

Relief of pain, spasms and nervous irritability; superior to opium or morphine for the relief of pain and spasm in the intestinal tract because it does not cause constipation (Milks).

Cantharides (Spanish Fly) (USP) — Vesicle irritation, partial sphincter paralysis, with dribbling of urine and teasing desire to urinate, accompanied with tenesmus (a painful spasm of the anal sphincter with an urgent desire to evacuate the bowel or bladder, involuntary straining, and the passage of but little fecal matter or urine) (Felter).

Useful mainly as an external blister; it is often serviceable in acute inflammation of the brain or the meninges and should be applied to the back of the poll; applied to the throat in acute laryngitis; as counter-irritation in cases of rheumatism of muscles and joints, placed over the diseased parts or nearby; serviceable in closing an open joint as a blister; close small umbilical hernias; used rarely internally, but may be of service in incontinence of the urine, and as a stimulant in cystitis and pyelitis (Milks).

Capsicum (Cayenne pepper, Red pepper) (USP) — Atonic dyspepsia (gastric indigestion; impaired gastric function), colic, with abdominal distension, congestive chills; cold extremities with small, weak pulse; tongue dry and harsh; buccal (pertaining to, adjacent to, or in the direction of the cheeks) and salivary secretions scanty, in fevers (Felter).

Internally, as a carminative (dilute well as it is very irritant to the mouth); externally as a counter-irritant but too powerful (mustard is better) (Milks).

GRAS; the extract/essential oil is also GRAS.

EMEA: All food producing animals (no provisions/restrictions).

Cardamom (Cardamom seeds) (USP) — Carminative (helps allay pain by causing the expulsion of gas from the alimentary canal) in flatulency; seeds are very warm, pungent and aromatic and form an agreeable addition to bitter infusions and other medicinal compounds (Felter).

GRAS; the extract/essential oil is also GRAS.

Cascara sagrada (Sacred bark, Chittem bark) (USP) — Constipation, due to neglect or to nervous and muscular atony of the intestinal tract; lesser ailments, depending solely upon constipation, with intestinal atony; an agreeable aromatic laxative (Felter).

Purgative (produces increased discharges from the bowels) and mild stomachic (stimulates functional activity of the stomach); efficient laxative for dogs and other small animals; to overcome chronic constipation due to torpor of the bowels (Milks).

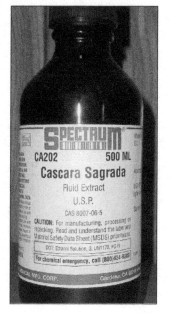

Cascarilla Bark (Sweet-wood tree) (USP) — Tonic and stimulant; dyspepsia, flatulency, chronic diarrhea, in debility attending chronic diseases, convalescence from acute diseases (Felter).

Astringent stomachic (causing contraction or constriction of stomach while stimulating its functional activity) (Milks).

The extract/essential oil is GRAS.

Castanea (Chestnut) (USP) — Convulsive cough; pertussis; possesses astringent and tonic properties for fever and ague (archaic term for general malaise probably due to malaria) (Felter).

Catechu (USP) — Strongly astringent; arrests mucous discharges when excessive; checks hemorrhages; chronic diarrhea, chronic catarrh (simple inflammation of mucous membrane), chronic dysentery (especially when combined with opium); fissure of the nipples (Felter).

Contains tannins, to which its action is due; better than tannic acid in diarrhea as their action is brought about more slowly and is more lasting (Milks).

Caulophyllum (USP) — See Chapter 10

Chamomile (USP) — Gastro-intestinal debility; weak, irritable stomach; flatus; colic; dysmenorrhea (abnormal menstruation) from cold; malarial affections; tonic and anti-spasmodic.

GRAS; the extract/essential oil is also GRAS.

Chaulmoogra — Worms, leprosy (*Mycobacteria leprae,* related to TB and Johne's).

Chenopodium (USP) — See Chapter 10

Chimaphila (Pipsissewa, Ground holly) (USP) — Atonic and debilitated states of the urinary organs, with scanty urine, but excessive voiding of mucus, muco-pus, or bloody muco-pus; offensive or non-offensive in character; chronic prostatitis (inflammation of the prostate) (Felter).

Very similar to buchu since the constituents are so similar (Milks).

The extract/essential oil is GRAS.

Cimicifuga (Black cohosh) (USP) — Muscular pains; uterine pains, with tenderness; false pains; irregular pains; rheumatism of the uterus; dysmenorrhea (abnormal menstruation); anti-rheumatic when the pulse is open, the skin not hot and constricted (Felter).

Astringent stomachic (causing contraction or constriction of stomach while stimulating its functional activity) (Milks).

EMEA: All food producing species; not for use in animals from which milk is produced for human consumption.

Cinchona (USP) — See Chapter 10

Cinnamon — Post-partum and other uterine hemorrhages with profuse flow and cold extremities; hematuria (urine contains blood or red blood cells), hemoptysis (bloody nose); stimulant, tonic, stomachic (stimulates functional activity of the stomach), carminative (helps allay pain by causing the expulsion of gas from the alimentary canal), and astringent (causing contraction or constriction of tissues) (Felter).

Carminative; disinfecting dental cavities and destroys nerves of carious teeth (Milks).

GRAS.

Coca (USP) — Labored and difficult breathing, with normal temperature; inordinate hunger and thirst (Felter).

Local anesthetic; mydriatic (causes dilatation of the pupil); respiratory stimulant (Milks).

Cod liver oil (USP) — Nutritive and alterative (Felter).

Valuable in any condition in which malnutrition is a prominent symptom; especially useful during convalescence from febrile diseases; improves the general health, increases the number of red blood cells, increases weight and accumulation of fat, and in growing animals favors the deposition and retention of calcium in the bones (Milks).

Coffee — Produces a mild, stimulating influence upon the organs of digestion, facilitating digestion, augmenting the biliary flow, and increasing peristalsis (gut activity), thus favoring a free action from the bowels; active diuretic (increases the secretion of urine), remedy for dropsies (edema, fluid build-up) of cardiac origin (Felter).

Slightly stimulates peristalsis, increases the appetite (in man), is a mild cholegogue (provokes the flow of bile) and laxative; decidedly diuretic without causing irritation; with marked dropsy (edema), a combination of digitalis and caffeine may be of more service than either alone (Milks); As an antidote to opium poisoning; its diuretic properties render it appropriate in aiding the absorption of pleuritic effusion (fluid build-up along the thoracic lining), ascites (fluid build-up in the peritoneal [belly] cavity) and dropsies, particularly of cardiac or renal origin, and in the hepatic form as well (Winslow).

The extract/essential oil is GRAS; caffeine is also GRAS.

Colchicum Root Bark (Meadow saffron) — Sedative, cathartic (quickens or increases evacuations from the bowels), diuretic (increases the secretion of urine) and emetic (produces vomiting); intestinal disturbances with gaseous accumulation, as in colic from intestinal irritation (Felter).

Conium (Hemlock, Poison hemlock, Spotted hemlock) (USP) — Neuralgic or rheumatic pains in the old and feeble; pain in stomach; pain of gastric ulcer; nervousness, restlessness (Felter).

Convallaria (Lilly of the valley) (USP) — Heart irregularities due to mechanical impediments; mitral insufficiency; dropsy (edema) of cardiac origin; palpitation and vehement heart action, with dyspnea (shortness of breath) and diminished arterial pressure; quickened pulse (Felter).

Cotton Root Bark (see Gossypium)

Digitalis (Foxglove) (USP) — Weak, rapid, irregular heart action, with low arterial tension; weak heart sounds; dusky; jugular pulsation, cough, and dyspnea (shortness of breath); edema (fluid build-up); anasarca (generalized infiltration of edema fluid into the subcutaneous connective tissue) with scanty, high-colored action; renal congestion (Felter).

Cardiac tonic/stimulant and diuretic (increases the secretion of urine) (Milks).

Dioscorea (Wild yam, Colic root, Rheumatism root) — Bilious colic (gall bladder pain); other forms of colic with spasmodic contractions; frequent small flatulent passages; colic, with tenderness on pressure; sharp abdominal pain, made worse by motion (Felter).

Echinacea — See Chapter 10

Ergot (USP) — See Chapter 10

Eriodictyon (Yerba santa) (USP) — Chronic asthma with cough, profuse expectoration, thickening of the bronchial mucous membrane, loss of appetite, impaired digestion, emaciation; cough with easy and abundant expectoration (Felter).

Eucalyptus (USP) — Sore throat with fetid odor; fetid and catarrhal (simple inflammation of the mucous membrane) states of the broncho-pulmonic tract; chronic catarrhal diarrhea; unhealthy fetid secretions of any part (Felter).
Mild irritant; used by inhalation in diseases of the upper respiratory tract; resembles turpentine in action (Milks); Carminative (helps allay pain by causing the expulsion of flatus from the alimentary canal) and antiseptic in chronic bronchitis; externally as an antiseptic and against flies (Udall).

Eupatorium (Boneset, Thoroughwort, Indian sage) (USP) — Pulse full and large, the current exhibiting little waves; skin full and hot with tendency to become moist, even during the progress of fever; deep-seated aching pains in muscles and periosteum (the lining which surrounds the bone) (Felter).

Fennel (USP) — Carminative (helps allay pain by causing the expulsion of flatus from the alimentary canal), stimulant, galactogogue (increases the secretion of milk), diuretic (increases the secretion of urine), diaphoretic (produces sweating); flatulent (gas) colic; amenorrhea (absence or abnormal cessation of the menses) and suppressed lactation; corrigent (pleasant carrier) of unpleasant medicines (Felter).
Carminative; (all carminatives) relieve pain by causing the expulsion of gas from the stomach and intestines, and to reduce griping pains (Milks).
GRAS; the extract/essential oil is also GRAS.

Fenugreek — As a vaginal wash to soothe irritation or inflammation; relieves uterine irritation and acts as an emmenagogue (stimulates menstruation); also to allay irritation of the throat and breathing passages (Felter).
GRAS.

Gamboge (Camboge, Gutti, Gutta gamba) (USP) — Drastic, hydrogogue cathartic, causing nausea, griping and copious watery stools, therefore used in dropsy (edema) in combination with squills; expulsion of tapeworm; dysmenorrhea (abnormal menstruation) (Felter).

Purgative (produces increased discharges from the bowels), cathartic; one ounce gamboge with ½ pound each of Epsom salts and common salt in 1 quart water make an efficient purgative in cattle (Milks).

Gaultheria (Wintergreen) — Cystic and prostatic irritation, inflammation of the urethra; undue sexual excitement, early renal (kidney) inflammation; stimulant, aromatic, and astringent; volatile oil used to make preparations agreeable (Felter).

Analgesic (alleviates pain), anti-pyretic (reduces body temperatures in fevers); carminative (helps allay pain by causing the expulsion of flatus from the alimentary canal) and counter-irritant (Milks).

Gelsemium (USP) — See Chapter 10

Gentian (Gentian root) (USP) — A powerful tonic, improves the appetite, strengthens digestion, gives more force to the circulation, and slightly elevates the heat of the body; contra-indicated with gastric irritation or inflammation. (Felter).

Simple stomachic; the most reliable for large animals; indicated in convalescence in febrile disease; with iron, as a tonic for large animals (Milks).

Geranium (USP) — Relaxed mucous tissues, with profuse, debilitating discharges; chronic diarrhea, with mucous discharges; chronic dysentery; diarrhea, with constant desire to evacuate the bowels; passive hemorrhages (Felter).

GRAS; the extract/essential oil is also GRAS.

Ginger (USP) — Loss of appetite; flatulence; borborygmus; spasmodic gastric and intestinal contractions; painful menstruation; acute colds; cool extremities (Felter).

Carminative (helps allay pain by causing the expulsion of flatus from the alimentary canal); often combined with sodium bicarbonate (baking soda) in atonic indigestion of ruminants and horses; as a carminative to expel flatus and prevent the griping pains associated with purgatives like aloe and magnesium sulphate (Milks).

GRAS; the extract/essential oil is also GRAS.

Glycerin (USP) — Stimulant, antiseptic, laxative and demulcent (soothes and protects irritated mucous membranes); emollient (used externally to mechanically soften and protect tissues); chapped nipples (small amount made with a few grains borax and rose-water); deafness due to hardened cerumen (applied locally) (Felter).

Emollient (when mixed with two or three parts water); injected rectally will stimulate peristalsis and cause expulsion of feces (Milks).

GRAS.

Glycyrrhiza/licorice root (USP) — Emollient (used externally to mechanically soften and protect tissues); demulcent (soothes and protects irritated mucous membranes), and nutritive; lessens irritation on mucous surfaces, being useful in coughs, catarrhs (simple irritation of mucous membranes), irritation of the urinary organs and pain in intestines due to diarrhea (Felter).

A very feeble drug; popular for coughs and colds as a demulcent; often used to cover the disagreeable taste of other drugs (Milks).

Gossypium/Cotton Root Bark (USP) — Uterine inertia (lack of force) during parturition (large doses). Menstrual delay; sexual lassitude; emmenogogue (stimulates menstruation), parturient (to aid in the birth of the young), and abortive (promotes uterine contraction more safely than ergot) (Felter).

Granatum (Pomegranate) (USP) — Teniacide and teniafuge for the destruction and expulsion of tapeworm (Felter).

Pomegranate and its alkaloids are used entirely as a remedy against tapeworms (Milks).

Guarana (Uabano, Uaranazeiro) (USP) — Diarrhea, leucorrhea (white or yellowish viscid fluid from vagina); diuretic (increases secretion of urine) (guaranine) (Felter).

Stimulant to the central nervous system, due to its caffeine content (Milks).

Gum Tragacanth (USP) — Demulcent (Felter).

Demulcent (soothes and protects irritated mucous membranes), and excipient (carrier) to hold emulsions, mixtures, etc (Milks).

Hamamelis (Witch hazel) (USP) — Tonic and astringent; diarrhea; excessive mucous discharges; pale mucous tissues (especially deep-red from venous engorgement, or deep blue from venous stasis); passive hemorrhages; muscular soreness and aching; poultice in painful swellings and inflammations (Felter).

Helleborus niger (Black hellebore, Christmas rose) — (Minute doses only); dropsy (edema), amenorrhea (absence or abnormal cessation of menses); discharges of mucus from bowels (Felter).

Hematoxylin (Logwood) (USP) — Tonic and unirritating astringent, less constipating than others; hemorrhages from the uterus, lungs and bowels, in old diarrheas and dysenteries (Felter).

Astringent for diarrheas of the young; less active than kino and krameria, but more pleasant on account of sweetish taste (Milks).

Humulus (Hops) (USP) — Tonic, hypnotic, febrifuge (decreases fever), antilithic (to aid in removing stones through diuresis) and anthelmintic (weakly so) (destroys or expels worms or prevents their development); dyspepsia (gastric indigestion) and fermentive dyspepsia (Felter).

Aromatic stomachic (stimulates functional activity of the stomach) (Milks).

Hydrastis (USP) — See Chapter 10

Hydrogen Peroxide (USP) — Antagonize septic conditions; applicable to all wounds as a local disinfectant and antiseptic; putrid surfaces and discharges call for it; low fevers; dyspnea (shortness of breath) attending chronic bronchitis; when lochia (post-partum dark uterine discharge) becomes putrid and the temperature leaps high, the ushering in of a puerperal septicemia (post-partum blood poisoning), no better cleansing agent than H_2O_2, with appropriate remedies it will greatly assist in preventing a fatal issue; purulent conjunctivitis (pus discharge stemming from inflammation of the conjunctiva) (Felter).

Particularly valuable in the preliminary treatment of septic wounds, abscess cavities, and fistulous (tunneling) tracts where it acts not only as a disinfectant but also mechanically removes exudates, blood clots, etc.; it should not be injected into deep cavities unless there is free exit for the gas to escape; very valuable wash for stomatitis (inflammation of the mucous membranes of the mouth), pharyngitis (inflammation of the mucous membrane and underlying parts of the pharynx), etc. (Milks).

Hyoscyamus (Henbane) (USP) — Sharp, dry cough, worse when lying down; muscular spasms; choking sensations; rapid and palpitating cardiac action; face warm and pupils dilated; fright, terror, restlessness (Felter).

The uses are the same as those of belladonna, but is often considered superior as a urinary sedative in the treatment of cystitis (inflammation of the urinary bladder) (Milks).

Hypericum (USP) — See Chapter 10

Ignatia (St. Ignatius bean) (USP) — General nervous atony; disposition to grieve; dysmenorrhea (abnormal menstruation), with colicky pains; sexual frigidity, impotence and sterility; coldness of extremities; muscular twitching, particularly face and eyelids; burning of the soles and feet (Felter).

Iodine (USP) — Very beneficial results in diseases involving the lymphatic structures; glandular obstructions; enlargement of the external absorbent glands; chronic enlargement of the liver, spleen, mammae, testes and uterus; leucorrhea (white or yellowish viscid discharge from the vagina), amenorrhea (absence or abnormal cessation of the menses), dysmenorrhea (abnormal menses), diseases of a hypertrophical (increased thickness) or cachectic state (depraved condition of general nutrition); locally, useful in chronic uterine engorgement, chronic indurations of the cervix uteri (uterine cervix), uterine ulcerations, leucorrhea; ringworm and corns (Felter).

Counter-irritant; cysts, enlarged bursas; goiter (hypertrophy of the thyroid glands); to reduce enlarged glands; parasiticide (ringworm); absorbent; antiseptic and stimulant for wounds; periodic opthalmia (moon blindness) of horses (Milks).

Ipecachuahana (Ipecac) (USP) — Emetic (induces vomiting); irritation, whether of stomach, bowels, nervous system or pulmonary tissues; active hemorrhages; acute bowel disorders with irritation; dyspnea (shortness of breath); irritative cough; hoarseness from cold; hypersecretion, with mucus rales (small doses) (Felter).

Externally — a powerful local irritant; internally — in small doses, stimulates the stomach, increases salivary and gastric secretions, and serves as a bitter (stimulates the gastro-intestinal mucosa without materially affecting the general system) or stomachic (increases the functional activity of the stomach); large doses are emetic and increase the secretion of bile (cholagogue) and intestinal mucus (purgative); very useful as an expectorant when the secretions are thick and scanty owing to its persistent action of increasing the bronchial mucus, which reflexively causes coughing (Milks).

Jaborandi/Pilocarpus (USP) — Source of pilocarpine; deficient secretion; marked dryness and heat of the skin and mucous tissues; pulse full, hard, sharp, and strong; muscular pain; muscular spasm; urine suppressed, of deep color and high specific gravity; elevated temperature, with deficient secretion; puffiness of tissues; marked restlessness; dry, harsh cough; edema; uremic poisoning; increased ocular tension; adapted to chiefly sthenic (strong and active) cases. (Felter).

To increase secretions of the gastrointestinal tract, combined with eserine, as in obstinate constipation or impactions in cattle and horses; weak ruminatoric (increase rumen activity); diaphoretic (produces sweat); to remove waste matter for the blood and in cardiac dropsy (edema); as a direct antidote to atropine (belladonna) — give in doses four times as great as the atropine. Contra-indicated in animals with heart or lung disease, colic where the heart is weak, pharyngitis (inflammation of the mucous membrane and underlying parts of the pharynx) or tetanus (lock-jaw) due to inability to swallow the large amounts of saliva produced; may produce abortion (Milks).

Juniper Oil — stimulating, carminative (helps allay pain by causing the expulsion of flatus from the alimentary canal) and diuretic (increases the secretion of urine); oil acts like copaiba in arresting mucous discharges (from urethra); with watermelon seeds, for ascites (accumulation of serous fluid in the peritoneal [belly] cavity); pyelitis (inflammation of the renal pelvis), pyelonephritis (inflammation of the renal pelvis and kidney itself with pus discharges), cystitis (inflammation of the urinary bladder); leucorrhea (white or yellowish viscid discharge from the vagina), cystorrhea (Felter).

Similar in action to turpentine; it is diuretic, expectorant (acts upon the pulmonary mucous membrane to increase or alter its secretions), carminative and stomachic (increases functional activity of the stomach); increases urine in passive congestion of the kidneys, resulting from chronic heart disease; in subacute or chronic diseases of the genito-urinary tract it is not generally regarded as serviceable as buchu, copaiba, or oil of sandalwood; most often used as an expectorant in chronic respiratory diseases (Milks).

GRAS.

Kamala (Spoonwood, Rottlera) (USP) — Tapeworm; round worm (Felter).

Irritant to the gastro-intestinal tract; strong purgative action (increases discharges from the bowels); action produced within 5 hours; about 50% effective in removing tenea from cats but less so in case of Dipylidium (Milks).

Kava kava — Neuralgia, particularly of the trifacial nerve; toothache, earache; ocular pain; anorexia (diminished appetite, aversion to food); chronic catarrhal inflammations (simple inflammation of a mucous membrane); vesical irritations; painful micturition; dysuria (Felter).

Kino (Gummi kino, Buja) (USP) — Pure and energetic astringent; can substitute for catechu; menorrhagia (increased menstrual flow); topical application in leucorrhea (white or yellowish viscid discharge from the vagina) and relaxed sore throat; suppress hemorrhage (Felter).

Kousso (Kooso, Cossoo, Cusso) (USP) — Purgative (increases discharges from the bowels) and anthelmintic; tapeworm (Felter).

Tapeworms (Milks).

Krameria (Rhatany) (USP) — Powerful astringent, with some slight tonic virtues; constipation with slight dyspeptic symptoms may be induced with immoderate doses; employed internally with advantage in menorrhagia (increased menstrual flow), hematemesis (bloody vomiting), passive hemorrhages, chronic diarrhea, leucorrhea (white or yellowish viscid discharge from the vagina), chronic mucous discharges; energetic styptic in epistaxis (causes contraction of blood vessels to check bleeding in bloody nose), hemorrhage from the cavity of an extracted tooth; application to spongy or bleeding gums (Felter).

Lactucarium (Lettuce opium) (USP) — Calmative and hypnotic, and as a substitute for opium, to which it is to be preferred in many instances because of its freedom from unpleasant after-effects, as constipation, excitement, etc; however, it is not to be considered equal in power to opium; insomnia; cough (Felter).

Laudanum/Tincture of opium (USP) — Possesses the medicinal virtues of opium, and may be used in all cases where the drug is indicated (Felter).

Lobelia (Indian tobacco) (USP) — Full, labored, doughy pulse; the blood moves with difficulty; angina pectoris cardiac neuralgia (heart pain); mucous accumulation in bronchae; rigidity of muscular tissue; rigid os uteri (uterine cervix), with doughy edges; rigid perineum or vaginal walls; nausea; as an emetic (induces vomiting) when tongue is coated heavily at base (Felter).

Resembles nicotine and coniine in action; emetic; with belladonna for asthma in dogs and heaves in horses; lobeline sulfate in ¼ grain doses to stimulate the rumen in cattle; weak ruminatoric (increases rumen activity); antispasmodic; antidote to strychnine poisoning (Milks).

Lycopodium (USP) — See Chapter 10

Male Fern (USP) — Expulsion of the tapeworm (Felter). Used for both types of tapeworms (Milks).

Matico (USP) — Aromatic bitter stimulant; bleeding from the lungs, stomach or kidneys; leucorrhea (white or yellowish viscid discharge from the vagina), chronic mucous discharges; dyspepsia (gastric indigestion; impaired gastric function); externally for arresting hemorrhages from wounds, leech-bites, etc.

Used as a genito-urinary stimulant and disinfectant; antiseptic to inhibit growth of bacteria and prevent decomposition of urine in the mucosa of the urinary tract; subacute or chronic pyelitis (inflammation of the renal pelvis), cystitis (inflammation of the urinary bladder) and urethritis (inflammation of the urethra) (Milks).

Mentha Piperita (Peppermint) (USP) — Gastrodynia (stomach pain), flatulent colic, cramps of the stomach and difficult digestion; equal parts of the essence and alcohol, used by atomization, relieve the cough of bronchitis and pneumonia (Felter).

Carminative to help relieve pain by causing the expulsion of gas from the stomach and intestines, and to reduce griping pains; valuable in colic and flatulence; externally good as an antiseptic (Milks).

GRAS.

Morphine and opium (USP) — Sudden, acute pain and pain from wounds, burns, scalds; peritonitis (inflammation of the belly lining), pleuritis (Inflammation of the chest lining), orchitis (inflammation of the testicles), metritis (inflammation of the uterus); severe and acute abdominal pains (Felter).

Used to relieve pain, overcome nervousness and excitability, lessen reflex excitability, diminish secretions, support the system; check distant hemorrhages (via sedation) (Milks).

Nux vomica (USP) — See Chapter 10

Opium tincture, camphorated/paregoric (USP) — A very valuable and useful opiate, which is efficient in allaying troublesome cough, nausea, slight gastric and intestinal pains; to cause sleep, and palliate diarrhea (Felter).

Pepo/pumpkin seed (USP) — Tapeworm; roundworm; ardor urinae.
The seed of the ordinary pumpkin is a very efficient and harmless vermicide (an agent to destroy parasitic worms) not only for tapeworm but for round worms as well. Sollman experimenting with anthelmintics found pepo highly efficient; he states that its active principle is soluble in water but destroyed by boiling. It has the advantage of being practically harmless for the higher animals (Winslow).

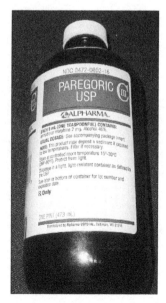

Physostigma/Calabar bean (USP) — Locally to induce contraction of the pupil in mydriasis or injuries to the eye; iritis (inflammation of the iris), corneal ulcers, iridal prolapse, and ocular inflammations; pulse feeble and tremulous, surface cool, extremities cold, and pupils contracted; or pupils dilated with small, rapid, tensive pulse; mental torpor; breathing difficult with sense of constriction (Felter).

Myotic; stimulates involuntary muscle, as in impaction of the bowels, to produce peristalsis; ruminatoric (increases rumen activity) in cattle, which may be repeated in an hour to advantage; acute bloating. Contra-indicated in over-distensions of the stomach and intestines on account of liability to rupture the walls by violent peristalsis; in spasmodic colic because it increases the convulsive contractions of the bowel, with consequent increase of pain; may cause abortion in pregnant animals; chronic bloating and chronic indigestion in cattle (Milks).

Phytolacca (USP) — See Chapter 10

Podophyllum (USP) — See Chapter 10

Polygonum (Water pepper, Smart weed) — Stimulant, diuretic (increases secretion of urine), emmenagogue (stimulates menstruation), antiseptic (antagonizes sepsis or putrefaction), diaphoretic (produces sweating) and vesicant; amenorrhea (absence or abnormal cessation of menses) (Felter).

Prunus virginiana (Wild Cherry) (USP) — Rapid, weak circulation; continual irritative cough, with profuse muco-purulent (combination of pus and mucous) expectoration; cardiac palpitation, from debility; dyspnea (shortness of breath); pyrexia (fever); loss of appetite; cardiac pain (Felter).
Extract/essential oil is GRAS.

Pulsatilla (USP) — See Chapter 10

Pyrethrum/Chrysanthemum cinerariaefolium (USP) — "Dalmatian insect powder" from Trieste; the powder destroys aphids, house flies and mosquitoes; powder must come into actual contact with the insect (Felter).
As a fly and insect spray (low toxicity for mammals) (Milks).
EMEA: All food producing species; for topical use only.

Quassia (Bitter wood, Bitter ash, Quassia wood) (USP) — Tonic, febrifuge (decreases fever), and anthelmintic (destroys or expels worms or prevents their development); remittent and intermittent fevers; dyspepsia (gastric indigestion; impaired gastric function), debility, worms; will remove ascarids (Felter).
Simple bitter; vermicide for pin worms — as a rectal infusion into the previously emptied rectum; does not contain tannin and may therefore be prescribed with iron (Milks).

Quebracho/Aspidosperma (USP) — Dyspnea (shortness of breath); stimulant to the pneumogastric nerve; disturbed relation between the pulmonic circulation and the action of the heart; asthma with emphysema (presence of air in the connective tissues of a part); asthmatic bronchitis; cough of la grippe (Felter).

Quercus alba (White oak) (USP) — Slightly tonic, powerfully astringent, and antiseptic; chronic diarrhea, chronic mucous discharges, passive hemorrhages; relaxation of mucous membranes, with unhealthy discharge; ulcerations, with spongy granulations; astringent injection for leucorrhea (white or yellowish viscid discharge from the vagina) (Felter).

Externally as an antiseptic; internally as a decoction in the strength of 1 ounce to 1 gallon of water as an injection in leucorrhea, pharyngitis, etc. (Milks).

Quinine/Cinchona (USP) — Periodicity, pulse soft and open, tongue moist; skin soft and moist, and nervous system free from irritation; intermittent and remittent fevers; periodic neuralgia (nerve pain); enfeebled innervation (weak nerve function); along with small doses of opium, its oxytocic power is increased; by its tonic and contractile action, it minimizes the danger of post-partum hemorrhage; in small doses for dyspepsia (impaired gastric function); small doses are a nervous and vascular stimulant; in large doses it is a sedative and muscular and cardiac depressant (Felter).

Powerful antiseptic, locally, in concentrations of 1:250 — 1:500; very toxic to the lower forms of animal (mobile organisms such as protozoa and ameba) and vegetable life; with urea hydrochloride as a local anesthetic at 0.25% in normal saline for all operations that can be done under cocaine; for oxyuris (pinworms) in a 1:2000 — 1:500 (0.5%) solution as a rectal injection; antiseptic for diseases of the eye (1-2%) such as suppurating conjunctivitis (formation of pus in association with inflammation of the conjunctiva), keratitis and ulcer of the cornea; internally: as a bitter, antipyretic (reduces fever), antimalarial, analgesic (alleviates pain) and antipyretic in colds, and in pneumonia (Milks).

Rhamnus cathartica (Buckthorn) — Powerfully cathartic (quickens or increases evacuations from the intestines) (Felter).

Rhubarb (USP) — Gastric irritation; irritative diarrhea with tenderness on pressure; sour smelling discharges; gastro-intestinal irritation with nervous debility, restlessness; light colored fecal discharges (Felter).

Small doses it acts as a bitter stomachic (stimulates functional activity of the stomach); in medium doses it is an astringent and serviceable in diarrhea of young animals; as a laxative in young animals suffering from constipation (Milks).

Rhus glabra (Smooth sumac, Pennsylvania sumac) (USP) — Relaxation of mucous tissues, with unhealthy discharges; mercurial ulcerations; spongy gums; ulcerative sore throat, with fetid discharges (author: essentially counteracts classic signs of mercury poisoning); flabbiness and ulceration of tissues (Felter).

Never used internally; as fluidextract diluted in 6-8 parts of water (tincture strength, author) is an efficient mouth wash in stomatitis and acute pharyngitis (Milks).

Rumex (Yellow dock) (USP) — Bad blood with chronic skin diseases; bubonic swellings; low deposits in glands and cellular tissues, and tendency to indolent ulcers; feeble recuperative power; irritative laryngo-tracheal cough; stubborn, dry, summer cough; chronic sore throat, with glandular enlargements and hypersecretion; cough, with dyspnea (Felter).

Ruta (Garden Rue) — Emmenagogue (stimulates menstruation), ecbolic (produces abortion), anthelmintic (destroys or expels worms or prevents their development), and antispasmodmic; its action is chiefly directed upon the uterus, and is capable of exciting menorrhagia (excessive menstrual bleeding), inflammation, and miscarriage; excellent vermifuge (agent to expel parasitic worms) (Felter).

Its actions depend on its volatile oil — ecbolic, emmenogogue but are dangerous to use since the irritation they set up on the gastro-intestinal tract may be fatal without producing abortion (Milks).

Sabina (USP) — See Chapter 10

Sanguinaria (Bloodroot) (USP) — Mucous membranes look red and irritable; burning and itching of mucous membranes, especially upper respiratory airways (Felter).

Santal (Red Sandal-wood, Red Saunders, Ruby wood) (USP) — Tonic and astringent (Felter).

Similar in action and use as Balsam of Copaiba: chronic discharges from the genito-urinary tract, pyelitis (inflammation of renal pelvis), cystitis (inflammation of the urinary bladder), urethritis (inflammation of the urethra) — but should not be given until the most acute symptoms have subsided since it is stimulant to the mucous membranes generally; subacute or chronic bronchitis (inflammation of the mucous membrane lining of the bronchi) with excessive purulent (pus) discharge, but generally inferior to guiacol (Milks).

Santonin (Levant wormseed) (USP) — To remove all kinds of intestinal worms but the tapeworm; retention of urine from atony; urethral irritation, with pain and scalding, accompanying uterine disorders; retention of urine in fevers (Felter).

Very efficient remedy for round worms but not so effective for tape or hook worms; drives the worms to the large intestine, which then should be removed by a purgative (Milks).

Sarsaparilla/Smilax (USP) — Alterative (a medicine used to modify nutrition so as to overcome morbid processes) (Felter).

Sassafras (USP) — A warm, aromatic stimulant, alterative (medicine used to modify nutrition so as to overcome morbid processes), diaphoretic (produces sweat) and diuretic (increases the secretion of urine); the oil, placed as drops on sugar, affords relief in the pain attending menstrual obstructions and that following parturition; also used in diseases of the kidney and the bladder (Felter).

Scoparius (Broom, Irish broom, Broom tops) (USP) — In large doses, emetic (produces vomiting) and cathartic (quickens or increases evacuations from the intestines); in small doses, diuretic (increases the secretion of urine); chronic dropsy (edema), especially dropsy of the thorax, combined with diseases of the lungs; never said to fail in increasing the flow of urine (Felter).
Used alone or with digitalis in cardiac dropsy (Milks).

Scutellaria (Skullcap, Madweed) (USP) — Tonic, nervine (calms the nervous system) and antispasmodic; intermittent fevers (combine with lycopus); nervous excitation, restlessness attending or following acute or chronic diseases; nervousness manifesting itself in muscular action; tremors; inability to control the voluntary muscles (Felter).

Senega (Senega snakeroot, Seneca snakeroot) (USP) — Relaxation of the respiratory mucous membranes and of the skin; cough deep and hoarse, with excessive secretion, mucous rales; stimulates secretions, especially as a sialagogue (increases salivary flow), expectorant (acts upon the pulmonary mucous membrane to increase or alter secretion), diuretic (increases the secretion of urine), diaphoretic (produces sweat), and emmenagogue (stimulates menstruation); chronic catarrh (simple inflammation of a mucous membrane) (Felter).

Senna (USP) — Wind or bilious colic (spasm of gall bladder); a laxative for non-inflammatory conditions of the intestinal tract; all forms of febrile diseases in which a laxative action is desired; excites increased peristaltic motion (intestinal wave action) (Felter).

Spigelia (Pinkroot, Carolina pink, Worm-grass) (USP) — An active and certain vermifuge (expels parasitic worms), especially in the young; useful in those conditions of the system caused by worms, which resemble febrile diseases (Felter).

A reasonably safe remedy for round worms but not terribly efficient; it should always be combined with a brisk cathartic (aloe or senna) (Milks).

Squills (USP) — Chronic cough, with scanty sputa; scanty, high colored urine; over-active kidneys with inability to retain urine; dropsy (edema), with no fever or inflammation; a general asthenic (weakness, adynamia; weakness or debility) condition (Felter).

Increase the force of the heart but decrease the rate and acts as a diuretic (similar to digitalis — increases secretion of urine); powerful expectorant for chronic bronchitis (Milks).

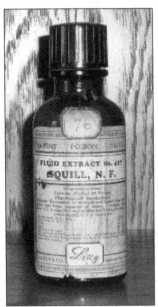

Stillingia (Queen's delight, Silver leaf, Yaw root) (USP) — Feeble tissues, with tardy removal of broken down material, and slow renewal of parts; mucous membranes, red and glistening, with scanty secretion; laryngeal irritation, with paroxysmal, hoarse, croupous cough; winter cough of irritation; periosteal (bone related) pain and tendency to form nodes (Felter).

Stramonium (Jimson weed, Thorn-apple, Stinkweed) — Delirium, furious, enraged, and destructive; restless, cannot rest in any position; seems to be fearful; pain, especially when superficial and localized; spasm, with pain; bloating and heat of face (Felter).

Action and use is practically identical to belladonna; especially valuable in asthma of dogs and heaves of horses; in the former case the leaves may be burned in the room, and will often prove of much benefit; stramonium is often preferred by horse dealers to belladonna for heaves in horses (Milks).

Strophanthus (USP) — Weak heart, due to muscular debility; muscular insufficiency; rapid pulse, with low blood pressure; cardiac pain, with dyspnea (shortness of breath) (Felter).

Circulatory stimulant, and diuretic (increases the secretion of urine), similar to digitalis; absorption and elimination more rapid (Udall).

Sumbul (Musk-root, Jatamansi) (USP) — Stimulant and tonic to the nervous system; low fevers, in gastric spasm, diarrhea, dysentery, leucorrhea (white or yellowish viscid discharge from the vagina), chronic bronchitis and other maladies accompanies by an asthenic (weakness, debility, adynamia) condition (Felter).

Tanacetum (Tansy) (USP) — Tonic, emmenogogue (stimulates menstruation), and diaphoretic (produces sweat); in small doses, cold infusion useful in convalescence form exhausting diseases, in dyspepsia (impaired gastric function), with troublesome flatulence (gas), jaundice and worms; warm infusion beneficial in intermittent fever, suppressed menstruation, tardy labor-pains; seeds efficient for worms; spray or inhalation strong tincture (1:4 -1:10) valuable in acute inflammations of throats and epidemic catarrh (simple inflammation of a mucous membrane) (Felter).

Ecbolic volatile oil (Milks).

Taraxacum (Dandelion) (USP) — Must be fresh, not dried root; Loss of appetite, hepatic (liver) sluggishness, and constipation; with cream of tartar, it's more diuretic (increases the secretion of urine) and laxative (acts mildly in opening or loosening the bowels); not for those apt to dyspepsia (impaired gastric function), flatulence (gas) and diarrhea (Felter).

Simple stomachic/bitter (stimulates functional activity of the stomach) (Milks).

The extract/essential oil is GRAS.

Thiosinamin (Black mustard volatile oil) — Gastic debility; a remedy for atonic (relaxed; without normal tone or tension) conditions only (Felter).

Thymol (Thymus vulgaris volatile oil) (USP) — Antiseptic (antagonizes sepsis or putrefaction) and disinfectant (substance with the power of destroying disease germs or the noxious properties of decaying organic matter [considered superior to carbolic acid]); destroys the vitality of organized and living ferments, it prevents the occurrence of putrefaction, and arrests it when it has commenced; in contact with putrid pus, it promptly removes any odor emanating from this substance, and keeps it without change until complete desiccation occurs; applied to wounds, ulcers, or abscesses, in the form of a weak solution, it promptly modifies their condition, and accelerates cicatrisation (the process of scar formation; the healing of a wound otherwise than by first intention); as a local application to prevent septicemia (blood poisoning) during the parturient period; employed in inhalations in bronchial and other affections of the respiratory organs, attended with gangrenous exhalations; it suppresses the gangrenous odor, and prevents the diseased bronchial surfaces and the abnormal secretions covering them, from undergoing butyric fermentation, thus checking a poisoning of the blood (Felter).

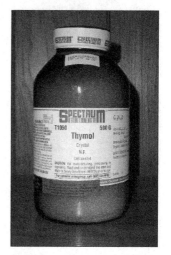

Externally, resembles action of phenol, but is less toxic and irritant; antiseptic for wounds, but its odor attracts flies and soon becomes distasteful to attendants; useful in fungus skin diseases; internally, as an intestinal antiseptic and a remedy against hookworms in dogs as well as treatment for tape and round worms (Milks); disinfectant and deodorant (Udall).

Extract/essential oil is GRAS.

Tiglii (Croton oil) (USP) — Powerful irritant and cathartic (quickens or increases evacuations from the intestines); obstinate constipation when no inflammation exists (Felter).

Tobacco (USP) — Parasiticide for killing mange, scab, lice and fleas; for sheep with full fleece, tobacco is one of the best curative and preventative agents in mange or scab; the Germans prescribe tobacco to stimulate peristalsis in ruminants, in doses of 2 ounces, with ½ lb. of common salt and 1 lb. of Glauber's salt for cattle; and for sheep, ½ oz., with 2 ounces of salt and 3 oz. of Glauber's salt. The decoction (1-2%) may be injected into the rectum of horses, in non-toxic quantities, to kill oxyurides and ascarids, and to excite peristalsis and relieve spasm in colic (Winslow). Useful as a parasiticide in mange of animals, especially sheep scab, although it is very efficient for lice, fleas, etc. of the large animals. As it is absorbed from the skin, larger amounts than can be given by mouth should not be applied. For mange in sheep and other animals, it is used in decoctions (2-5%). Law's dip is as follows: Tobacco, 16 pounds; oil of tar, 3 pints; soda ash, 20 pounds; soft soap, 4 pounds; water, 50 gallons. Sufficient for 50 sheep. Steep the tobacco for ½ hour; strain, and add the other ingredients at 70° F (Milks).

Tonga — Hoarseness and aphonia (loss of voice as a result of disease or injury of the organ of speech), with burning and constriction of throat, and thin, nasal discharge; intensely sore throat, with bleeding and fetor; the tongue red and sensitive (Felter).

Triticum (Couch grass, Quick grass, Dog grass) (USP) — Irritation of the urinary apparatus; pain in the back; frequent and difficult or painful urination; gravelly deposits in the urine; catarrhal (simple inflammation of a mucous membrane) and purulent (pus) discharges from urethra.

Turpentine (Oleoresin of *Pinus* species) (USP) — Topically are irritants, occasioning heat, redness and inflammation of the skin; internally, act on mucous tissues, lessening excessive morbid secretions; stimulant to the general system, quickening pulse, increasing surface temperature and causing sensation of warmth in the stomach; anthelmintic (Felter).

Turpentine (oil of) (USP) — Germicide (destroys parasites), antiseptic antagonizes sepsis or putrefaction), deodorant (conceals or destroys foul odors), and expectorant (acts upon the pulmonary mucous membrane to increase or alter its secretions); kills yeast and renders vaccine virus innocuous; winter cough, chronic bronchitis; secondary bronchial affections with heavy mucous secretions; relieves flatulent dyspepsia (gaseous impairment of stomach function); applied to wounds, ulcers, burns, it acts as a deodorant (conceals or destroys foul odors) and protective (an agent to protect the part to which it is applied), favoring rapid healing (Felter).

Carminative (helps allay pain by causing the expulsion of flatus from the alimentary canal); anthelmintic; diuretic (increases the secretion of urine) (Milks).

Ustilago (USP) — See Chapter 10

Uva ursi (Bearberry) (USP) — The effects depend entirely upon its stimulant, astringent, and tonic powers, though, in smaller doses, it tends to relieve chronic irritation of the bladder; as an astringent, chronic diarrhea and dysentery; retardation of putrescent changes in the urine; relaxation of urinary tract, with pain and mucous or bloody secretions; chronic vesical irritation, with pain, tenesmus (painful spasm of the anal sphincter with an urgent desire to evacuate the bowel or bladder, involuntary straining, and the passage of but little fecal matter or urine), and catarrhal discharges (simple inflammation of a mucous membrane) (Felter).

Diuretic (increases the secretion of urine) and astringent; uses and action as buchu (Udall).

Valerian — Excites the cerebro-spinal system; large doses cause mental excitement, visual illusions, giddiness, restlessness, agitation, and frequently nausea; cases requiring are those evidencing enfeebled cerebral circulation, there is despondency and depression; in properly selected cases it relieves irritability and pain, and favors rest and sleep (Felter).

Veratrum viride (USP) — See Chapter 10

Viburnum prunifolium (Black haw) (USP) — Uterine irritability, and hyperesthesia (heightened sensibility to pain); threatened abortion; uterine colic; dysmenorrhrea (abnormal menses), with deficient menses; severe lumbar and bearing down pains; cramp-like, expulsive menstrual pain; intermittent, painful contractions of the pelvic tissues; after-pains and false pains of pregnancy (Felter).

To quiet the uterus in threatened abortion (Milks).

Vinegar/acetum — An agreeable cooling drink in fevers, especially when tongue is red; diminishes inordinate vascular action, allays thirst, neutralizes excess alkali, and increases urinary discharge; in putrid diseases it acts as an antiseptic; in urinary affections with a white sediment mainly of calcium and ammoniaco-magnesium phosphate, it has been recommended; in dysentery, vinegar saturated with common house salt has been very beneficial; local application for external inflammations, contusions, severe injuries to joints, swellings, etc (Felter).

Whiskey/alcohol — Prostration, with soft, feeble pulse, hurried respiration, and irregular heart action; prostration, with dry tongue (Felter).

Wintergreen Oil (see Gaultheria)

Xanthoxylum (Prickly ash) (USP) — Atonicity of the nervous system (larger doses); capillary engorgement, sluggish circulation, tympanites (tight pinging sounds due to gas) in bowel complaints, intestinal and gastric torpor (with deficient secretion), dryness of the mucous membrane of mouth, flatulent colic, uterine cramps; for painful bowel disorders, the berries are preferred.
Extract/essential oil is GRAS.

Zea/cornsilk (USP) — Diuretic (increases the secretion of urine) and slightly anodyne (diminishes sensibility to pain), especially useful in urinary troubles associated with renal or cardiac disorders (dropsy/edema); calculi, gravel; cystic (urinary bladder) irritation due to phosphatic and uric acid concretions, and in both acute and chronic inflammations of the bladder; pyelitis (inflammation of the renal pelvis) , catarrh (inflammation of mucous membrane) of the bladder and urine retention; its diuretic (increases secretion of urine) action is largely due to its tonic action upon the heart an blood vessels.

One compound, garlic (Allium sativum), is not on the listing given by Fish, but is in *King's American Dispensatory* (and is not part of the common Homeopathic materia medica). There is a perennial interest in it among farmers and it will therefore receive its due attention here. Garlic has been listed in the National Formulary (NF) since at least 2002.

Garlic (USP) — *Allium sativum.* Nat. Ord. — Lilliaceae. The bulbs contain cysteine sulfoxides and gamma-glutamylcysteines with alliin (S-allylcysteine sulfoxide) being the most abundant cysteine sulfoxide — which is itself biosynthesized from its parent compound, S-allylcysteine. Alliin is the most abundant sulfur compound in unbruised garlic; whole garlic contains between 6-14 mg. of alliin per gram (fresh weight). Once bruised, the enzyme alliinase acts on it within 10 seconds, converting all alliin to allicin (a thiosufinate).The total allicin yield has been calculated to about 2.5-5.1 mg/gm of fresh, crushed garlic or about 5-20 mg per clove. Allicin and its transformational products account for most of garlic's medicinal properties. However, allicin is unstable and begins reacting with itself to form more stable compounds; it has a half-life of 2-16 hours at room temperature. Refrigeration will increase the half-life by a factor of 20. Alliin and alliinase can survive drying and processing, but freeze-drying retains the highest levels. In order to deliver allicin to the body, supplements need to have an enteric coating to by-pass the stomach and be active in the small intestine for absorption. Garlic oil (steam distilled) is devoid of allicin, but it does contain other sulfides which can be similar to using whole garlic, but be careful in choosing brands since high quality oils can have more than 50 times the sulfide content of inferior ones. Allicin and its derivatives have been shown to be antimicrobial. Peer reviewed studies show these compounds toxic to *Staph aureus, Salmonella typhimurium, Clostridium botulinum, Candida albicans* and amoebas which cause dysentery. Both alliin and allicin have been shown to be anti-oxidant, in a dose-dependent fashion (with excessive doses being pro-oxidant); however, S-allylcysteine and allicin are both better antioxidants than alliin. It is considered GRAS (Generally Recognized As Safe) in the U.S. (Ganora).

Stimulant, diuretic (increases the secretion of urine), expectorant (acts upon the pulmonary mucous membrane to increase or alter its secretions) and rubefacient (causes irritation and redness of the skin), all due to its volatile oil; beneficially used in coughs, catarrhal (simple inflammation of a mucous membrane affections), hoarseness and worms; applied along the spinal column and over the chest, in the form of poultice, it is very useful in pneumonia; the juice can be made into a syrup with sugar for coughs and pulmonary affections (Felter).

Table 10: Plant Medicine Doses as Used by Conventional Veterinarians

(Fish 1930, p. 8-41 — abridged; see the indications and uses of the remedies immediately preceeding this listing)

T. = Tincture   **C/T.** = Compound tincture (mixture)   **F.E.** = Fluidextract
H&C = Horses and cattle   **Sh&Sw** = Sheep and swine
All doses are in cc's (ml's) to be given orally, unless otherwise noted ("gr" = grains)

	H&C	Sh&Sw
Aconite T.	2-6	.25-1
Aloe	8-40	4-15
Areca nut	15-30	2-6
Arnica .T	15-30	4-8
Asafetida T.	60-120	8-15
Balsam Copaiba	15-60	2-8
Balsam Peru	15-60	4-8
Balsam Tolu	15-60	2-4
Belladonna leaves T.	15-30	4-8
Bryonia T.	15-30	2-4
Buchu leaves F.E.	30-60	2-4
Caffeine citrate	1-2	.25-.5
Calendula T.	15-30	4-8
Calumba T.	60-120	12-24
Cannabis Indica T.	15-45	2-4
Cantharides	1-2	0.3-1
Capsicum	4-8	0.6-2
Cardamom T.	60-90	12-24
Cascara sagrada F.E.	8-45	0.6-4
Cascarilla Bark F.E.	15-30	4-8
Castanea F.E.	30-60	8-15
Castor Oil	500	60-120
Catechu C/T.	30-60	8-15
Chamomile	30-60	4-8
Chaulmoogra Oil	2-12	0.3-2
Chenopodium Oil	6-12	0.6-1.3
Chimaphila F.E.	30-60	4-15
Cimicifuga T.	30-90	8-15
CinchonaBark C/T.	60-120	15-30
Cinnamon Oil	2-6	0.3-0.6
Coca F.E.	30-120	15-30
Cocaine HCl	.3-.6 gr	.03-.1gr
Cod Liver Oil	60-120	15-30
Codeine	0.4-2.0gr	.03-.2gr
Colchicum Root T.	15-45	4-6
Conium F.E.	4-8	0.6-1.3
Convallaria	4-8	0.6-1.3
Cotton Root Bark	15-60	4-8
Digitalis T.	12-24	3-10
Dioscorea F.E.	8-24	2-4
Echinacea F.E.	4-15	2-4
Ergot T.	15-60	4-15
Eriodictyon F.E.	15-60	2-8
Eucalyptus Oil	8-15	1.3-3.3
Fennel	30-60	8-12
Fenugreek	30-60	8-12
Gamboge	15-30	1.3-4
Gaultheria Oil	8-30	2-8
Gelsemium T.	15-60	4-12
Gentian F.E.	15-30	4-8
Geranium F.E.	8-30	2-4
Glycerin	30-60	8-15
Glycyrrhiza	15-60	4-15
Gossypium F.E.	8-30	2-8
Granatum F.E.	15-30	4-12
Guarana F.E.	8-30	2-4
Gum Tragacanth	60-90	15-30
Hamamelis F.E.	30-60	8-15
Helleborus Niger F.E.	4-15	0.6-2
Hematoxylin F.E.	15-45	6-12
Humulus T.	30-120	4-15
Hydrastis F.E.	8-30	4-8
Hydrastis Glycerite	8-30	4-15
Hydrastis T.	30-60	4-15
H2O2 3%	15-60	4-15
Hyoscamus T.	30-90	8-15
Ignatia F.E.	2-4	1.3-2.6
Iodine T.	8-15	1.3-2.6
Ipecac F.E.	4-8	1-2
Jaborandi F.E.	8-15	2-4
Juniper Oil	4-8	0.6-2
Kamala F.E.	15-30	4-12
Kava kava F.E.	15-30	4-12
Kino F.E.	30-60	4-12
Kousso F.E.	15-60	4-12
Krameria F.E.	15-30	4-8
Lactucarium F.E.	8-30	1-4
Laudanum	15-60	4-15
Lobelia T.	30-60	4-12
Male Fern F.E.	12-24	4-8
Matico T.	30-60	8-15
Mentha piper. oil	1-2	0.3-0.6
Morphine	.2-.6gr	<0.13gr
Myrrh T.	8-15	4-8
Nicotine	.001-.006	.001-.002
Nux vomica T.	4-24	1.3-2.6
Opium T. (paregoric)	60-120	15-30
Pepo (Pumpkinseed)	-----	15-60
Physostigma F.E.	1-2	.13-.25
Phytolacca F.E.	4-8	1.3-3
Pichi F.E	8-24	2-4
Pilocarpus F.E.	8-15	2-4
Pipsissewa F.E.	30-60	4-15
Pomegranate	30-60	4-12
Polygonum F.E.	15-30	4-8
Prunus Virginian F.E.	15-60	4-8
Pulsatilla F.E.	2-8	0.3-0.6
Pyrethrum	15-30	2-6
Quassia F.E.	30-60	8-15
Quercus Alba F.E.	15-30	4-8
Quinine (antipyretic)	8-15	1.3-2.6
Rhamnus F.E.	30-60	4-8
Rhubarb F.E.	30-60	4
Rhus glab. F.E.	15-30	4-8
Rumex F.E.	30-60	4-8
Ruta F.E.	15-30	15-30
Ruta Oil	2-4	.13-.6
Sabina F.E.	30-60	2-4
Sabina Oil	8-15	0.5-1
Sanguinaria F.E.	4-24	0.6-2
Santal F.E.	15-60	8-12
Santal Oil	4-12	0.6-3
Santonin	15-30	4-8
Sarsaparilla F.E.	30-60	4-8
Sassafrass F.E.	30-60	4-8
Sassafrass Oil	2-8	0.3-0.6
Scoparius F.E	15-30	4-8
Scutellaria F.E.	15-30	4-12
Senega F.E.	4-15	1-2
Senna F.E.	120-150	30-60
Serpentaria F.E.	15-30	2-4
Spigelia F.E.	4-30	2-4
Squill F.E.	4-8	0.3-1.3
Squill T.	24-48	6-12
Stillingia F.E.	4-30	2-8
Stillingia T.	15-45	4-8
Stramonium F.E.	1.3-4	0.3-0.6
Stramonium T.	4-8	0.6-2
Strophanthus T.	4-15	0.3-1.3
Sumbul F.E.	8-24	1-4
Sumbul T.	15-30	2-8
Tanacetum Oil	1.3-4	.13-.4
Taraxacum F.E.	30-60	8-15
Terebene	8-24	2-4
Thiosimanin	2-4	0.5-1
Thymol	2-8	0.3-2
Tiglii Oil	1-2	0.3-0.6
Tonga F.E.	8-30	2-4
Triticum F.E.	30-60	8-24
Turpentine Oil		
(carminative)	30-60	4-15
(anthelmintic)	60-120	15-30
Ustilago F.E.	15-60	2-4
Uva ursi F.E.	60-120	8-15
Valerian F.E.	30-60	4-8
Valerian Oil	2-4	0.6-1.3
Veratrum viride F.E.	2-4	1.3-2
Veratrum viride T.	8-12	2.6-4
Viburnum Prun. F.E.	30-120	8-15
Vinegar	30-120	2-8
Whiskey	60-120	30-60
Wintergreen Oil	8-30	2-8
Xanthoxylum F.E.	15-60	4-12
Zea F.E.	30-60	8-15
Zingiber F.E.	8-30	4-8
Zingiber T.	30-60	8-15

Classification of Medicines According to Their Physiologic Actions

(Fish, p.156, abridged)

Anthelmintics (remedy for destroying or expelling worms or to prevent their development):
Aloes (enema), aspidium, chenopodium, koussein, oil turpentine, extract male fern, pelletierine tannate, pumpkin seed, quassia infusion, sodium chloride, santonin, sodium santoninate, spigelia, thymol

Antipyretics (a medicine to reduce body temperatures in fevers):
Acetanilid, benzoic acid, salicylic acid, aconite tincture, aspirin, phenacetin, quinine and salts, resorcin, salicylic acid, veratrum viride tincture

Carminitives (a remedy which helps allay pain by causing the expulsion of flatus from the alimentary canal):
Anise, asafetida, calumba, capsicum, cardamom, caraway, cascarilla, chamomile, cinchona, cinnamon, cloves, gentian, ginger, nutmeg, nux vomica, oil cajeput, oil mustard, orange peel, pepper, pimenta, quassia, sassafras, serpentaria

Galactogogues (an agent to increase the secretion of milk):
Lactic acid, castor oil (topically), extract malt, jaborandi, pilocarpine hydrochloride

Gastric tonics (a medicine promoting nutrition and giving tone to the system):
Alkalies: before meals, aromatics, berberine carbonate, bismuth salts, bitters, carminitives, hydrastis, nux vomica, quassin

Oxytocics (ecbolics) (an agent to aid or produce parturition):
Cotton root bark, ergot, hydrastine, hydrastine hydrochloride, pennyroyal, quinine, rue, savine

Tonics, General:
Vegetable: Bitters, berberine carbonate, cinchona alkaloids and salts, cod-liver oil, eucalyptus, hydrastis, quassin, salicin

General Actions of Drugs

(Winslow, pp.19-58)

(See dosage Table 10 from P.A. Fish (page 166), if the following are not found within recipes in the treatment chapters; doses for minerals not stated later in the recipes need to be obtained from pertinent texts listed in the bibliography.)

If using any of the following drugs, they should be feed-grade (*i.e.,* linseed oil, castor oil, etc.).

Drugs Acting on the Digestive Organs

Stomachics (a drug to stimulate functional activity of the stomach):
Bitters — Gentian, calumba, quassia, hydrastis, taraxacum
Aromatic bitters — Cascarilla, chamomile, serpentaria
Aromatics — Coriander, capsicum, pepper, ginger, cardamom, fennel, fenugreek, anise, calamus, mustard, spearmint, peppermint, alcohol

Antacids: Sodium carbonate, sodium bicarbonate, potassium carbonate, potassium bicarbonate, solution of potash, magnesia, magnesium carbonate, calcium carbonate (chalk), solution of lime (lime water)

Antiseptics (an agent antagonizing sepsis or putrefaction): Creosote, creolin, bismuth subnitrite, bismuth subcarbonate, bismuth subsalicylate, sodium sulphite, bisulphate and hydrosulphite, hydrogen dioxide (hydrogen peroxide)

Emetics (a medicine to produce vomiting) *[author: cows do not vomit, unless there is severe pathology of the fore-stomachs]*:
Specific — Apomorphine, lobeline, morphine, senega, squills
Mixed — Tartar emetic, ipecac, copper sulphate, zinc sulphate
Local — Tepid water, mustard, salt, alum

Gastric sedatives and anti-emetics (an agent which allays vomiting): Ice, hot water, bismuth subcarbonate, bismuth subnitrate, carbon dioxide, morphine, menthol, creosote, aconite, belladonna, hyoscyamus, cocaine, cerium oxalate, lime water, minute doses of: arsenic, ipecac, alcohol, iodine, silver nitrate; chloroform, chloral, bromides, nitrites, brandy and champagne

Laxatives (a medicine acting mildly in opening or loosening the bowels): Olive oil, cottonseed oil, magnesia, sulfur, nux vomica, small dose: linseed oil, castor oil; liquid petrolatum and other mineral oils (act as a mechanical lubricant and not absorbed)

Simple purgatives (a medicine to produce increased discharges from the bowels): Aloe, linseed oil, castor oil, rhubarb, senna, cascara sagrada, frangula, bryonia

Drastic purgatives: Croton oil, colocynth, gamboges, scammony, jalap, elaterium

Saline purgatives: Magnesium sulphate, sodium sulphate, sodium phosphate, potassium bitartrate

Direct cholagogues (a drug provoking the flow of bile): Sodium salicylate*, podophyllum*, aloes, rhubarb, colchicum, sodium sulphate, sodium phosphate*, ipecac, euonymus, nitro-hydrochloric acid*

have been found by clinical evidence to be most active

Drugs Acting on the Circulation

Drugs acting upon the blood

Hematinics: Iron and its salts, copper salts, potassium permanganate, manganese dioxide

Drugs acting on the heart:

Increase **the** *force* **of the heart beat:** Digitalis, adrenalin (epinephrine), squill, physostigmine, strophanthus, sparteine

Increase **the** *rate* **of the heart-beats:** Belladonna, atropine, hyoscyamus, stramonium, cocaine

Increase **the** *force and rate* **of heart-beats:** Alcohol, chloroform, ether, ammonia, ammonium carbonate, strychnine, caffeine, quinine, arsenic

Decrease **the** *force and rate* **of the heart-beats:** Aconite, veratrum viride, ergot, antimony salts

Drugs Acting on the Blood Vessels

Systemically to contract vessels: Cocaine, ergot, atropine, digitalis, strophanthus, squill, sparteine, strychnine, hamamelis, hydrastis, physostigmine, adrenalin (epinephrine)

Systemically to dilate vessels: Amyl nitrate, nitroglycerin, spirit of nitrous ether, alcohol, salicylates, ether, chloroform, thyroid secretion, chloral, aconite, opium, secondary action of belladonna, hyoscyamus, stramonium

Drugs Influencing the Brain

Cerebral excitants

Camphor, caffeine, quinine, cocaine

Cerebral depressants

Anodynes by reason of their action on the brain (*anodyne* — an agent which diminishes sensibility to pain): Codeine, morphine, opium, alcohol, anesthetics, chloral, cannabis indica, gelsemium, bromides

Narcotics (a powerful remedy causing stupor): Opium and its derivatives, alcohol, anesthetics, chloral, cannabis indica, belladonna, stramonium, hyoscyamus

Hypnotics: Opium, morphine, chloral, bromides, cannabis indica

General anesthetics: (an agent used to produce insensibility to pain): Ether, chloroform, nitrous oxide

Stimulate the motor centers: Strychnine, atropine, physostigmine

Depress the motor centers: The bromides, chloral, alcohol, anesthetics

Drugs Acting on the Spinal Cord

Stimulate motor cells of inferior cornua: Strychnine, brucine, thebaine, ammonia, ergot

Depress the motor cells of inferior cornua: Physostigmine, bromides, ergot, gelsemium, emetine, turpentine, saponin, chloral, morphine, apomorphine, alcohol, ether, chloroform, camphor, nicotine, veratrine, salts of: magnesium, sodium, potassium, lithium, antimony, zinc, silver

Drugs Acting on the Nerves

Influence peripheral sensory nerve-endings

Stimulate: Counter-irritants

Depress — local anodynes: Aconite, menthol, carbolic acid, atropine, morphine, chloral, prussic acid, sodium bicarbonate, veratrine, heat, cold

Depress — local anesthetic: Cocaine, eucaine, stovaine, novocaine, holocaine, cold, ether spray

Influence motor nerve-terminations

Stimulate: Strychnine, pilocarpine, aconite, nicotine

Depress: Curare, conium, atropine, amyl nitrate, cocaine, camphor, lobeline, nicotine (and many others)

Drugs Acting on Nerves of the Special Senses

Act on the eye

Mydriatics — (a drug causing dilatation of the pupil, paralyze 3rd nerve endings): Atropine, belladonna, homatropine, hyoscyamine, hyoscine, scopolamine, gelsemine; cocaine stimulates sympathetic endings.

Myotics — (a drug causing contraction of the pupil)

acting locally (stimulate 3rd nerve endings): Physostigmine, pilocarpine

acting centrally: Anesthetics, opium

Drugs Acting on the Respiratory Organs

Expectorants (a medicine to act upon the pulmonary mucous membrane to increase or alter its secretions)

Increase secretion: Apomorphine, potassium iodide, ipecac, pilocarpine, ammonium chloride, squill, camphor, balsams, sulfur, tar, turpentine, terpin hydrate, terebene, volatile oils

Decrease secretion: Belladonna, hyoscyamus, stramonium, opium, volatile oils (first increase, then decrease secretions)

Altering the nutrition of bronchial mucous membrane: Cod-liver oil, sulfur, potassium iodide

Exerting an antiseptic action: Turpentine, terebene, terpine hydrate, balsam of Peru, balsam of Tolu, cubebs, copaiba, tar, ammoniacum

Locally stimulating and antiseptic to mucous membranes: Eucalyptol, guiacol, creosote

Drugs stimulating the respiratory centers: Strychnine, atropine, caffeine, cocaine, belladonna, hyoscyamus, stramonium, strong ammonia

Drugs depressing the respiratory centers: Morphine, codeine, chloral, bromides

Drugs relaxing spasm of the bronchial muscular tunic and relieving cough:
Locally: White of egg, linseed tea, syrups, mucilage, external counter-irritation and heat
Systemically: Opium, codeine, hyoscyamus, stramonium, cannabis indica, nitrites, chloral, bromides, chloroform, phenacetin, adrenalin (epinephrine)

Drugs allaying spasm and cough: Opium with belladonna

Drugs Acting on the Urinary Organs

Diuretics (a drug to increase the secretion of urine):
   Increase the glomerular fluid: Water, potassium acetate, citrate, bitartrate; digitalis, squill and strophanthus (when the circulation is poor); caffeine

Stimulating renal cells or lessening absorption from tubular cells, or both:
   Caffeine, theobromine; volatile oils, resins or aromatics such as buchu, juniper, turpentine, cantharides; glucosides such as scoparin and asparagin; all salts, glucose and alkalies

Urinary antiseptics: Benzoic acid, boric acid, methylene blue, salicylic acid, salol, buchu, copaiba, cubebs, volatile oils

Urinary sedatives: Hyoscyamus, belladonna, opium, alkalies (with an acid urine)

Drugs Acting on the Sexual Organs

Influencing chiefly the male generative organs

Aphrodisiacs:
   Direct aphrodisiacs: Strychnine, phosphorus, alcohol (act on centers); cantharides (local irritant); yohimbine (causes congestion of the sexual organs)
   Indirect aphrodisiacs: Iron, strychnine, arsenic, full diet

Anaphrodisiacs (depress sexual activity): Opium, bromides, purgatives, nauseants, bleeding (venesection), spare diet

Influencing the female sexual organs:

Emmenagogues: (a drug to stimulate menstruation):
   Direct: Savin, rue, cantharides (irritants); ergot
   Indirect: Aloe (purgative); iron, arsenic, strychnine, full diet (in debility)

Ecbolics (oxytocics) (a drug to produce abortion): Ergot, quinine, hydrastis, savin, corn smut, cotton root bark

Restraining uterine contractions: Cannabis indica, bromides, chloral, anesthetics

Influencing milk secretion, increase flow of milk (galactogogues):
   Pilocarpine, leaves of the castor oil plant, alcohol, full diet; rubbing the udder with nettles (Udall, 1922)

Drugs Influencing Metabolism

Tonics (impossible to define precisely): Tonics improve the general nutrition and health, and generally understood to refer to drugs promoting appetite and digestion (gentian); the state of the blood (hematinics, as iron); the condition of some organs (heart, as digitalis; nervines, strychnine). Tonics are indicated in the treatment of debility (general or special) and anemia.

Drugs Influence Body Heat

Diminish metabolism: Quinine

Dilate superficial vessels: Salicylic acid, alcohol, nitrous ether, opium and ipecac

Depress circulation: Aconite, veratrum, digitalis, antimony and venesection (bleeding)

Anti-pyretics (a medicine to reduce body temperatures in fevers): Acetanilid, antipyrin, phenacetin

Drugs Influencing the Skin

Dilate superficial vessels: Cantharides, iodine, mustard, capsicum, croton oil, oil of turpentine and other volatile oils, camphor, heat

Contract superficial vessels: Hamamelis, ergot, hydrastis, cocaine, tannic acid and drugs containing it, cold, and mineral salts

Styptics (hemostatics) (an agent causing contraction of blood vessels to check bleeding): Ferric alum, ferric chloride and subsulphate; adrenaline

Emollients (a substance used externally to mechanically soften and protect tissues): Lard, petrolatum, cacao butter, olive oil, cottonseed oil, lanolin

Demulcents (a mucilaginous or oily substance to soothe and protect irritated mucous membranes): Acacia, linseed infusion or tea, licorice, syrup, molasses, honey, glycerin, white of egg, milk, starch, sweet oil

Diaphoretics (a medicine to produce sweating): Pilocarpine, alcohol, opium, ipecac, aconite, camphor, external heat, blankets

Anhidrotics (a medicine to inhibit sweating): Atropine, belladonna, hyoscyamus, stramonium, nux vomica, quinine, salicylic acid (locally), cold externally

Drugs Which Destroy Microorganisms & Parasites

Disinfectants or germicides

Disinfectant — a substance with the power of destroying disease germs or the noxious properties of decaying organic matter

Germicide — An agent to destroy parasites: Lime, chlorinated lime, chlorine, heat

Antiseptics (external/surgery): Tincture of iodine, alcohol, hydrogen peroxide, potassium permanganate, zinc chloride, zinc sulphate, iodoform, salicylic acid, boric acid, thymol, balsam of Peru

Antiseptics (internal): Bismuth salicylate, bismuth subnitrite, quinine, volatile oils

Anthelmintics removing tapeworms (anthelmintic — a remedy for destroying or expelling worms or to prevent their development): Aspidium (horse and dog), oil of turpentine, kousso, areca nut (sheep), pumpkin seed, aloe, linseed/cottonseed/or castor oil

Anthelmintics removing ascarids:
Horses: Oil of turpentine, copper sulphate, carbon disulphide
Dogs: Santonin, spigelia, oil of chenopodium

Anthelmintics removing pinworms: Oil of chenopodium, thymol and cathartics (orally), salt, lime solution, quassia, iron salts, phenol, tannic acid, oil of turpentine (enema)

Anthelmintics removing strongyles: Thymol, oil of chenopodium, turpentine, copper sulphate, chloroform

Anthelmintics removing bots (gastrophilus spp.): Carbon disulphide

Anthelmintics against lungworms: Turpentine (intratracheal injections; author: only to be done by a veterinarian!)
Chloroform (injections in nostrils)

Vermifuges (to expel dead parasites from bowels after anthelmintics): Aloe and oil

Antiparasitics (skin) (a substance that destroys or drives away insects):
Against fungi of ringworm (Trichophyton spp.): Tincture of iodine, creosote, chrysarobin ointment, cantharides, croton oil, salicylic acid, boric acid, thymol
Against ray fungi of lumpyjaw (Actinomyces): Tincture of iodine, potassium iodide, iodoform, copper sulphate
Against fungi of thrush or aptha, sporadic aphthous stomatitis: Boric acid, potassium chlorate, potassium permanganate, alum, salicylic acid

Against mites of scab, itch or mange: sulfur, lime-sulfur dips, tar, crude petroleum, Peruvian balsam, salicylic acid, cantharides

Against lice: Staphisagria, oil of tar, Peruvian balsam, oil of anise, tobacco, pyrethrum, creosote preparations

Against fleas: pyrethrum, oil of anise, creosote preparations

Agents Acting Upon Micro-Organisms (germicides & antiseptics) (Ellingwood, 1919)

Iodoform, aristol (thymoliodide), boric acid, creosote, guaiacol, thymol (essential oil), potassium permanganate, hydrogen peroxide, sulfur, sulfurous acid, sulfur dioxide, bismuth salicylate

Since antibiotics are not allowed to be used in organic farming, it is of utmost importance to know what materials may help in acting as disinfectants, antiseptics, antimicrobials and germicides. The list "Agents Acting Upon Micro-Organisms" should be of help. It is from Finley Ellingwood's superb book, *American Materia Medica, Therapeutics and Pharmacognosy,* which is a gold mine of information about the use of Eclectic remedies.

Drugs Excreted by the Mammary Gland

(Fish 1930, 176)

Acid, boric; acid, salicylic; aloe; atropine; bromine and its compounds; chloroform; copper and its salt; croton; ether; iodine and its compounds; potassium and antimonium tartrate; rhubarb; sodii sulph.; turpentine

(Winslow 1919, 49)

Opium; all volatile oils; purgative salts; rhubarb; senna: castor oil; scammony; jalap; iodine; potassium iodide; antimony; zinc; iron; bismuth; neutral salts; ammonia; acids; sulfur; atropine; copper; colchicum; euphorbium; ergot; salicylic acid; veratrine; strychnine; croton oil; aloe; turpentine

For current European information and regulation *see: www.emea.europa.eu*

The website is a partnership of participating European countries set up to show what the withholding times are for both chemotherapeutic synthetic medicines and plant medicines.

Table 11: Potential Toxicities

Plant (as % of body weight, or plant part)	Amount to ingest to be toxic	Toxic principle
Asclepias (milkweed)	0.1%-0.5% of BW	galitoxin
Cicuta maculata (water hemlock)	8 ounces	cicutoxin
Corydalis (fitweed)	2% of BW/clinical signs; 5% BW/death	isoquinoline
Aesculus glabra (horse chestnut)	0.5-1.0% of BW	aesculin
Eupatorium rugosum (white snakeroot)	1-10% of BW (cumulative dosage)	trematone
Atropa belladonna (deadly nightshade)	Poisoning is uncommon	atropine
Conium maculatum (poison hemlock)	1% of BW	coniine
Delphinium (larkspur)	0.5-1.5% of BW	polycyclic diterpene
Lobelia (cardinal flower, giant lobelia)	Consumption of enough is rare	lobeline
Lupinus (lupine, bluebonnet)	1% or less of BW	lupinine
Lathyrus (wild pea, vetchling)	Unlikly unless cumulative	amino propionitrile
Colchicum autumnale (autumn crocus)	Mostly in bulbs	colchicine
Euphorbia (spurges)	1% of plant material	ingenol, phorbol
Phytolacca americana (pokeweed)	All parts toxic, taproot most	phytolaccin
Podophyllum peltatum (mayapple)	root, green fruit, leaves	podophyllin
Ranunculus (buttercup)	bulbous roots	protoanemonin
Ricinus communis (castor bean)	one seed	ricin
Xanthium strumarium (cocklebur)	1% BW of cotelydons and seeds	carboxytractyloside
Senecio (tansy ragwort, groundsel)	Flowers, leaves and stems (not roots)	pyrrolizidines
Symphytum (comfrey)	Risk to humans, not animals	pyrrolizidines
Kalmia (laurel, lambkill, calfkill)	3 cc nectar/kg BW or 2% BW of leaves	andromedotoxins
Rhododendron (rhododendron, azalea)	3 cc nectar/kg BW or 2% BW of leaves	andromedotoxins
Convallaria majalis (lily-of-the-valley)	Flowers, leaves, seeds; not fleshy berries	Digitalis glycosides
Apocynum (dogbane)	All parts	Digitalis glycosides
Digitalis purpurea (foxglove)	All parts of the plant	Digitalis glycosides
Aconitum napellus (monkshood)	All parts of the plant	Aconitine
Allium (garlic, onions)	Bulb	N-propyl disulphide
Hypericum perforatum (St. Johnswort)	All parts of the plant	Hypericin
Ergot from Claviceps (aka Secale)	Entire fungus	Lysergic acid

Source: Toxicology, *Gary Osweiler, Lippincott Williams & Wilkins, Philadelphia, 1996.*

Chapter 12

Homeopathic Remedy Guide for Dairy Barns

Since I dispense many homeopathic remedies, a standard chart is presented here with remedies and their indications. This is somewhat parallel to the *Comparative Materia Medica* and the *Materia Medica of Plant Medicines* but is a lot more abbreviated. The information shown is how I usually prescribe and label them when dispensed. What is presented here is how they are commonly used by dairy farmers, as currently taught by leading classical veterinary homeopaths. I was taught in this manner while I was a herdsman, and although I view potency of remedies in a different light now, I still carry a full array of homeopathic remedies in various potencies to meet the needs and requests of farmers.

Homeopathic remedies are delicate and need certain storage considerations. They should not be stored in direct sunlight, they should not be stored near any substance that has a strong odor, and they should not be stored near any strong electric/magnetic machinery. Thus, your homeopathic remedies are best kept in a cabinet away from the vacuum pump or refrigerator. Do not keep them near essential oils or strongly aromatic tinctures (like garlic).

Homeopathic remedies are taken note of during regulatory inspections. Properly labeled remedies are required by law (in Pennsylvania) when used for food producing animals. A properly labeled homeopathic remedy will have all the following: **the name of the remedy, what it is to be used for, the dosing frequency, any withdrawal times, the expiration date, the veterinarian's signature, and date of signature.** The above is true for any medicine that needs a veterinarian to dispense it in food producing animals. Although no homeopathic remedies are technically allowed for animal use (due to no clinical trials having been performed for FDA requirements), their production is still protected by the Homeopathic Pharmacopeia of the United States, passed in the first half of the 20th century. This being the case, they still can be obtained over the counter (OTC), except certain remedies of certain strengths which require a doctor's prescription. However, the moment when an OTC homeopathic remedy enters a food producing animal, the rules change, simply because it involves a food producing animal. Therefore, to be acceptable for regulatory inspection, a label is required on the homeopathic bottle in order for it to sit on the shelf in the dairy barn. In Pennsylvania, if remedy bottles are in a box or kit, it is not necessary to label each one **IF** you have a *materia medica* inside the box. The following listing is to fulfill that purpose if needed. Therefore, if needed, make a copy of it, and place it in your homeopathic box in the barn. Pennsylvania was the first state to allow homeopathic remedies to be in barns **IF** properly labeled (or accompanied by *materia medica)*. This was in response to the federal PMO regulation of 1997 which mandated that homeopathics, aloe products, colostrum-whey products (in bottles with rubber stopper and metal ring) and conventional antibiotic powders used topically on hooves need to have an appropriate veterinary label or the farmer will be debited 7 points on regulatory inspections. It is only a matter of time until other states begin to enforce the federal PMO. Hopefully, properly labeled products will be the answer in other states as well. There are no withdrawal times associated with homeopathic remedies. If homeopathic remedies are being used for mastitis treatment, withhold the bad quarter from your tank, as it will affect your bulk tank count. However, the other three quarters, if of good quality, could go into the tank because of no antibiotic residues.

The following is a list of commonly used remedies found on dairy farms, some basic indications and their dosage and frequency of administration. They should be either placed directly into the mouth or into the vulva to give good contact with mucous membranes. This is the preferred way to dose an animal for an individual problem. If dosing a herd, remedies may be put into the water system provided there is no blacklight or hydrogen peroxide being used to improve water quality.

Please remember the following basic rules when using homeopathic remedies with dairy cows. Always try to recognize 3-4 guiding symptoms prior to selecting a remedy. Use a certain remedy only as long as the symptoms indicate its use — if the symptoms change then change remedies to fit the new symptom picture. Start with low potency and proceed to high potency if you are seeing a resolution to the problem — in other words, use high potency to "finish the problem off." Use of high potency requires that you are certain that the chosen remedy is the remedy for the case, through studying the totality of symptoms, otherwise there will be no effect (like missing the bullseye in a dart game). In addition, when using high potency the remedy does not need to be given very frequently (depends on remedy and patient condition, however). Using the multi-potency homeochords of some of the plant derived remedies will help to minimize mistakes. If using a combination of remedies at one time, make sure each one fits the symptom picture and complements the other ones it is with. Lastly, remember that it is fine to use homeopathic remedies in combination with other supportive measures (like IV fluids and plant medicines if needed).

The multi-potency homeochords are shown when the homeopathic remedy indications are so similar to the Eclectic indications that it seems more rational to get small yet real material amounts of the remedy administered while simultaneously also giving the energetic aspect of the remedy for the homeopathic indications. This also allows the animal to select the concentration or energy level of the remedy to which it can respond best. Bear in mind that not all remedies are suitable for homeochord use as some remedies and nosodes act differently at low and high potency (i.e., sulph and hepar and folliculinum).

ABL (Apis/Bryonia/Lycopodium) — fresh cow with swellings, swollen udder, slow kidney function, fluid build-up.
   30C — 10 pellets 2-3 times daily for 3 days.

Aconite — any sudden onset of: fever, cough, mastitis, fright; if the animal has sweated, it is too late for aconite (perhaps Belladonna may work).
   Homeochord — 10 pellets every 20-30 minutes as needed.
   30C — 10 pellets every 20-30 minutes as needed.
   200C — 10 pellets every 20-30 minutes as needed.

Antimonium tart — moist weak cough; gradually going backwards, excess mucus but difficult to clear; lack of thirst; eyes may have mucus around perimeter.
   30C — 10 pellets 3 times daily for 5 days.

Apis — swellings, hard quarter, udder edema, right-sided cysts.
   30C — 10 pellets 3 times daily for 3-5 days.
   200C — 10 pellets 2-3 times daily for 3-5 days.

Arnica — pain, bruising, trauma, mild bleeds, pink milk, sore muscles.
   Homeochord — 10 pellets every couple hours as needed.
   30C — 10 pellets every couple hours as needed.

Arsenicum — foul diarrhea, watery mastitis, toxic, chilly, sips at water.
   30C — 10 pellets 3-4 times daily for 3-4 days.

Belladonna — fever, hot mastitis, heat stroke, dry mouth, bounding heart.
   Homeochord — 10 pellets every 30-60 minutes as needed.
   200C — 10 pellets every 30-60 minutes as needed.

Bellis perennis — deep bruising of muscle and soft tissue (after surgery).
   12X — 10 pellets every few hours as needed.

Berberis — kidney problems, pyelonephritis (pus in the kidneys), painful lower
   back; liver congestion. Symptoms swing in extremes (thirst then not at all).
   Homeochord — 10 pellets 3-4 times daily for 3-4 days.

Bryonia — right-sided mastitis, lays on bad quarter, slow moving cow, respiratory involvement.
   Homeochord — 10 pellets 3-4 times daily as needed.
   30C — 10 pellets 3-4 times daily for 3-4 days.

BUB — (Bryonia/Urtica/Belladonna) fever, moist lung sounds, belly edema,
   large udder, fresh with mastitis.
   200C — 10 pellets 3 times daily for 3-5 days.

Calc carb — slow blocky cows, peaceful, short limbed, large joints; mastitis.
   Use in dry cows to help prevent milk fever.
   30C — 10 pellets twice daily for 5 days.
   200C — 10 pellets 1-2 times daily for 3-5 days.

Calc fluor — bone lesions, lumpy jaw, and wooden tongue.
   30C — 10 pellets twice daily for 7-10 days.

Calc phos — dairy type cows, mastitis, after milk fever.
   30C — 10 pellets twice daily for 5 days.

Cantharis — cystitis (inflammation of the bladder) with frequent painful
   urination with reddish urine.
   Homeochord — 10 pellets 4 times daily for 3-4 days.

Carbo veg — bloat, cold, collapse.
   30C — 10 pellets every 30-60 minutes as needed.
   200C — 10 pellets every 20-30 minutes as needed.

Caulophyllum (mother tincture — Ø) — to strengthen contractions,
   retained placenta, cow not clean. 10 pellets 4 times daily for 3-5 days.
   30C — prepare uterus for labor, retained placenta.
   Before labor — 10 pellets once daily for 7 days.

Causticum — weak, sagging ligaments and musculature; non-pregnant uterus hangs over pelvic brim too far. Uterine inertia during labor.
6C — 10 pellets 1-2 times a week for 2-3 weeks.

Chelidonium — sluggish digestion, slow to eat, lethargy.
Homeochord — 10 pellets 3-4 times daily as needed.
12C — 10 pellets 3-4 times daily for 3-5 days.

China — fluid loss from diarrhea, bloat, symptoms that come and go periodically.
Homeochord — 10 pellets 3 times daily as needed.
6X — 10 pellets 3 times daily for 5 days.

Colchicum — rumen bloat, dysentery with trapped gas and straining; will not eat; worse from moving.
3X — 10 pellets every 15-30 minutes as needed.

Colocynthus — colic, rumen bloat; diarrhea forcibly expelled; calf scours; better from movement.
3X — 10 pellets every 15-30 minutes as needed.

Conium mac — persistent calving paralysis, pinched nerve.
Homeochord — 10 pellets 2-3 times daily for 3 days.

Drosera — dry cough, upper respiratory conditions; viral pneumonia in calves.
8C — 10 pellets 3-4 times daily for 3-4 days.

Echinacea — acute septic infection; metritis; bites; lacerations.
Homeochord — 10 drops 4-6 times daily for 2-3 days.

Ferrum phos — fever comes on slower than aconite; reddish throat; cough.
6X — 10 pellets 3-4 times daily for 3-4 days.

Folliculinum — help ovulate near heat/at breeding.
6X — 10 pellets 3 times daily for 2-3 days.

Gelsemium — weakness, flu-like symptoms; muscle tremors; depressed; coliform mastitis.
Homeochord — 10 pellets 4 times daily as needed.

Graphites — slimy, wet, sticky oozing discharge; udder rot, foot rot.
30C — 10 pellets twice daily for 7 days.

Gunpowder — deep hoof abscess with joint involvement.
30C — 10 pellets 3 times daily for 7-10 days.

Hairy wart nosode — hairy wart/strawberry heel prevention and treatment (with appropriate attention to hoof).
30C — 10 pellets twice daily for 10-14 days.

Hepar sulph 10X — painful abscesses not yet open.
10 pellets 3 times daily for 5-8 days.

Hepar sulph 200C — painful abscesses, open and draining; mastitis, foul smelling; stops suppuration; pinkeye; uterus discharge.
10 pellets 2-3 times daily for 5-8 days.

Hydrastis — foul smelling yellowish discharges from uterus; sluggish digestion.
Homeochord — 10 pellets 3 times daily for 3-5 days.

Hypericum — pinched nerve/calving paralysis; pinched teat end.
Homeochord — 10 pellets every couple hours as needed
30C — 10 pellets every couple hours as needed.
200C — 10 pellets every couple hours for 3-4 days.

Iodium — shriveled ovaries, skinny cow, eats and milks well.
6C — 10 pellets twice daily for 10 days.

Ipecac — pink milk; coccidia; hemorrhages with gushing blood.
30C — 10 pellets 3 times daily as needed for pink milk.

Lachesis — left-sided cysts, septic mastitis, septic metritis, reddish watery discharges.
30C — 10 pellets 2-3 times daily for 3-5 days.
200C — 10 pellets 2-4 times daily for 3-5 days.

Ledum — puncture wounds, insect bites, cold at area; milk allergy; potential tetanus.
30C — 10 pellets 3-4 times daily for 2-5 days.

Lillium tigrinum — nymphomania due to cystic follicle on ovary.
30C — 10 pellets twice daily for 5 days.

Lycopodium — ketosis, poor digestion.
200C — 10 pellets 3-4 times daily as needed.
1M — 10 pellets 2-3 times daily for 3-4 days.

Mag phos — muscle twitching, grass tetany.
4C — every hour as needed.

Merc corr — diarrhea, slimy and blood stained; intestinal sloughing and straining.
6C — 10 pellets 4 times daily as needed.

Nat mur — fluid build-up, cysts, udder edema.
30C — 10 pellets 4 times daily for 5 days.
200C — 10 pellets 1-2 times daily for 4 days.

Nux vomica — off feed, constipated, mild bloat, strains, colic.
Homeochord — 10 pellets every hour as needed.

Ovarian — reproductive health, help cow to cycle.
6X or 5C — 10 pellets twice daily for 5 days.

Phos — fever, inflammation, mastitis, minor bleeds, is sick but looks fine.
30C — 10 pellets 3-4 times daily for 3-5 days.
200C — 10 pellets at least twice daily for 3-5 days.

Phytolacca — mastitis, swollen glands, sore throat, cow fights stripping.
Homeochord — 10 pellets 3-4 times daily as needed.

Podophyllum— effortless green pipestream diarrhea.
6C — 10 pellets 3-4 times daily as needed.

Pulsatilla — pus discharges (no odor), mastitis, runny nose, no heat.
Homeochord — 10 pellets every 4 hours for 3-5 days.
30C — 10 pellets 2-3 times daily for 3-5 days.
200C — 10 pellets 1-2 times daily for 3-5 days.

Pyrogen — bad uterus, fevers, mastitis, sick cow.
30C — 10 pellets 4 times daily for 3-5 days.
1M — 10 pellets at least twice daily for 3-5 days.

Rhus tox — arthritis, feels better after movement.
200C — 10 pellets twice daily for 7 days.

Ruta grav — sprains and tendonitis; increase tone of uterine contractions.
6C — 10 pellets twice daily for 2-3 weeks.

Sabina — bloody uterine discharge, cow not clean; tendency to abortion; persistent bleeding after calving.
Homeochord — 10 pellets 3 times daily as needed.
30C — 10 pellets 3-4 times daily for 3-5 days.
200C — 10 pellets twice daily for 5 days.

Sepia — reproductive health; helps regulate cycle; uterine discharge after calving.
30C — 10 pellets twice daily for 5 days.
200C — 10 pellets once daily for 1-5 days.

Silica — chronic discharge, softens scar tissue, mastitis.
30C — 10 pellets twice daily for 7-10 days.
200C — 10 pellets once daily for 10-14 days.

Spongia — harsh dry cough, burning throat with difficult breathing.
30C — 10 pellets every few hours as needed.

SSC — mid-lactation mastitis.
30C — 10 pellets 3-4 times daily as needed.
200C — 10 pellets 2-3 times daily as needed.

Staph nosode — mastitis and high somatic cell count.
    30C — 10 pellets 3-4 times daily for 3-5 days.
    200C — 10 pellets twice daily for 3-5 days.

Strep nosode — mastitis and high somatic cell count.
    30C — 10 pellets 3-4 times daily for 3-5 days.
    200C — 10 pellets twice daily for 3-5 days.

Staph/Strep nosode — mastitis and high somatic cell count.
    30C — 10 pellets 3-4 times daily for 3-5 days.
    200C — 10 pellets twice daily for 3-5 days.

Sulfur — dry scaly skin, chronic nasal discharge with scabs on nose, mastitis, to finish up problems.
    30C — 10 pellets 3-4 times daily for 5-10 days.
    200C — 10 pellets twice daily for 7 days.

Symphytum— helps bone healing after fracture has been corrected.
    6C — 10 pellets 1-2 times daily for 2-3 weeks.

Thuja — warty cauliflower growths which bleed easily; ill effects of vaccination.
    30C — 10 pellets twice daily for 7-10 days.

Urtica urens — milk letdown and increased urine flow; better from warmth.
    12C — 10 pellets 3 times daily as needed or 1 hour before milking.

Ustilago — fluid uterine discharge, especially if dark and into lactation.
    Homeochord — 10 pellets 3 times daily for 5-7 days.
    30C — 10 pellets twice daily for 7-10 days.

Constitutional Remedies

These remedies are ones that resonate with the animal's character and traits. One must take into account the animal's physical appearance, outward behaviors and essential energy. A constitutional remedy will help re-vitalize the whole patient. It doesn't matter so much what the current problem is because the remedy will act on the animal in its totality and will thus be nudged towards healing. This also gives you time to pick the precise remedy for the current situation. Constitutionals are good for long-standing, deep-seated problems but also as quick, time-buying remedies. If you feel correct in choosing the constitutional remedy for a particular animal, give a high potency as these will act deeply on the animal's most profound level of "self." Essentially, almost any remedy could be a constitutional remedy; however, this would take a great deal of studying to know the "character" of each remedy well enough. This topic is more for an intermediate or advanced level homeopathic user, but I will mention a few here to give the reader an idea of what's involved with respect to cattle. It is interesting that constitutional remedies are often times originally from a mineral substance — which fits with my earlier noted thought that minerals, being so slow to form in the earth, work more deeply in the body when used homeopathically. The following list is a composite, referenced from the works of Day, Gibson, Macleod and Sheaffer (see Bibliography). Modalities, such as "worse from" or "better from" refer to the animal's response to the stimulus mentioned. And although a listed symptom may seem the opposite of the one just preceding it, take note if there is a modifier (such as periodicity, appearance, etc.) as these symptoms would have been noted during a remedy's original proving.

Calc carb:

For a beefy looking animal; more blocky and square; muscled; tends to be naturally over-conditioned; slower moving; not bothered by much; peaceful and content; can work with her and she won't care; fleshy udder, maybe doesn't milk out so well; tendency to be obese, flabby, soft; lethargic; stays put; good cud chewer; fatness without fitness; possibly lean neck with bigger belly; poor circulation; generally wide pupils; can be almost catatonic; slow, dull, uninterested, timid and shuns any form of mental effort; internal heat with external chilliness; cold is often felt in patches; normally hungry; may eat dirt, chalk, etc. — manifesting impaired calcium assimilation; thirst may be strong, especially for cold water; cornea may tend to ulceration and photophobia; liable to colds; chronic nasal discharge, thick, with crusts; stubborn constipation with clay colored manure; recurrent diarrhea, worse in afternoon; manure often has undigested feed; enlarged lymph nodes possible; long periods of heat and strong; udder may bag up slightly before heat; right ovary pain; leucorrhea is a thick discharge.

Can help induce resorption of abscesses in deep tissues (like hardware). Possibly good in worms.

Worse in cold air, draft, change from warm to cold; wet weather; contact with cold water.

Better from warm dry weather; lying on side of lesion; and when constipated.

Follow-up progression:
Belladonna — Calc carb — Pulsatilla, Rhus
Calc carb — Lyco — Sulfur
Don't follow *Bryonia* with *Calc carb*

Calc phos:

For the classical looking dairy cow, looks like a wedge of cheese with point at head and widening out towards the hook bones; tends to be a lean conditioned animal; cow is usually a heavy milker; doesn't like to be bothered — just likes to be left alone to make milk; perhaps high-strung; may be kicky, especially when trying to work with her; udder milks out well; thin, emaciated, tall in general; nervous, sensitive, timid, bad temper, wants to be left alone; contrary — when outside wants to be in, when in wants to go out; desire of change of scene; tendency to go off alone; strains or stress bring on illness; less appetite near cycling; left sided symptoms; clear thin fluid runs from nose in cold room, may be blood stained; belly aches after eating; relief from burping and passing gas; leucorrhea both day and night; resembles egg white, worse in morning; increased sex drive.

Worse from cold, wet and changes of weather; east winds; melting snow; movement or exertion.

Better from warm dry weather; when resting.

Complementary — *China, Nat mur, Ruta, Sulfur.*

Nux vomica:

Spare, lean, thin, quick, active, nervous, irritable, but hearty and full of life; movements are brisk and jerky; everything is done rapidly (walking, eating, etc.); tense, over-anxious, jittery, scheming, highly irritable, fiery temperment; malicious, feels better after a good blow-up; over-sensitive to touch, pain, noise, odors; permanently chilly, hugs the fire; hates the wind; appetite is unreliable; feels better by taking food, but only in a small amount; dysphagia, food down esophagus and then regurgitates; loosening of teeth; flatulent distention with colic is worsened by eating and drinking; tendency to jaundice with increased liver size; morning diarrhea with slimy stool with difficult passage; diarrhea may alternate with constipation; stools generally hard; too much green food; rumenal stasis; plant poisoning; ravenous hunger, especially about a day before an attack of dyspepsia; stomach very sensitive to pressure; sore abdominal walls; constipation with frequent ineffectual urging, incomplete; irregular peristaltic action, passing small amounts each time; shallow respiration.

Worse in cold, dry, windy weather; touch; worse in AM after eating.

Better from damp, wet weather; after a nap.

Best if given in evening.

Phos:

The animal likes attention; likes to be near others; likes drinking water, and lots of it; scared of thunder and other sudden loud noises; bleeds rather easily; pneumonia of rapid onset; hard, dry cough with rust expectorant; hepatitis; ketosis; superficial small hemorrhages, pink milk; mastitis with fever and thirst; gums ulcerated and bleed easily; thirsty; tender stomach with colic; stools pale and clay colored; urine is brown with reddish sediment; weak and rapid pulse; ascending paralysis and trembling of muscles; acute gastric ulcer;

enlarged lymph nodes near neck; mastitis with abscess formation; doesn't know she is sick; epistaxis with small hemorrhages; bleeds easily; swelling and necrosis of lower jaw; easily bleeding gums; thirst for very cold water; tongue dry, smooth and red; weakness after stool; suppuration of udder with or without fistulous tracks; increased respiratory rate — but shallow; trembles; heat between shoulder blades; small wounds bleed very much.

Worse with cold; before thunderstorm; lying on left side — can only lie on right side; chilly every evening; worse from touch, warm food or drink; changes in weather.

Better in dark, lying on right side, cold food, cold open air, washing with cold water. Sleep.

Complements: *Arsenicum, Lycopodium, Silica, Sepia*
Incompatible: *Causticum*

Pulsatilla:

For the curious, but cautious cow; wants to be friendly but is timid; she likes to be with the group (herd); will approach you but you can't approach her; a bit on edge when working with her; tends to be a clean animal; mild, gentle, yielding. Changeable; fond of company, averse to solitude; more timid than anxious; likes sympathy; slow, phlegmatic; may become irritable and touchy, but not violent; desires and absorbs affection, but doesn't return it; seeks open air; discharges are thick, bland, and yellowish-green; symptoms ever changing; thirstless, peevish, chilly; thick bland discharges from the ears; thick, profuse, yellow, bland discharges from the eyes; lids inflamed and sticky; styes; yellow mucus from the nose, abundant in AM, smells bad as if old; dry mouth without thirst; averse to warm food or drink; flatulence; thirstlessness; no two stools alike; above normal manure production; increased urine output, worse when lying down; loose cough in AM with copious expectoration; pressure upon chest; chilly extremities; fever with chilliness; external heat is intolerable; thick green, creamy non-irritating uterine discharge; slow cycles; abnormal presentations during labor; uterine inertia and feeble labor pains; thick muco-purulent discharges from all mucus membranes; ovarian hypofunction and retained placenta.

Silica is the "chronic" of *Pulsatilla*.

Sepia:

Sad, silent, solitary, lacking zeal; wants to be left alone but not forgotten; sallow, sweating, sagging, weak, trembly, apathetic; feels cold even in a warm room; tall and angular or puffy with tendency to portliness; sweats not accompanied by thirst (like Apis); tendency to abortion; venous congestion; flatulent abdomen; liver sore and painful; relieved by lying on right side; gaunt, but not relieved by taking food; manure dry and hard; almost constant dribble of manure; red, adhesive sand in urine; leucorrhea yellow or yellow-greenish — irritating; cycle late and not much shown; increased maternal instinct; discharges post-partum of all kinds; regulates entire cycle; use routinely; prolapsed uterus (post-replacement) and vagina; estrogen insufficiency; uterine straining; great tenderness of cervix; dryness of vagina; persistent sterility; coldness between shoulders; restlessness in all limbs; joints seem weak and unreliable; shivering with thirst; ringworm; takes cold easily.

Worse from sitting still, before cycle, forenoon and dusk; dull cloudy weather; before thunder.

Better from violent movement in open air; enjoys facing wind; afternoon.

Sepia OK with *Nat mur.*

Sepia not good with *Lachesis.*

Sulfur:

For the cow which tends to be dirty, caked with manure often; lies in the muddy stream or mud itself; poor, rough hair coat in general; not a cow you would like others to see; doesn't mind being sloppy; associated with burning; exhibits an earthward tendency; down to earth; physical awareness; hair tends to be coarse, lusterless and not well groomed; skin is dry, rough, scaly with tendency to eruptions, sores and pustules; obvious redness at all orifices of body; all discharges tend to excoriate the skin; averse to water and washing and to standing for any length of time; likes open air; averse to hot stuffy atmosphere; thirst is marked, will drink a lot of water; mucus membrane discharge will cause redness, rawness and soreness of surrounding skin surfaces; periodicity — symptoms recur with regularity; sterility; irregularly irregular cycles; puerperal fever; teats cracked; marked weakness of paraspinal muscles, resulting in chronic swayback; itching of skin; scratching feels good, but then burns; aggravated by heat; skin problems alternate with other problems; wounds slow to heal; sepsis common and tends to chronicity; ulcers become indolent; sensitive to water; standing long periods causes distress.

Worse on waking; aggravation at noon and night.

Better from dry warm weather and lying on right side.

Can counter serious after effects of vaccination. *Aconite* for acute, *Sulfur* to prevent recurrence. Good in acute disease that is slow in clearing up or in chronic illness. If vague symptoms — *Sulfur* may bring to surface more definite symptoms from which to then treat more specifically.

Chapter 13
General Considerations For Treatments

It is necessary to take a "multi-prong" approach to dealing with dairy cows in a wise, holistic manner. The multi-prong approach takes into account the practical realities of the farm's physical environment, the cow's current condition and the willingness to tinker with various medicinal remedies. Moreover, the three fundamental components of whether or not an animal will become sick also must be considered. Prior to any treatments, the *stresses* a cow is facing, the *environment* in which she lives and her *nutritional* plane must be taken into account. These three factors will also help to determine what the kinds of treatments that are needed (natural or conventional) as well as her prognosis (predicted outcome). For instance, a cow just fresh with twins and a retained placenta, very skinny, in a dirty environment (calved in a damp box stall) comes down with coliform mastitis and concurrent high fever. The internal stresses upon her are enormous (having recently calved and suffering from an infected uterus), her immediate environment is probably excessively moist, she is in obvious negative energy balance and she is feverish due to the coliform infection which is probably working its way into her entire system, and she is rapidly becoming dehydrated, acidotic and possibly toxic. Obviously, the cow is quite sick. We need to make her feel better quickly and effectively. We need to stabilize her with appropriate fluids, pump her stomach for nutritional advantage, bed her in a clean/dry place (and keep it that way) as well as needing to work on the uterine infection and the inflamed udder with its mastitis. (See specific treatments for uterus and mastitis). But what if the same cow calved out on pasture in a clean area, was in good body condition coming into lactation, and had been vaccinated against coliform mastitis (vaccination will reduce the severity of symptoms)? Now we can concentrate more on just the uterine infection, her negative energy balance and have the farmer strip her out — instead of first needing to stabilize a very sick cow. The sick cow described is not fictitious — it is unfortunately a very real life situation that can happen any day of the week on any dairy farm. The conditions that set the cow up for such a disaster cannot be changed once the cow is sick — *we have to deal with the case as it presents itself.* Her prognosis likely depends on her condition *prior* to becoming sick. Obviously the vaccinated cow with the retained placenta has a better prognosis than the other one.

For those seeking a "quick fix" or want simple-minded input-substitution, the learning curve regarding natural treatments will be steep. Input-substitution is a term met with derision among truly holistic farmers for treatments that are essentially "band-aids" intended to simply replace conventional products. Input-substitution would be, for example, a natural product which claims (or is implied) against worms to substitute for a conventional wormer or a natural product for ketosis which substitutes for conventional propylene glycol without taking into account other changes that are needed. In the case of worms, you will find many natural treatments in the treatment section under that listing. However, to truly treat parasitic worms also requires increased attention to grazing management and calf rearing hygiene — not just one treatment of ivermectin or a natural substitute. We need to know the biological life cycle of a parasite and break it in various places in the environment as well as in the animal. Fly problems would be another example. Simply substituting a natural fly spray composed of essential oils instead of a standard pyrethrin/piperonyl butoxide spray is but one part of fly control/reduction. The core issue in reducing fly populations is to reduce moisture in manure and in the air and on the animals themselves (see section on Flies).

Natural treatments should be viewed as part of an overall treatment plan — *not simply stand-alone treatments.* They *can* be stand-alone treatments, but never count on it. Although skeptics of natural treatments may claim that it is then not possible to evaluate whether or not a specific natural treatment *by itself* is responsible for a cure, we in the holistic realm should not be daunted by such remarks, for our way of looking at things is *to see the overall picture.*

We must strive to see all the pieces of the puzzle, not only concentrating on single treatment "silver bullet" effects. For if we do revel in the rarely seen "silver bullet" effect of a single natural treatment which happens to have a strong beneficial input-substitution effect, we may become too easily blind to potential side-effects of such a treatment — side effects either within the single animal or quietly invading the whole farm system. If a natural treatment does cause an obvious beneficial effect, we should note it but remain alert to potential side effects. The well known phrase, "for every action there is a reaction" is a basic truth when it comes to biological systems. We in the alternative realm should heartily embrace multi-prong approaches, for by spreading out the burden to cause a specific beneficial effect, we are lessening the chances that a harmful organism can figure out how to get around a single input. Hence, the idea of input-substitution is misguided and any beneficial effect of a specific input-substitution treatment will most likely be overcome in a matter of time by organisms' adaptive abilities (as seen in the continual need to revise and devise new insecticides which quickly lose their efficacy upon rapidly generated populations of crop insects). Hopefully in previous chapters the reader has been given ample reason to accept the idea of a multi-prong approach to tackling problems over the input-substitution approach. However, there are times when a simple input-substitution may be temporarily needed, but rectifying underlying causes to problems is obviously paramount to lend any credibility to the well known term "sustainable agriculture." By the way, sustainable agriculture is not simply a farm enterprise which remains in business for the long term due to continuous profitability. While this is preferable to not being in business at all, sustainable agriculture, in my opinion, is agriculture which enhances the local agro-ecology, personal growth, optimizes animal welfare and is increasingly self-sufficient — as well as profitable.

In the following treatment section, many options presented have themselves a combination of ingredients or a sequence of events rather than a single active ingredient solution. There are two reasons for this, one from a historical viewpoint, the other from a holistic viewpoint. Historically, in the Regular school, combinations of ingredients were used simply because there were no known "magic bullets" at the time and mixtures were probably more effective. Holistically, medicinals are combined to complement various aspects of the ingredients and to therefore arrive at a biologically (and/or energetically) balanced treatment. The Regular school mixtures had more of a "sledgehammer effect" while the Eclectics (and present day Western herbalists) strived to help the patient by specific medication or by using a biologically synergistic mixture. This is in direct contrast to classical homeopaths who aim to use one single highly diluted and succussed remedy at a time and conventional doctors who use pharmaceutical products with single active ingredients, respectively.

A fundamental aspect of treatment — and one that cannot be emphasized enough — is to first **correct any occurrence of fever, dehydration and impaired circulation.** This is definitely part of a multi-prong approach — perhaps the most fundamental step. One of the most important goals in treating a dairy cow is to get her feeling good enough to begin eating once again (with the goal of resuming production quickly). That is the ultimate goal — for if a cow is off-feed, she will only begin to become well again if she begins eating. And the most common reason a cow is "not eating quite right" or is entirely off-feed is due to fever, often with secondary dehydration and impaired circulation. Too many books on complementary and alternative veterinary medicine (CAVM) overlook this fact entirely. It doesn't matter what therapy is being written about — acupuncture, homeopathy, or medicinal herbs — there is near unanimous omission of fluid (electrolyte) therapy for rehydration. Fever is definitely addressed in many therapies, but fluids for rehydration are not. How can any therapy ever be fully utilized if the cow is lacking in essential fluids in the circulatory system? As a practitioner I cannot see how folks go about using whatever remedy, even conventional, without fluid therapy if it is needed. Farmers are quite willing to let the veterinarian give IV fluids or pump a cow's stomach with 5-10 gallons of a nutritional slurry to "jump start" the rumen bugs. Practitioners, especially ones promoting alternative therapies, would be wise to point out that animals that need IV fluids *before* commencing on a course of natural treatments. It is just common sense that an animal's circulatory system — *the most fundamental biological system that sustains daily life* — needs to be in correct functioning order. It is basic cardiology that the heart only pumps what is brought to it. If there is too little blood in the system, there needs to be given a blood transfusion (which is easy in cows) or, as is more commonly done, give 4-5 liters of IV fluids (lactated ringers solution, hypertonic saline, calcium, dextrose, physiologic saline, etc.). When I'm called out to an animal that is droopy and has been given various natural remedies with no success, it is often due to dehydration and poor circulation which has yet to be addressed. Once fluids are given, they almost always perk up (and stay perked up). And as the saying goes, one gallon of IV fluids is worth ten gallons of fluids pumped into the stomach. This is definitely true. However, to do both is even better!

Fevers are both good and bad. Fevers indicate that the cow is actively fighting an infection somewhere. But, I cannot think of much good in a pregnant cow having a fever of 104° F for a few days and letting the fever run its course just to be "natural" about everything. There is in all likelihood a good chance that she will abort, secondary to the fever (if the cause of a pending abortion is not the source of fever itself). And, except for fresh cow problems (which there are many), the entire herd is usually pregnant in various stages. In addition, as mentioned earlier, the whole idea of a dairy farm is to produce milk

— when feverish, a cow usually doesn't eat and drops in milk production. I believe these two reasons are good enough to treat fevers, let alone the idea to make a cow simply feel better. I generally don't treat fevers in cows that are not pregnant who appear well and are eating yet have a fever (*i.e.,* a fever was incidentally noticed upon rectal palpation during a routine herd check). There are definitely many natural treatments that deal with fevers. It cannot be emphasized enough that if we can get a cow to feel better and resume eating, then there is a much greater chance that she will become well again without reverting to prohibited materials for certified organic use.

To be sure, these treatment guidelines are not meant to substitute for appropriate conventional medicines *if needed.* Conventional medicines are extremely valuable if used conservatively with care and wisdom. Conventional medicines include such treatments as antibiotics, anti-inflammatories, antipyretics, anthelmintics, etc. However, I think an animal will be stronger in the end if it can quickly get over an illness on its own with holistic approaches that support the entire animal. If cows are constantly being treated with conventional drugs at the slightest sign of illness, this is akin to always being on crutches and not sincerely trying to walk on one's own when perhaps it is possible and more strengthening to do so. Often times we just need a nudge in the right direction and we're back on track again. The same is true with using natural treatments in dairy cows. In my opinion, given two products, I will always choose the one which allows the animal to heal itself. However, it should be stated that at times an antibiotic may be needed, and if it is, it should be used before it is too late when absolutely nothing will save the life of the animal. Antibiotics are a wonderful discovery and many readers of this work probably know someone whose life may have been saved due to the judicious use of an antibiotic. What I object to is the routine dispensing and administration of antibiotics at the slightest sign of illness or even before illness — and especially the revolting practice of continually feeding youngstock medicated feeds (that's an artificial crutch to prop up high density animal farming if there ever was one!). Otherwise, an antibiotic *used occasionally and at the right time* can be a wonderful treatment. I certainly carry a small supply of antibiotics with me, although much less than other livestock veterinarians.

When using natural therapies, the animal caregiver must have, or develop, a heightened awareness for the smallest changes in the animals' health picture. ***You must jump on the problem before permanent changes can take hold within the animal.*** You really need to be committed to this concept. Also, your veterinarian will be able to better help you if you have him/her involved early on, rather than when all else has failed. It should also be mentioned that the intent or will of the care provider to truly want to help the animal will probably also help the patient move in the direction of becoming well again. Although this concept is in stark opposition to the objective and detached scienctific method, I believe that a good heartfelt intention of the care provider is very important. In my opinion there is absolutely nothing wrong with a caring and compassionate person trying with all their heart and energy to enable an animal to become well again. I have much more sympathy for a farmer who stays up late tending to a sick cow than for a farmer who has patience only to give one shot of medicine and demands it to work quickly. In whatever way one can "feel" or "sense" which natural remedy may be the correct one if it seems to be the right one, try it. (No one is born a scientist, but we are all born with the ability to feel on many different levels.) Go with your gut feeling if it is strong. This will become almost habit for those folks that really develop their skills and senses for natural remedies. For those just beginning, it is my sincerest desire to help you get a "foot in the door" with natural treatment remedies until such a point where you, too, feel comfortable enough to be able to pick and choose remedies of your liking for each individual case. Ultimately, you must pick and choose the cases with which you want to use natural treatments. Just remember to do things in a timely manner. Farmers are notorious for being optimistic, almost to a fault. Do not "wait and see" what happens in a day or two if an animal is "not quite right." Investigate the situation and begin remedial action promptly.

Administering Remedies

You must be certain that the cow or calf receives its medicine. Therefore you need to give the actual dose in the mouth or, alternatively, possibly into the vulva. While the vulvar mucous membranes can absorb material, it can do so only with small volumes (which may not adequately deliver what the cow needs for proper dosage). Alcohol based tinctures should *not* be given in the vulva due to potential irritation and burning of the sensitive mucosal lining. Both are mucus membranes, and as mentioned earlier, the mucus membranes are an entry way to the body and immune system (see pg. 103-104). By giving the remedies directly to the individual animal, this also keeps you in touch with the animal throughout its treatments, allowing you to detect changes. Unlike conventional therapies, which are based on relatively large doses, the Eclectic tinctures and homeopathic remedies rely somewhat more on the frequency of dosing to stimulate the animal's vitality or to directly oppose the disease process. Obviously, by stimulating the animal's vitality and opposing the disease process at the same time would seem the best of both worlds — thus another reason to consider using low potency homeopathic remedies, Eclectic tinctures or homeochords. Conditions with rapid onset (see the problem at next milking or earlier) generally need very frequent and/or strong treatments while longer standing cases call for dosing less frequently but for a longer period of time. In homeopathy *the remedy should change when the symptoms change* in order to fit the new symptom picture — or stop giving the remedy altogether if no longer needed. In Eclectic medicine, the idea is the same, except that some of the energetic alkaloids definitely need to be stopped when the desired effect is obtained (*i.e.*, aconite to drop fever by reducing the heart rate — usually after 4-5 doses at most). That is why it is so important to *take note of the slightest changes* in an animal's condition. Some people put the remedies upon the grain feed on the floor while others put them in the animal's water bowl. Neither method is good. Instead, *put them directly in the mouth or into the vulva.* Putting remedies in water bowls does have its place, especially when dosing a group of animals. To administer homeopathic doses to a group of animals, you can add 50 pellets of each remedy to a gallon of spring water, shake and dissolve for one hour, then shake again, then add the gallon of medicated water to 50 gallons of water. Repeat this daily for 5-7 days. Make sure there is no scent of chlorine in the water, nor much hydrogen peroxide activity, nor outright algae scum — in other words, a recently cleaned water tub with fresh water.

Oral Fluids

It is at times necessary to administer a medication or supplement via the mouth. If these supplements are in liquid form, careful attention is needed to give the fluid correctly. If given too hastily or carelessly, it is quite possible to get the fluid into the lungs and cause aspiration pneumonia. I see the negative results of improper drenching far too often. Please read the following section carefully.

Adult cows:

To give a cow a liquid, a dosing syringe or hand-held pump with nozzle should be used. First restrain the cow so full control of head is maintained. Place one arm over the muzzle of the cow and with that hand pull the mouth open using your fingertips by raising the nose *slightly* with your arm. Keep the cow's head roughly parallel to the floor — *avoid at all costs having the nose point high up into the air* as this will allow fluids to get more easily into the windpipe and cause aspiration pneumonia. Once the cow's mouth is slightly open, insert the nozzle of the delivery syringe near the back of the lips and deliver a small amount of fluid. Then stop, allowing the cow to either swallow or cough it out. You cannot force a cow to swallow, it's as simple as that. But by delivering small quantities at a time, chances are best for correct swallowing to take place. If the cow is carrying on, coughing and thrashing, STOP. You are probably getting fluid into the windpipe (trachea). Have a vet show you proper technique if you are unsure. Treating apiration pneumonia is generally not rewarding, needing high doses of antibiotics, anti-inflammatories, and antipyretics only to send the cow for salvage a few weeks later when the antibiotic withdrawal time has been observed. Drenching correctly is an art that many people do correctly, if having patience and a firm but gentle touch. If possible try to use gel tubes with a "caulking gun" delivery system. You will have much better control over the amount delivered due to the clicking mechanism. Give a few clicks, stop and allow the cow to swallow, then a few more clicks and so on. When finished drenching a cow correctly it is likely you will be "wearing" some of the drench solution because cows will reject, if allowed, some of the drench material. This is OK.

Calves:

Calves often need fluid replacement therapy due to losses associated with diarrhea. It is simple to fill the bottle with electrolytes and allow the calf to suck on the nipple. Easy enough. If calves are dehydrated due to scours, it is advisable to give them *less fluid per feeding but feed more frequently,* avoiding "slug" feeding. Feeding four times daily in these instances is usually sufficient. Do not take the calf off of milk completely, it has nutrients that straight electrolytes do not have. Instead feed half a bottle of milk in AM then half a bottle of electrolytes at noon, then half bottle milk at PM milking, then half bottle of electrolytes at end of evening.

If, however, the calf cannot suck from the nipple, either call your vet to evaluate the calf, or you can try force feeding it yourself. There are calf feeder bags with an attached tube available at any feed store. These can be truly beneficial if used correctly or can quickly kill a calf if used incorrectly. The trick is to pass the tube gently down the left side of the calf's mouth into the throat (avoiding the windpipe). At the end of the tube there is a large bead-like structure that allows it to only enter the throat. If straddling the calf and looking down upon the calf's head, the left side is also your left side. The idea is to remember that the word swallow has two "ll's" (swallow), so pass the tube left of center as you straddle the calf and look down upon the calf's head. Gently insert the beaded end with attached tube into the mouth underneath the left nostril. After you pass it about 6 inches, you will feel a slight resistance, keep advancing it forward about 12-15 inches; then, and only then, open the valve to allow fluid to run in. Keep the bag only somewhat higher than the calf's head. Once the bag is emptied, clamp the line or bend the line shut and then extract the tube. If the fluid accidentally got into the lungs, the calf will die in about 1-2 minutes. If not, you've done it correctly. The first time is without doubt quite stressful for the animal care giver. But afterwards you will have some confidence and with experience even more. Calves that don't suck their first colostrum can in a likewise manner be tube fed to ensure adequate colostrum intake within the first two hours of life.

Many treatments can be bought commercially in pill form. Also, many home remedies can be placed into 1oz. gelatin capsules (porcine derived) to be given orally. Be careful when administering any kind of large pill boluses (especially to calves). Balling gun injuries can cause a terrible infection at the back of the mouth/top of the throat (necrotic laryngitis) that will give breathing trouble later and/or abscesses which are difficult to treat, even conventionally. Therefore, when giving a bolus, lay the end of the balling gun just behind the hump of the tongue, then push. Close mouth and massage throat to stimulate swallowing. Ideally, any pill should be lubricated to help the cow swallow. *When the cow licks its nose, it has swallowed the pill.*

Intravenous (IV) Fluids

Always warm-up the bottles to body temperature in a bucket of warm water. The only time you might not want to warm up a fluid bottle is if an animal has a high fever (*e.g.,* heat stroke) and you want to cool the animal rapidly. IV fluids are a quick and easy way to deliver fluids and medicines directly into the bloodstream if needed. To be certain, I rely heavily on IV fluid therapy when I am called out to see sick cows. As already mentioned, a dehydrated cow will not respond as well to whatever other therapeutic measures are taken as compared to a cow which has a good, normal circulatory system. It is critical to give fluids if needed. To assess extent of dehydration, remember to pinch the eyebrow and see if it stays "tented up." If it does, the cow (calf) needs fluid therapy. There is a much more immediate and beneficial effect with IV fluids than only pumping fluids into a rumen or drenching cows as just discussed.

I am administering an IV bottle.

A) Jugular vein (for down cow)

This is definitely my preferred site for fresh cows down with milk fever. Have the cow's head in a halter and tie the lead of the halter to just above the cow's hock in a position that you normally see a cow sleeping. Look for the groove along the neck. The vein is always in the valley there, somewhere. Use your thumb to put pressure on this groove close to where the neck meets the chest. Or use a baling twine as a tourniquet to see the vein. You should see the vein rise. Do this a few times to see the position of the vein. Tap your finger on the "raised" vein to confirm your placement. With the needle pointing towards

My wife, Becky, giving an IV bottle.

the udder, make a quick thrust into the vein. Remember — going through the skin takes a good, hard thrust as you are basically going through leather. If in it, you should see a continuous flow of blood. Hold the needle hub as you attach the IV line. If the cow moves around or you are not sure that you're still in the vein, lower the bottle to below the entry point of the needle. You should see blood coming immediately back into the line if you're still in the vein. This is easiest if you use a clear plastic line (versus a tan one). Use of a 14 gauge, 2 inch disposable needle is best for ease of penetrating the skin and placement of needle in the vein. The color and flow of the blood can indicate the condition of the cow. Normal blood will easily flow out of the IV needle and look like a velvet red. If the blood is very dark and flows weakly out of the needle, there is serious circulatory deficiency and poor oxygenation. Giving a cow 3-4 liters of fluids will usually change a very dark blood to a brighter blood as the circulation and oxygenation improves. Usually adding 200cc 8.4% sodium bicarb into the IV mix will also help.

B) Jugular vein (for cow that is standing)

Put a halter on the cow's head. Tie the cow's nose tight to the stall divider in front of her, having the cow's head below the tie-stall bar and nose pointing down to the side divider. Using a nose-lead (nose-tong) along with a halter will also help to keep the head from moving about. The jugular vein should rise due to gravity from the head pointing downward to the side. With a quick thrust, jab the needle into the vein, pointing it towards the udder. If you have an automatic head locking system, the job is a little easier. Get the cow's head into the head-lock, put a halter on and tie the head off to the side. Scan the neck for the vein and plunge the needle in towards the udder. If continual blood is draining out, hook-up your IV line. Check to make sure you are in the vein by periodically lowering the bottle below where the needle is inserted and seeing if blood immediately backs up into the line. If not, reposition the needle, otherwise an abscess will develop. If giving calcium to a "wobbly" cow (with just a touch of milk fever), holding the bottle no higher than the backbone is generally a safe rule and running in the calcium between 5-10 minutes is safe as well. When taking the needle out, vigorously rub the IV site to help stop blood dripping.

C) Milk vein

This is the site that I routinely use for standing cows. If careful, you won't create a problem. If not careful, you will really regret it, for abscesses at this site are *very bad* for the cow. Have someone jack the cow's tail straight up, or use a kicker over the hook bones (I've never used a can't kick bar but it may be OK). Stand on the left side of the cow — the side upon which you would lean the right side of your body as you face the head of the cow. Make sure the site on the milk vein is clean — extremely important!! Clean the site with alcohol. No going through manure! Then slap the vein a bunch of times so the actual insertion thrust of the needle won't take the cow by surprise. Aim the needle towards the udder. Punch the needle in *at a 45 degree angle*. A good stream of blood should immediately be seen. If not, ever so slightly rearrange the inserted needle until you see the stream. It is then OK to hook up your IV line by holding the needle hub while you connect the line. *Check to make sure that you're in the vein by lowering the bottle below the needle site to see blood coming back in the line.* Also, hold the line against the rib area with your right hand to give it support while you hold the bottle with your left hand. When done, quickly remove the needle, and *immediately push up against the vein with the palm of your hand for at least 30 seconds*. Then check for any dripping for another half minute. If so, again apply pressure until no drips. If the cow is uncomfortable and carrying on during the IV, you are likely not in the vein. *Immediately* check if you are in the vein or not by lowering the bottle lower then the needle. There should be blood coming back into the IV line immediately. If not, take the needle out and push up against the vein while rubbing it vigorously to disperse the fluid which was incorrectly deposited under the skin next to the vein (which causes irritation to the cow).

Physical Exam & History

When I pick up the phone and a farmer says to me, "Doc, I got a sick cow," my first question almost always is: "How long is she fresh?" Why is this my first question? Because most problems in dairy cows occur in the peri-parturient time — that time span roughly 2 weeks before calving to 3 weeks after calving. This is a major stress period in the life of a cow due to her internal hormonal changes to get ready for calving and her dietary changes as the farmer starts her getting used to the lactating cow ration. It is simply an incredibly challenging time period for the cow, the calf and the farmer. Metabolic problems like milk fever (hypocalcemia) and ketosis (acetonemia and hypoglycemia) are either prevented or enabled in this time period, mainly due to the feeding strategy and the cow's body condition. Obstetric problems easily give rise to

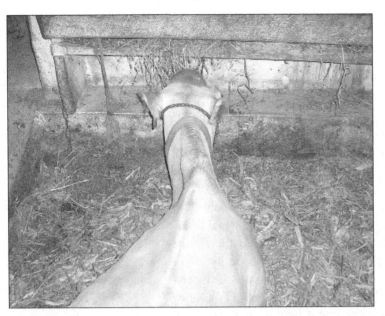

S curve in neck denoting low potasssium.

retained placenta and intra-uterine infections, which in turn can too easily lead to a displaced abomasum ("twisted stomach") and/or generalized septicemia. Many sets of problems can be predicted almost based entirely on a cow's stage of lactation or an animal's age.

Obtaining a complete history is crucial to accurate diagnosis, even when a physical exam has just been done. One of my bovine professors in veterinary school, Dr. Robert Whitlock, really drummed into our heads the term "history, history, history!" Since the animals cannot talk, we veterinarians need to have the gaps filled in by the animals' caretaker, the farmer. Getting a complete history can make for a really interesting tale, sometimes including nearly the entire life story of the cow — especially if she's had a few extra problems in her life time as compared to her herd mates. Other times, even with thorough questioning, I don't get much in the way of filling in the blanks.

In any event, check the cow *at minimum* for temperature, respiratory rate, heart rate, rumen contractions, pings (for a twisted stomach), and CMT plating of all 4 quarters. I check for hydration/dehydration by pinching the eyebrow and seeing if it tents up, and if so, for how long. Looking in the mouth at the color of the oral mucous membranes can reveal a lot about the circulatory system (red, pink, splotchy pink, pale). Also, simply observing whether her eyes are bright and glistening or if they are dull and depressed is important. Feeling their nose to see if it is moist (as it should be) and feeling their ears to check for coolness or warmth is also critical in assessing circulatory integrity. It is paramount to correct circulatory problems if they exist.

By applying gentle pressure at acupuncture trigger points (see Acupuncture Charts), it is possible to gain even further information about an animal's condition or for a veterinarian to help confirm a diagnosis. This I have found true when checking the bladder meridian for respiratory problems near the withers, the kidney area, the rumen points and some of the points for cystic ovaries and mastitis. Although I use therapeutic acupuncture mainly for calving paralysis, I have had success with it in the treatment of cysts and non-functioning rumen.

As mentioned earlier, many alternative therapies come with their own diagnostic work-ups and terminology. As far as I am concerned, I stay very close to standard physical examination but take into account my own observations of the cow's behavior or general character as used in homeopathic work-ups as well as occasionally using the acupuncture trigger points. Some conditions are so obvious (like milk fever) that a complete, thorough physical exam and history will be wasting time when it is obvious the cow is nearly comatose due to extremely low blood levels of calcium. Other times, with vague standard physical symptoms, a thorough questioning of the farmer regarding history and the cow's character as well as picking up more subtle signs or signals is required on my part.

One last point, always look at the cows' hooves throughout the herd. Those hooves will tell you a lot about the rumen health of the cows. It's always enlightening to a farmer if I observe an obvious line on all the hooves of many of the cows somewhere below the hairline and parallel to it. Seeing an obvious line is evidence of a herd-wide major stress in the ration. It takes about 8-12 months for a hoof to grow out (from hairline to floor), so it is possible to trace back the approximate time of the feed stress. The line indicates an irregular hoof growth at that time which is usually due to a ration which has induced acidosis in the rumens of the cows.

For those interested in natural treatments, steer clear of lightning quick veterinarians if you can help it. Try to work with a vet that likes to explain the situation and gives options for treatment. Definitely try to work with a vet that will teach you how to become a better herdsman. I guess part of the reason for taking the time to write this book is that I do like to teach and explain procedures as I do when working on a cow or a herd problem. I hope the following chapters allow farmers to better understand some of the common (and not so common) problems that are encountered as well as show treatments from my clinical experience and those that were the mainstay of treatments from earlier generations of veterinarians.

The Medicine Cabinet

No matter how well a farmer bases cow health on good soil and pasture nutrition, there will always be instances where individual cow treatment is needed. Even on the very best of farms, where it has taken years to get the soil and pastures to be in complete synergy for the cows' health, there will be some cows that simply need intense individual attention (usually right around calving time). Even with the most highly successful graziers, there can be temporary "break downs" when there's a string of problems for a couple months. Therefore, every farmer should have some basic supplies on hand. This enables the farmer to jump on problems as early as possible and in so doing become more self-reliant. The following is a list of necessary items for any animal caregiver to have "at-the-ready:"

1) Thermometer

2) Herbal tinctures (see Agri-Dynamics, Inc. in Resource Contacts)

3) Homeopathic remedies (see Washington Homeopathic Products in Resource Contacts)

4) Vitamin A, D, E; vitamin B complex; vitamin B12; vitamin C; vitamin E & selenium

5) Colostrum-whey products (both injectible and oral pills)

6) Laxative boluses — magnesium oxide/magnesium hydroxide ("pink pills")

7) Pill gun

8) Syringes (3cc, 12cc, 20cc, 35cc, 60cc)

9) Needles (14, 18 and 20 gauge, 1-2 inch length)

10) Magnets

11) Alcohol and alcohol pads

12) Calving chains/straps

13) Butane powered de-horner

14) Calendula-echinacea ointment (for teat abrasions)

15) Teat dilators

16) Come-alongs and beam hooks

17) Calf electrolyte replacement and calf tube feeder

18) Adult cow electrolytes (calcium, dextrose, hypertonic saline, etc.)

19) Probiotic pills

20) Ketosis tubes

21) Calcium gel tubes (calcium propionate — not calcium chloride)

22) Infusion pipettes

23) Calendula tincture for uterine infusion

24) Mineral oil or vegetable oil

25) Epinephrine

26) Apple cider vinegar

27) Molasses

28) IV line (clear is preferred over tan colored)

29) Electric prod (better than a pitch fork!)

30) Ketone and pH strips

Chapter 14
Treatments
Part 1
General Problems

Treatments that immediately follow the disease description and that are sequentially numbered are the ones I commonly use or recommend. The generic conventional treatment will immediately follow my list of common treatments; this is to let the reader know what the vet may recommend. Treatments which cite the author, year and page referenced are ones that I may or may not have used, but are presented to give the reader as many possible options for the disease entity. This approach should give the reader a fair amount of options from which to choose.

Note — there will be compounds in the treatment that were not covered in Chapters 10, 11 and 12. It is simply not possible to cover all compounds within a book which is meant to have practical value to a farmer. For those interested in learning more about certain compounds named, please refer to the bibliography.

In the treatments shown, the tinctures as used by the Eclectics will be designated by the symbol "Ø." This Ø also denotes a homeopathic "mother tincture," which is the only strength possible for veterinarians to still obtain for the energetic alkaloids (aconite, belladonna, bryonia, gelsemium, hyoscyamus, nux vomica, phytolacca, veratrum viride, etc.). It is simply not possible to obtain the official USP fluidextracts anymore. The reader may need to consult the tables for conversions of old apothecary language (such as "drams") in order to convert to modern metric values if needed ("cc" or "ml"). In addition, it is up to the reader to locate and obtain materials, some of which may be easy to procure, while others may not be (see Agri-Dynamics, Inc. and others in the Resource Contacts). Words in *italic* print are homeopathic remedies, and the dosing frequency is shown as: *the number of doses a day times the number of days given.* For example: *"Sepia 2 x 3"* would mean "Homeopathic Sepia twice daily for three days." Normally, 10 homeopathic pellets are given at each dose to an adult cow. Calves should get 5 pellets per dose.

Special Note: Agri-Farmacy, LLC, a company owned and operated by Jerry Brunetti and Dr. Karreman, was established due to request by farmers to provide easy access to natural products mentioned in the 1st edition of this book. See resource contacts in the appendix for contact information.
See Recommended Remedies for a listing of commercially available products.

General Doses (Fish, 1930)

> Cow will take $1\frac{1}{2}$-2X that of a horse
>
> Sheep will take $\frac{1}{3}$ that of a horse
>
> Hogs will take $\frac{1}{8}$-$\frac{1}{10}$ that of a horse

Frequency of dosing tinctures (crude drugs)

> To decrease fever (Aconite, Spirits of Niter, Potassium nitrate, Potassium Chlorate): every 2 hours (severe); 3-4 times/day (mild)
>
> Purgatives: every 20-30 hrs (severe); every 40-48 hrs (mild)
>
> Tonics: 1-3 times per day
>
> Stimulants: every 2-6 hrs
>
> Uterine contractions: every 30 minutes
>
> Anodynes (pain relievers): every 30 minutes

To convert fluidextract dose to tincture dose, for alkaloids, multiply by 10

Hypodermics of alkaloids are given usually at one-half the dose by mouth. Intravenous doses one-half to two-thirds of the hypodermic dose.

FEVER (>102.5° F or 39.5° C) & Heat Stroke

Taking the temperature of a cow that "doesn't seem right" is a basic requirement of being a good cow person. So often I will be called to see a cow that is "off-feed" with the farmer thinking it is a twisted stomach or constipation when the cause for not eating is a high fever. A fever is not necessarily bad, it is simply a sign that the animal is trying to fight off an infection of one sort or another. In one sense fevers are good, as it shows that the cow is able to rally in response to the presence of an abnormal occurrence intruding upon her system. It adjusts her internal temperature to make the environment unfavorable to the invading pathogen. But do not let fevers go un-checked. Low-grade fevers usually indicate an abscess or pus accumulation somewhere while high fevers usually indicate a viral or severe bacterial infection. Generally any fever will diminish a cow's appetite enough for the farmer to notice. It is the intensity of the feverish state that determines which measures need to be taken. A high body temperature that remains unchecked has the potential to cause a pregnant cow to abort. Is the abortion due to the fever or is there an underlying cause giving rise to the fever? Determining the cause of a fever is definitely the goal, but until then there are some worthwhile remedies to help your cow feel better. If a fever can be brought down to within normal limits, the cow will often begin eating again, which will in turn boost her strength and she may well return to health. **By all means do not let a fever continue for more than a day or two without consulting with your vet.** If an animal looks really ill but has a "normal" temperature (100.5-102.5), her temperature may actually be falling downward through the normal range into the sub-normal range. If this is suspected, and a rapid heart rate is observed, the prognosis is very poor as this can indicate shock or toxemia.

In the case of heat stroke, it is absolutely necessary to hose down the cow with cold water, usually for about 20 minutes. Most cows, if standing, will stay put and not even need to be tied since they enjoy the water upon them. When hosing down a cow, do not forget to hose the back of the head since that is where the temperature regulation of the brain is located. Cows with heat stroke will commonly have a temperature from 106-108. Above 108, brain damage starts occurring and even if prompt hosing is carried out, chances begin to lessen that the cow will recover. It is usually apparent when heat stroke has occurred, especially during a hot stretch. This condition may be associated with recently fresh cows having some age and having also a low blood calcium level (*i.e.*, milk fever). Treat for the milk fever first, then try to get her up, and begin to hose down regardless. Some may need to be "floated" (put in a water tank as part of therapy to help rise her up). After initial treatment, however, when this is needed the chances are not good that she will go on to be a milk cow any longer — perhaps beef in a few weeks, but not dairy. Generally speaking, a cow will never fully recover when her temperature goes above 108° F. The best outcomes in heat stroke cases are those cows which begin eating and drinking right after the emergency therapy. Therapy should not only be hosing down the cow, but also 3-4 liters of IV fluids such as lactated ringers solution for obvious dehydration. Homeopathic belladonna (low or high potency) given every 30-60 minutes would also be indicated for heat stroke. Conventional therapy for heat stroke would use flunixin, furosemide, dexamethasone, and possibly an antibiotic, as well as hydrotherapy. The dexamethasone and antibiotic are prohibited for organic livestock.

The normal temperature of a dairy cow ranges between 100.5 and 102.5 F.

Aconite Ø (3-5cc) orally in mouth. This could be repeated every 1-2 hours for a few times during the first day. This combination is especially for a cow with a bounding heart rate and fever and bright pink gums and inflammation (can give IV in dextrose).

Aconite Ø for fever with weak pulse.

Belladonna Ø for inflammation with dilated pupils and quiet.

Gelsemium Ø for fever in a touchy cow that is mean-tempered.

1) *Aconite* homeochord every 15 minutes for 3-4 doses for sudden onset, followed by *Belladonna* homeochord given every hour for 3-4 doses.

2) *Pyrogenium* 4 x 3, especially for a fever due to a retained placenta or a "sloppy" uterus in general; also in fever associated with mastitis.

3) *Phos* 30C or 200C every couple hours as needed for those cows that have fever but look otherwise normal — especially those cows with a possible mild virus as seen by moistness ringing the eyes.

4) Tincture of garlic, echinacea, goldenseal, wild indigo and barberry; 15-20 cc orally 2-3 times daily.

5) Powdered garlic, ginseng, echinacea, wild indigo and barberry; two 1 oz. capsules, 2-3 times daily.

6) Vitamin C, 250-500 cc, IV once daily or 5 cc/100 lbs IM (no more than 30cc per site)

7) Aspirin, 3 cow size pills, 3 times daily for 1-2 days.

8) Observe the cow's eyes — are they sunken or in any other way different? If sunken the cow most likely needs fluids electrolytes IV to counteract dehydration. This may mean giving 2-4 liters of lactated ringers solution, 1-2 liters of hypertonic saline and 200-300 cc 8.4% sodium bicarbonate. *Sunken eyes are a bad sign — call vet — don't wait until tomorrow.*

Conventional: Non-steroidal anti-inflammatories (*e.g.,* flunixin).

(Dadd, 1897)

Lemonbalm, wandering milkweed, thoroughwort or lady slipper:

2 oz. of any with 2 qts. boiling water. When cool, strain and add a wineglass of honey (p.185).

If putrid or malignant, add small quantity capsicum and charcoal (p.186).

Ad libitum chamomile tea (p.186).

Combination:

Capsicum, powdered	1 tsp.
Bloodroot, powdered	1 oz.
Cinnamon, powdered	$^1/_2$ oz.
Thoroughwort or valerian	2 oz.
Boiling water	1 gallon

When cold, strain and give 1 qt. every 2 hrs. (p.186)

(Waterman, 1925)
Fever due to pneumonia and bronchitis:

Epsom salts 3-4 oz. daily
Fresh water   as needed
Box stall, well ventilated, well lit; blanket & rub legs
Mustard or liniment applications freely (p.371)

Internally:

Alcohol	$1^3/_4$ oz.
Water	1 pt.

Can repeat in 30-60 minutes and again in $1^1/_2$-2 hrs.

Fluidextract Aconite	15 drops
Fluidextract Belladonna	$1^1/_2$ tsp.

Can repeat in 2 hrs.

1st two days: mix —

Aconite	$1^3/_4$ drams
NH4Cl	$2^1/_2$ oz.
Saltpeter	2 oz.
Alcohol	8 oz.
Water	2 pts.

Mix; give 4 oz. 4-6 times daily (p.86)

After 2 days:

Fluidextract digitalis	$4^1/_2$-6 drams
Fluidextract belladonna	$^3/_4$-1 oz.
NH4Cl	$1^3/_4$-2 oz.
Saltpeter	2-3 oz.
Alcohol	6-8 oz.
Water	$1^1/_2$-2 pts.

Mix; 2 oz. 4-6 times in 24 hours.

If high fever or not responding to above:
1^1/$_2$ dram aspirin in 3 oz. alcohol 2-3 times in 24 hrs. OR
^{3}/$_4$-1 dram doses of quinine 3 times per day

Upon improvement:
Fluidextract Nux vomica   1tsp.
Fluidextract Gentian   2 tsp.
Mix in a little water, give 3 times per day (p.87)

(Herbal, 1995)
Fevers associated with following symptoms:
Angelica root: with coughs, bronchitis, pleurisy (p.90)
Blue vervain: with liver involvement (p.93)
Boneset herb: with upper respiratory tract infection + flu (p.94)
Catnip: colds-flu, esp. acute bronchitis (increased perspiration) (p.96)
Ephedra: lung involvement (p.101)
Meadowsweet herb: has aspirin like compounds (p.112)
Myrrh gum: upper respiratory tract infection & oral infection & glandular
    infection (p.113)
Peppermint leaf: colds + flu (p.116)
Chinese skullcap: thirst, cough, expectoration, snot; has antimicrobial,
    antipyretic and anti-inflammatory properties (p.121)
Willow bark: a source of natural aspirin compounds (p.126)

(Nuzzi, 1992)
Peppermint tea + distilled water (p.93)
Elder/Peppermint/Yarrow tea (equal parts)
Infection Formula (esp. for pneumonia; strong "antibiotic"):
    Boneset, Cayenne fruit, Chammomile flower, Echinacea root, Oregon
    grape root, Osha root, Red root, Usna lichen (p.54)
AntiViral Formula: Boneset, Chaparral, Echinacea root, Ginger root,
    Lomatium root, Osha root, St. Johnswort, Tronadore, Usnea lichen
    (p.54)

(Fish, 1930)
Aconite, tincture
Quinine & salts
Veratrum viride, tincture (p.150)
Quinine (for catarrhal fever) (p.90)

TABLE 12: Summary of the Action of Eclectic Tinctures for Fever (Ellingwood)

	ACONITE	GELSEMIUM	VERATRUM	BRYONIA
When	At onset, and during fevers in sthenic cases	In sthenic cases, with nervous irritation and spasm	Only in sthenic cases, with rapid heart action	In sthenia or asthenia
Age	Young and middle age, less prompt in the aged	Young; in full doses for strong adults	For strong adults; at parturition	Any time in life
General	All acute fevers and inflammations	Acute cerebral engorgement, nervous irritability	Threatened local engorgement	Inflammations of serous or synovial membranes
Heart and pulse	H-strong and rapid; P-rapid, quick, hard	H-strong, irritable, violent in action; not for bad heart	H-strong and rapid; P-full, large, hard; not for bad heart	H-may be weak or strong. Not a heart depressant; P-quick
Temperature	Always present in acute cases	Usually present, of a nervous type	Usually present; temp. not as important as heart	Fever present
Skin and eyes	Skin hot/dry; eyes bright	Skin hot/dry; eyes bright, pupils contracted	Skin may be cool	Usually hot, either moist or dry
Secretions	May be abruptly suppressed	Usually suppressed	Skin and kidneys usually free in action	Usually deficient
Pain	General distress; local pain in inflamed organs	Severe headache with extreme restlessness	Pain may be local; may be bursting headache	Local soreness on pressure; general muscle aching

(M.R.C.V.S., 1914)
Camphor, tartarized antimony, acid tartrate of potash — 1 dram of each
Nitre   2 drams
Aniseeds   4 drams
Glycerin, enough to make a ball (p.195)

(Dun, 1910)
Salicylic acid and sodium salicylate
Aconite
Quinine (in intermittent fevers) (p.763)
Benzoic acid (makes Friar's Balsam) (p.650)

(Winslow, 1919)
Cinchona and its alkaloids (p.351)
Quinine (p.359)
Salicylates (p.363)
Methyl salicylate from wintergreen; Oil of Birch;
Oil of Wintergreen
11 parts methyl salicylate (98%) = 10 parts salicylic acid (p.364)

COW NOT FINISHING FEED/Rumen atony/ Slow rumen

Check for fever
1) Powdered ginger, gentian, cayenne, sodium bicarb and caffeine; two 1 oz. capsules, twice daily.
2) Tincture of ginger, gentian, nux vomica and fennel Ø 15-20 cc in mouth 3 times daily (can give IV in dextrose).
3) Probiotic pastes or yogurt drench.
4) Reduce grain, feed only topnotch hay ("candy hay"), no silage; allow to graze.
5) Get vet to tube rumen with a nutritional slurry.
6) Magnet, if never given before.
7) If cow is fresh about one to two weeks, consider twisted stomach (especially if cow has a bad uterus, a fever, and is also ketotic).
8) Bottle of dextrose and calcium IV, especially if a heavy producer.
9) Tincture of red root, celandine, milk thistle, dandelion, oregon grape berry 10-15 cc in mouth 3 times daily for liver and digestive sluggishness (especially if possible moldy feed); can give it as an IV in bottle of dextrose. Also consider botulism (cow's tongue can be pulled out and cow doesn't retract it), if moldy baleage.
10) *Carbo veg* for gassy rumen and cool skin.
11) Ruminant laxatives (magnesium oxide/pink pills) for 3-4 milkings.
12) Aloe vera juice; 8-40 cc orally, 2-3 times daily.

Decision Tree for Off-Feed Cow

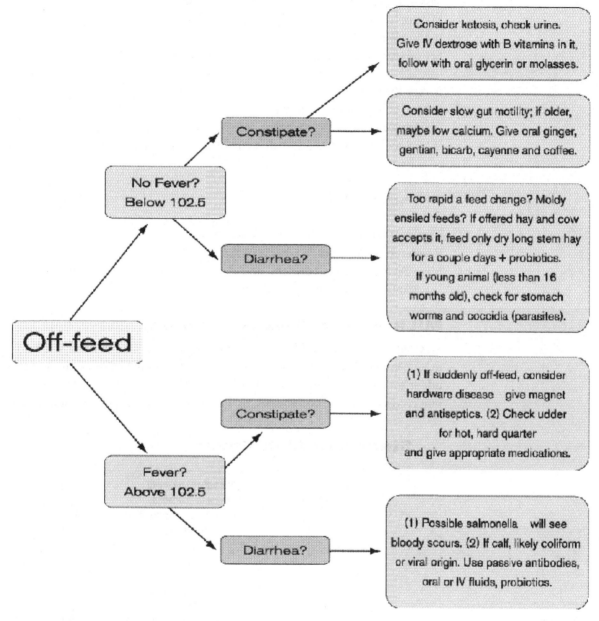

13) 5cc vitamin B12 at acupuncture points: BL-23 (bilateral), BL 41-01, 43-01, BL-18 and LV 14.

14) Nux vomica Ø IV, 20cc, in bottle of dextrose.

Conventional: Probiotics, laxatives, stomach tube with probiotic slurry.

(Dadd, 1897)

Make a mucilaginous drink of slippery elm, or marshmallows, and give a pint every two hours. *If impacted rumen,* try:

Sassafras (Laurus sassafras)	1 oz.
Mayapple (Podophyllum peltatum)	4 drams
Boiling water	2 qts.

Let mixture cool and give a pint every four hours (p.192)

If very thirsty,

| Lemon balm (Melissa officinalis) | 2 oz. |
| Boiling water | 2 qts. |

When cool, strain and add half a tsp. of cream of tartar; give half a pint every 2 hours

goldenseal, powdered	1 oz.
Caraway	2 oz.
Cream of tartar	$^{1}/_{2}$ oz.
Powdered poplar bark	2 oz.

Mix. Divide into six powders, and give one every four hours in a sufficient quantity of chamomile tea (p.166)

MILK FEVER (Hypocalcemia)

Cold ears, weak, wobbly, and staying down are cardinal signs. This is a derangement of calcium mobilization, when the demands for calcium to go to udder for milk outstrip the ability of the bones to release it. To reduce occurrence, cows in dry period should not be fed anything with appreciable calcium

Stages of Milk Fever

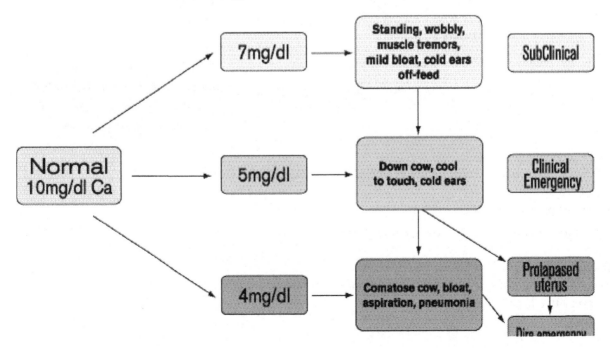

— this will "trick the bones" into sensing that there's a lot of calcium and not allow them to prepare to release it when really needed (when milked out). Additionally, diets in the dry period should contain no more than 2% potassium (K), as this will also induce milk fever. To diagnose if a cow this is mildly hypocalcemic, pull up on the skin at her withers and see if she opens her mouth.

SubClinical Milk Fever Syndrome

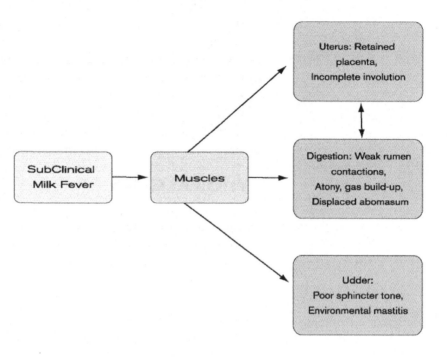

Clinical

1) Bottle of calcium in the vein, slowly (approx. 5-10 minutes is usually safe); it is always safest to keep the bottle no higher than the backbone. Do not give if cow is toxic (dime-slit pupils, blood shot eyes); when toxic, calcium can adversely affect the heart more easily. Record in healthchart as "electrolytes if organic."

2) Bottle or a tube of oral calcium. Caution: Giving oral medicine to a milk fever cow that is down or flat out is dangerous because her swallowing is weakened, and you can cause aspiration pneumonia. The tubes are best for standing, wobbly cows. A tube can be given after a bottle in the vein, to help maintain calcium levels high in the bloodstream. Just make sure the oral preparation is not made with calcium chloride as this can burn the esophagus (throat), especially if two doses are given. Try to have calcium propionate or calcium oxide preparations on hand.

3) *Calc phos* alternating with *Mag phos* (especially if muscles are twitching).

4) Calcium-rich feed after initial treatment above.

5) Consider boosting phosphorus if calcium treatments are not working (also consider other diseases). The phosphite form of phosphorus which is in most calcium combination products does not become available biologically as does the phosphate form. Either drench the cow with sodium phosphate or ask vet to administer an IV solution (1 Fleet® enema bottle added into a bottle of hypertonic saline, given fairly slowly).

Conventional: Calcium IV or under the skin (use calcium gluconate 23% if injecting under the skin); oral solution of calcium *if* the cow is not flat out.

Prevention of milk fever:

1) Grass hay in the dry period (or alfalfa with anionic salts).

2) 2 oz. of apple cider vinegar twice daily in the dry period, 2 weeks before freshening.

3) Give a tube of calcium oxide or calcium propionate gel about 12 hours before calving and another just after calving (after giving a bucket or two of water).

4) *Calc phos* 30C twice a week for three weeks prior to calving.

5) After the calf is out, offer the cow as much warm water as she wants; spiking it with a little apple cider vinegar and molasses can be done.

HYPOKALEMIA

Deficient blood potassium is sometimes met with in practice. Typically, the cow will not have responded to standard calcium therapy or will have been on the corticosteroid, isoflupredone. The cow will often have an "S" look to its neck and not be able to rise. Therapy with IV CMPK or other fluids with supplemented potassium will usually correct the situation quickly, although a repeated IV administration of the electrolytes may be needed. This condition can be induced if a farmer is administering isoflupredone (a steroid) too often for pinched nerve or ketosis.

KETOSIS (Acetonemia)

This often happens as milk production really starts to climb, about 1-3 weeks post-calving. If ketosis occurs earlier, it often can be secondary to a digestive upset like a twisted stomach. Primary ketosis indicates a liver problem and may be seen with cattle over-conditioned at calving. In sheep and goats, ketosis can occur before lambing/kidding and is known as Pregnancy Toxemia. If pregnant cattle are ketotic, beware of twinning and not being fed enough to keep up with the demands of pregnancy. Symptoms usually include a cow eating her hay and ensiled feeds, but not her grain, which is actually what she needs for quick energy. Also, she'll usually have firm manure and her behavior will be somewhat aloof and may seem sleepy. Occasionally, they will "strut" as they walk, holding their head a bit cocked high as well. Checking the urine for ketones is simple. Some folks can also smell the ketotic scent in the cow's breath or milk. Not all people can, it is genetically pre-programmed.

1) Tincture of Red root, celandine, milk thistle, dandelion, oregon grape berry Ø, 10 cc 3 times daily as needed (can be given with IV dextrose).

2) *Lycopodium* and *Chelidonium* homeochord 4-6/day as needed.

3) Vitamin B complex, 15cc in muscle, once daily for 3 days.

4) Niacin boluses, 2 at each milking for 6 milkings.

5) Feed the cow whatever she likes best and will eat a lot. Pour molasses onto her feed.

Ketosis

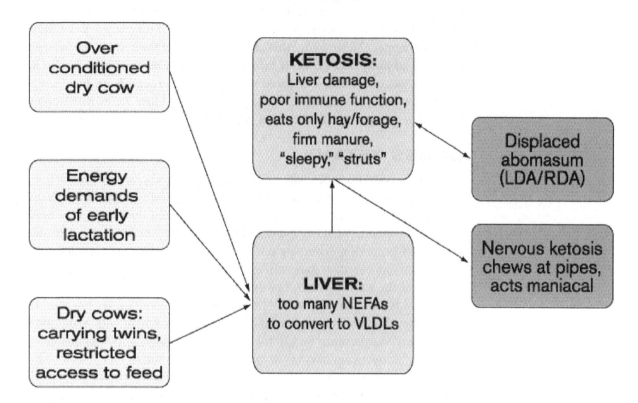

6) 50:50 molasses and apple cider vinegar mix; 8 oz. orally twice daily as needed.

7) Glycerin, 1 cup for 3-4 milkings.

8) Dextrose in vein, 1 or 2 bottles. Do not put under the skin or in muscle, it's very irritating and skin may slough.

9) Consider twisted stomach.

Conventional: Dextrose with dexamethasone (and possible insulin for chronic non-responding cases).

NERVOUS KETOSIS

This condition is simply "regular" ketosis gone on too long without any treatment. Symptoms are the same as above but then the cow will begin to chew the pipes, chew at herself, lick at her feed but not take any in, be clumsy on her feet, eventually fall down and continue her biting and chewing behavior. The prime rule-out for this would be rabies. Fortunately, rabies in dairy cows is very rare in my area, but seeing a cow with nervous ketosis one must keep in mind rabies — especially if she does not respond to treatment and/or remains down and can't get up. The treatment is the same as for "regular" ketosis, but follow-up treatment with glucose promoting substances is more heavily required until she starts eating more normally again. Rarely do I see nervous ketosis in association with a twisted stomach (displaced abomasum). If it does happen, I have found that after treatment for the ketosis, the twist will generally go away on its own. If I do a surgery on a nervous ketosis cow, her prognosis is guarded (as compared to an excellent prognosis on most simple left sided twisted stomachs with secondary ketosis.) Once a cow with nervous ketosis goes down, it is unlikely that she will get back up.

GRASS TETANY

This is a condition seen when cows are grazing lush pastures usually in the spring but sometimes in the autumn when lush regrowth occurs. It is a relative imbalance of magnesium in the cow's bloodstream and the levels in the brain. The result is a somewhat nervous and hyper-alert type of behavior, often staggering and becoming extremely difficult to restrain. They will lay down and then when approached they will get up. This is repeated until complete restraint of the animal is accomplished. A tip off that it is grass tetany will be the animal is a first or second calf heifer, thus putting milk fever lower on the possibilities. Treatment is to give an IV milk fever treatment with a solution of calcium and magnesium. Be careful with solutions high in potassium as well due to the bloodstream having relatively more potassium in this instance when magnesium is low. Occasionally a cow will collapse due to grass tetany, and the only way you or a vet will be able to decide it is grass tetany is due to the history of the cows grazing lush pasture, especially small grains like rye, wheat and oats. These cows may start convulsing, being on their side "paddling" with their legs and with irregular eye blinking. Either you or the vet should give the IV treatment, however, being very careful to listen to the heart with a stethoscope to make sure of regular beats while treating her. The best thing to do is call the vet and give *Mag phos* and *Calc phos* every 5-10 minutes while you're waiting. For follow-up after initial IV treatment, magnesium hydroxide is beneficial ("pink pills") or magnesium sulfate (epsom salts).

Conventional: Cal-Phos #2 IV (electrolytes — OK for organic).

FAT COWS/FATTY LIVER

Cows that are over-conditioned (excess weight) are potential disasters. The reason is that when the cow freshens, her body demands major shifts in body reserves and metabolism to accommodate milk production. Normally, cows will "milk the fat off their back" and that is OK. They will drop in body condition as milk production peaks at about 45-90 days in milk; they will then replenish their body reserves as they reduce milk production further into lactation. If they are too fat, however, at freshening, the liver becomes overloaded with fat as the fat is mobilized from body reserves. This creates a fatty liver, which severely hinders the normal liver function of detoxifying blood and many other enzymatic processes. The key word for this condition is *prevention*. Most grazing herds, fortunately, don't have this problem. It usually occurs in herds heavily fed corn silage and concentrates, especially during the dry period. I would rather have a moderately lean cow freshening than a fat cow freshening because the lean cow is already drawing off her reserves somewhat while the fat cow's catabolism (breakdown of muscle and protein to supply needs) is totally "on vacation." Don't get me wrong, I always want to have the cows in the correct body condition, but I think lean cows would be in potentially less trouble beginning lactation than overly fat cows. Research has shown that choline can help mobilize fat through the liver. The product Reassure™ made by Balchem is an excellent product for herds that are chronically fat. However, due to its formulation, it is not allowed for certified organic herds. Organic herds could use injectible choline in the form of Methaplex™, 20 cc in muscle once daily for 5 days before freshening. Apple cider vinegar would probably help as will by giving acetic acid to the rumen, which its microbes can utilize immediately. It will also gently acidify the system and perhaps help bones to release needed minerals into the blood stream so they are available upon commencement of lactation (also see Milk Fever prevention). Another possible treatment are liver protectants herbs (*i.e.,* Phytonic) 10 cc orally once daily for 7-10 days just prior to anticipated freshening date.

FAT NECROSIS/FAT SAPONIFICATION

A rare and infrequently diagnosed disease among cattle is fat necrosis of the omentum. The omentum is the sling which holds the intestines in place. When cows graze fescue with endophyte toxins, the usually soft fat that resides in the abdomen may become rock hard and slowly strangulate some part of the loops of bowel. The cow then dies. Usually it is further into lactation as the condition takes time to accumulate. Once a few cows die in a similar fashion, the farmer usually opts to have a cow under go a complete necropsy (after death analysis). If there is fat necrosis, even a quick and dirty field necropsy will easily reveal this. Always make sure that fescue, if using it in a pasture rotation, is endophyte free.

IMPACTED RUMEN

This may be a primary problem due to not enough water intake and lots of hay being eaten or a secondary problem due to slow movement of the entire digestive tract. An impacted omasum can give similar signs, however, an impacted omasum can usually only be diagnosed by exclusion of other possible syndromes whereas an impacted rumen can be detected and diagnosed when doing a rectal on the cow (treatment would be similar). Sometimes 1st

Very painful cow (peritonitis).

calf heifers will not know how to drink from a water bowl correctly and become dehydrated (treatment: place a 5 gallon bucket of water in front of her and see if she greedily drinks it down). It is fairly easy to feel an impacted rumen — go to the rumen side and push in deeply just behind the ribs and a bit high up (near the triangle). It will feel very firm like cement; it will feel this way also if a rectal exam is done and the rumen felt internally. Check a few other cows happily eating to feel a mushier or doughy type rumen, then check the slow eating cow again.

Check for fever

1) Powdered ginger, gentian, cayenne, sodium bicarb and caffeine; 2 capsules twice daily as needed.
2) Tincture of ginger gentian, nux vomica and fennel; 15-20 cc orally 2-3 times daily, especially if constipated and straining (can give IV in dextrose).
3) Magnesium oxide pills (pink pills), 3 pills twice daily as needed.
4) Mineral oil, 1 quart twice daily, as needed. Add a flavoring agent to the oil so the cow can taste it so as to not inhale it into the windpipe — turpentine or ginger work well.
5) Probiotic *(Lactobacillus)* products to stilmulate lethargic rumen bugs.
6) Fresh green grass or alfalfa if possible.
7) If rapid decline in milk and low grade fever, consider hardware and give a magnet.

Conventional: Laxatives.

(Waterman, 1925)

Epsom salts	1½-2 lbs.
Sodium bicarbonate (baking soda)	1 oz.
Ginger	1 oz.

Dissolve in 2 qts. of lukewarm water and give as a drench (p.383)

Follow up with:

Aromatic spirits of ammonia	1 oz.
Sulfuric ether	$^1/_2$ oz.
Fluidextract of jaborandi	2 drams
Fluidextract of calibar bean	½ dram
Fluidextract of nux vomica	$1^1/_2$ drams
Fluidextract of belladonna	1 dram
Water to make	1 pint

Shake; give as one dose and repeat three to four times a day until animal is relieved

If in great pain, give 5 grains of morphine (or in modern times use butorphanol) three times a day. If still impacted and constipated in 36 hrs. repeat the dose of Epsom salts, and if after 36 hrs. longer, the mass is still unmoved, give 1 lb. Epsom salts along with 2-4 drams of gamboge, or $^1/_2$-1 dram of croton oil; if the oil is used, give it in a pint of raw linseed oil. Give the animal all the water it wants and wet the food given to keep it moist. If still no better, consider having a rumenotomy done by a veterinarian to physically remove the impacted rumen contents.

HUMPED BACK and/or GRINDING TEETH

This is how cows may indicate abdominal pain. Pain may be due to hardware, ulcer, impacted rumen or omasum, peritonitis (belly infection), intestinal obstruction, perforated ulcer, pyelonephritis (kidney infection), etc. Hardware is typically seen as a cow going abruptly off-feed, major drop in milk, firm and fibrous manure, hump back, and low grade fever (T = 102.7°-103°F). Ulcers in cows may be due to stress near calving time, "hot rations" (1st calf heifers will develop laminitis — you may see shifting leg lameness), leukosis (enzootic bovine leucosis *a.k.a.* cancer) or other causes. Observe feed intake and color of manure. Black, tarry manure may indicate a bleeding ulcer. A cow may stamp her feet occasionally, yet infrequently, over a long period of time in order to relieve ulcer pain. A cow down with milk fever will also occasionally grind her teeth (but not always).

Cows milling around a burn pile — source of hardware (above). Close up of burn pile and hardware (below).

Check for fever

1) Magnet.

2) *Arnica* 4 x 3 for pain. Avoid aspirin when seeing black, tarry manure.

3) *Phos* alternating with *Calendula* 4 x 3; for internal bleeding.

4) Tincture of garlic, echinacea, goldenseal, wild indigo and barberry; Ø, 15-20cc in mouth 3 times daily for 3-5 days.

5) No grain, only long-stem hays such as timothy/orchard grass for 3-4 days. Also oats ("Quaker Oats") 2 cups twice daily with molasses on it.

Conventional: Non-steroidals anti-inflammatory, analgesic *(e.g.,* flunixin), antibiotic, magnet, +/-bicarb.

DISPLACED ABOMASUM ("DA" or "TWISTED STOMACH")

Call the vet if you suspect a twist. You can try by carefully listening on your own first (you don't have to have a stethoscope but it will help amplify the sound). Place your ear upon the cow's left ribcage and flick your finger sharply against the space between the ribs. If you hear a tin can-like "ping," that is what a displaced abomasum sounds like. On the left side of the cow, this is usually what it is about 95% of the time (a bloated rumen would account for the other 5%). On the right side, there are more reasons for "pings." However, if the "ping" on the right is over a large area and extends forward quite a ways and towards the elbow, it is possible that a right-sided displaced abomasum is present. A right-sided ping found only behind the last rib and below the "short ribs" is common and can be heard even in a cow milking a hundred pounds. Cows with diarrhea also will have a ping back here.

A common occurrence on modern dairy farms these days (especially those that feed heavy amounts of corn silage early in lactation and push the cows hard) is to see a DA within the first two weeks of lactation. The "perfect set up" equation for a DA goes like this: retained placenta/metritis + off-feed + ketosis + feeding lots of corn silage + keeping cow tied in stall = twisted stomach. I can almost guarantee it. From experience, I can say that a hard calving with a first calf heifer, which is then introduced to the fresh cow ration to make lots of milk, usually ends up with a twisted stomach. The best way to prevent the need for potential surgery is to GRAZE the animal. She will get exercise (to help drain the lymph away from the uterine vessels) as well as getting fresh feed (not ensiled). At least, allow her to walk around freely most of the day and bulk her up with HAY, not corn silage.

If electing to try to correct it medically with the suggestions below, remember the twist may recur once you resume pushing grain and silage. Most twists occur soon after calving because there is space in the abdomen after expulsion of the calf. It is therefore important to keep the cow's rumen bulked up with roughage (hay) if this condition is to be prevented. The manure consistency associated with a DA can range anywhere from stiff and firm to normal or loose. Although a right-sided DA can become deadly if let go, they actually respond better than a left-sided DA to the medical treatments below *if it occurs only a day or two after calving*. A cow with a large ping on the right, looking depressed with dull eyes, not eating a thing and not passing manure needs to be seen immediately by your veterinarian, *do not delay!*

Although most DA's occur within about 3 weeks of calving, they certainly can occur later in lactation, although much less commonly. Actually, any ruminant of any age or sex can develop a displaced abomasum. In my experience, farmers aren't as quick to think about a possible DA later on and will let it go for a few days. Because the rumen has been at maximal capacity for a while already in lactation for a while, the abomasum tends to float up onto the right side at this stage more often than not. I have already had to do DA surgery on cows eight months pregnant! Fortunately, they were successful because the cow was still comfortably standing. It is when there is a right-sided DA present and the cow does not want to stand for longer than a couple of minutes for examination before lying down again when it has become too late to do anything for her. A right-sided twist is potentially fatal. However, this is the rare instance. Usually by keeping a watchful eye on your cows you can detect slight changes in eating patterns and the treatments below may help.

1) Absolutely no silage, a little grain and lots of your best, sweetest hay.

2) Exercise/jog the animal; this helps the stagnant gut move around.

3) Tincture of ginger, gentian, nux vomica and fennel Ø, 15-20 cc in mouth 3-4 times daily (can give IV in bottle of dextrose).

4) Powdered ginger, gentian, cayenne, sodium bicarbonate and caffeine, 2 capsules twice daily as needed.

5) *Nux vomica* also can be used with above.

6) Give laxatives (magnesium oxide/pink pills) to induce diarrhea for 2 days (increase gut motility).

7) Bottle of calcium with electrolytes IV, especially if third calf cow or older.

8) Cast cow onto right side, then roll her and stand her up; may correct it, at least temporarily. Basically, when standing behind the cow, make her fall clockwise and then roll her over on her back and get her up. Everything is done clockwise when standing behind her. (Note: rolling a cow may give a right sided twist.)

9) Acupuncture: 5cc vitamin B12 at BL 43-01.

Conventional: Surgery.

If a surgery is needed, free gas in the abdomen (pneumoperitoneum) may occur as a consequence of standing surgery. Once the trapped air eventually makes its way out through the system, feed intake will increase greatly. Sometimes an abscess will develop following surgery done to correct a DA from the underside of the cow (ventral paramedian approach). Abscesses underneath don't worry me, for gravity will help them open and drain. Once open, keep it clean with a germicide spray and clean bedding area.

TWISTED CECUM

This condition is about as rare as mesenteric torsion (see Colic), but the prognosis for recuperation and productivity are much better if a cecal torsion is diagnosed. Cows don't usually show as many colicky signs with a dilated or twisted cecum as compared to a mesenteric torsion. They will, however, look bloated and full on both sides. They will be completely off-feed. Upon rectal palpation, a large balloon filled organ can be palpated directly below the rectum. This is usually gas filled with liquid manure at the bottom. A large ping may be heard on the right side, behind the ribcage. Usually there will be no manure on the sleeve after palpation. Like the colicky cow needing surgery, the heart rate, relative dehydration and duration of condition will indicate surgery or slaughter. I have had one cow (certified organic) with a twisted cecum go to surgery at the U. of Penn veterinary school and they did an excellent job without using any antibiotics. The cow recovered and went on to be productive for a few lactations longer. One note regarding cecal torsions, it is thought that *Salmonella* may be a contributory cause of this sporadic condition. If so, the cow may be a carrier, and it could theoretically occur again. I have, however, yet to see that happen.

COLIC

Cows show colic by shifting their weight on their back legs from side to side. I have yet to see a cow "roll" like a horse. Colic in cows is not good. Although it occasionally occurs in a pregnant cow as a result of the calf's intra-uterine movements, it usually indicates an abnormality occurring somewhere in the small intestine. If the cow also looks very "full" or bloated on both sides, looks uneasy and is shifting its weight on its back legs or is kicking at its belly and laying down and getting up again, call the vet without delay. Sometimes a cow that begins to kick furiously at her belly (without any sign of bloat) will be passing a kidney stone. The episode will last a very short time, about 20-30 minutes and be finished by the time the vet arrives. Sometimes with these kinds of signs it will be an impending case of diarrhea. However, most cows with diarrhea do not exhibit colic. More often than not (and fortunately it is rare to see colic in a cow) it is due to a small bowel obstruction or a mesenteric torsion. This is when the mesentery that the intestines are embedded in becomes twisted with resultant impaired circulation. A nearly pathgnomonic symptom of mesenteric torsion is shown by the cow dipping her right hook bone and leaning downward with her belly. This can also be diagnosed by a veterinarian upon rectal palpation by feeling tight bands and loops of bowel in the abdomen, when normally none are felt. If a diagnosis of mesenteric torsion is arrived at, there are only two options: surgery or slaughter, either one to be done without delay. I have occasionally pumped in a gallon of mineral oil into a colicky cow at night if I have felt a "tight band" in the abdomen, when normally none are felt. A tight band can be indicative of an early mesenteric torsion but also possibly a gut which is totally atonic (non-moving) with gas filled loops of bowel. Generally the heart rate and behavior of the cow will aid in diagnosing the problem. If there is no manure on the rectal sleeve, that is a bad sign. If there is diarrhea in the rectum, it may be an odd case of gastrointestinal upset (as seen with bad feed, especially bad ensiled feed). In any event, these kinds of cows need careful attention over the course of 12-24 hours time. In addition to the mineral oil, using a gut stimulant like the ginger, gentian and nux vomica Ø mixture would be of value. Allowing the cow outside to exercise and eat green grass would also be of benefit. If absolutely no improvement is seen in a cow with colic after 6-8 hours, call the vet in for sure.

1) Tincture of ginger, gentain, nux vomica and fennel, 20 cc orally 2-3 times daily.

2) Peppermint Ø, 20 cc every couple hours if gassy.

3) Allow cow to be outside or moving around somewhere; allow to graze or bring fresh greens to her.

4) Stomach tube with mineral oil or carefully drench with mineral oil which has been given a mint flavor.

(Dadd, 1897)

Powdered aniseed	$^1/_2$ tsp
Powdered cinnamon	$^1/_2$ tsp.
Powdered asafetida	$^1/_2$ tsp.
Gruel of slippery elm	2 qts.
Oil of aniseed	20 drops

Give as one dose (p.166-167)
Peppermint tea and brisk friction upon the stomach and bowels (p.167)

GRAIN OVERLOAD (ACUTE RUMEN ACIDOSIS)

Fortunately this condition is fairly rare. Occasionally I'll be called out to see a cow that got into the grain cart or grain bin. Like most diseases, the earlier the intervention, the better the outcome — this is especially true in grain overload. One of the problems with grain overload is that, although it is not difficult to get a cow over it if it is early, the consequences may surface in a few months in the form of laminitis/founder (seen as sole ulcers of the hoof).

If the animal still looks bright and alert, not dehydrated, not too much rumen bloating and no scours, try the following:

1) Activated charcoal, to adsorb toxins in the gut.

2) Mineral oil, to coat the gut so toxins will not permeate it.

3) Antacid pills (pink pills) with magnesium oxide to help off-set the probable drop in pH caused by a large slug of grain.

4) Sodium bicarbonate (baking soda), a handful, also in order to counteract the low rumen pH.

Conventional: Same as above.

If the cow is known to have eaten a large amount of grain and she is depressed with cold ears, scours and is still standing, massive IV fluid therapy will be needed in the form of many liters of lactated ringer's solution with sterile injectible sodium bicarbonate to correct the obvious metabolic acidosis. It is also necessary to keep the cow away from water for about 8-12 hours, as water can accelerate the acidosis conditions in the rumen. However, giving activated charcoal and a gallon of mineral oil would still be good. Depending on the circumstances, a rumenotomy can be performed to empty out the rumen of the offending grain. This is not a common procedure for most veterinarians, but it certainly could be done. Once the animal is down, depressed, grinding teeth and has a really watery diarrhea, the prognosis is grave (in other words the cow will probably die despite treatment attempts). Again, in suspected rumen acidosis, time is of the essence!

SUBACUTE RUMEN ACIDOSIS (SARA)

Unfortunately this condition is far too common and, in my opinion, is the root cause of many digestive and hoof problems in modern dairy cattle. It basically arises out of the practice of feeding as energy dense a ration as is possible, wth minimal fiber. It would be most likely found in large herds fed total mixed rations, and compounded by walking on concrete continuously. However, it must be said that grazing cows can also experience sub acute rumen acidosis, especially in the spring when the pastures are lush with very low fiber levels while also feeding grain with little hay. (However, the subsequent effects may not be as harsh on a grazed animal as that in a confinement system.) The most common scenario is that of cows being pushed fast and hard for prolonged maximal production. A consequence of feeding such acidosis inducing rations is that of laminitis (founder). This is seen in sole ulcers a few months later. Interestingly, it is often easy to see when a herd of cows had a nutritional stress by looking for a line of different hoof growth. This line will be seen on all four hooves and on many cows. If it is only seen on one cow, then only that particular cow had a severe stress (perhaps an illness or really rough freshening time). Another thing to look for is erosion of the back of the heel away from the hair line with resulting continual peeling away of successive layers of new heel growth. These can become infected due to environmental dirt; however, the original cause of the separation of heel from hairline is still due to too high an energy rich ration compared to fiber (or possibly mineral imbalances). A few insightful nutritionists theorize that herds which experience a high rate of strawberry heel probably are fed acidosis causing rations. As a veterinarian, I can understand this being possible, especially when considering that the normal circulation to the hoof is hindered in laminitis due to rumen acidosis and this impaired circulation weakens the hoof-hairline (coronary band) integrity, allowing bacteria in the environment to colonize the area. The result is hairy hoof wart/strawberry heel. There's really not much treatment for laminitis and sole ulcers, mainly due to the fact that the damage is done, occurring previously. However, if these symptoms are observed, a frank discussion about acidosis with the nutritionist is in order.

BLOAT (archaic: Hoven, Blown, Blasted)

This can be a life-threatening emergency. If the animal has obvious difficulty breathing, you don't have much time (sometimes only minutes). You need to make a quick stab with a short, sharp pointed knife into the upper-left flank ("near side" in horseman terms) just behind the last rib and just below the short ribs. Usually the rumen is obviously distended and ballooned out in this area. A sharp ping may be heard in the area if the rumen is really stretched tight. But normally it will have a "pong" sound when not extremely life threatening. Call your vet! There are two kinds of bloat, either free-gas or frothy. Free-gas bloat can be relieved by passing a stomach tube and allowing gas to escape. If the tube is passed correctly and no gas escapes but very fine, rumen color foam is observed, chances are that the cow may have frothy bloat. Frothy bloat occurs mainly on cattle put on lush, legume pasture. It may take a couple of days for the symptoms of frothy bloat to show. Free-gas bloat is usually a single animal problem whereas frothy bloat usually involves more than one animal. Calves will sometimes get bloat as well. This condition in calves can come on quickly and kill them in a few hours if you are not aware. Calfhood bloat is usually a single animal condition, yet it can repeatedly affect a calf for a few months of life and then go away as mysteriously as it came. Some adult cows will bloat every so often. This is not good as it usually indicates that the nerve which stimulates normal turn-over of the rumen is damaged. This is usually caused by an abscess, often the end result of a previous bout of "hardware." If not a dire emergency, after relieving gas build-up, try:

1) Give 1 pint vegetable oil; repeat in 20 minutes. Force animal to walk around.

2) Tincture of ginger, gentian, nux vomica and fennel.

3) *China* every 15 minutes until bloat subsides; can alternate with *Carbo veg.*

4) *Colchicum* every 10-15 minutes if kicking at belly and looking back at flank.

5) *Colocynthus* for cramping, arched back, looking back at flank.

6) Walk/jog the animal to help work out gas.

7) Poloxalene (Therabloat) may be needed for the frothy type (OK for organic).

8) Place a gag at the back of the mouth and tied behind the ears. This will promote saliva production (as the animal tries to work the gag away with its tongue) which has bicarbonates. This can help reduce amount of frothy bloat.

Conventional: If pasture bloat — poloxalene; free gas bloat — stomach tube to release pressure and mineral oil.

(Waterman 1925)

Aromatic Spirits of Ammonia	1 oz.
Fluidextract of jaborandi	4 drams
Fluidextract of Calabar bean	1 dram
Fluidextract of belladonna	1 dram
Hyposulphite of soda	4 oz.
Water, to make	1 pint

Shake; give as one dose, and repeat in 30 minutes if necessary

If these items are not on hand, give a good dose of ginger or an ounce of turpentine in a pint of mineral oil. After the crisis passes, follow up with $1^1/_2$ lbs. Epsom salts and 2 tbs. ginger in 2 qts. of water. (p. 381)

(Dadd, 1897)

Cardamom seeds	1oz.
Fennel seeds	1oz.
Powdered charcoal	1 tbsp.
Boiling water	2 qts.

Let the mixture stand until sufficiently cool; then strain and administer in pint doses, every ten minutes

Clyster (enema):

Powdered lobelia	2oz.
Powdered charcoal	6oz.
Common salt	1 tbsp.
Boiling water	2 qts.

When cool, strain, and inject (rectally)

To help the animal after the crisis,

Marshmallows	2 oz.

| Linseed | 1 oz. |
| Boiling water | 2 qts. |

Set the mixture near the fire and allow it to macerate for a short time; after straining through a sieve or coarse cloth, it may be given and repeated at discretion (p.144)

ADULT DIARRHEA

There can be many causes of diarrhea. Among the most common in adult cattle would be simple indigestion from bad feed, winter dysentery, *Salmonella*, Johne's disease, BVD, clostridia or intestinal parasites, etc. Until diagnosis by a vet, remove all grain and silage. Have cow only on a grassy dry hay diet (some oats are OK). Check dehydration by squeezing skin along neck or eyebrow and seeing how long it takes to snap back. The longer it takes, the worse the dehydration. Also monitor eyes for how sunken they are and how bloodshot the whites are — the worse, the more serious.

Calf with mouth sores, possible BVD case.

Simple indigestion from eating some spoiled feed, especially ensiled feeds, is a common reason for diarrhea. It is usually self-limiting, in that once the offending material is passed out, the cow resumes normal activity. Conservative feeding (hay only), and some probiotics usually are enough to get the cow going.

Winter dysentery is a condition that farmers see yearly in late autumn or early winter. It affects many animals at once, especially first calf heifers. It is thought to be caused by a virus, although no vaccines have been made yet. It is not deadly and with conservative feeding, most cows will come through it fine in a few days. Cows generally continue eating, although with reduced appetite and reduced milk.

Salmonella is a potentially fatal condition. Cows usually run high fevers, have a watery and somewhat bloody, fetid explosive diarrhea, won't eat a thing, drop off in milk, and look really sick. As the condition develops, parts of the intestinal lining may slough and be passed out externally. *Salmonella* is bacteria which may become ingested through spoiled feed or by contact with it from manure. A cow sick with *Salmonella* may need antibiotic treatment to save her, although

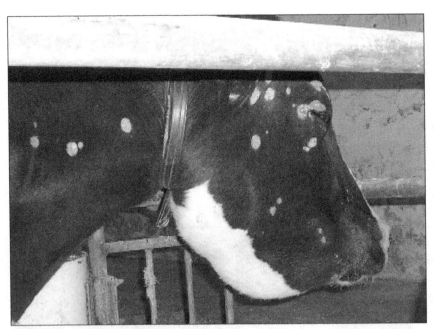

Cow with bottle jaw and ringworm.

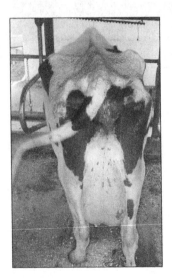

Johnes disease and poor body condition.

this may cause her to become a carrier. Also, certain strains of *Salmonella* are transmittable to humans, so care must be taken not to drink bulk tank milk if there is a *Salmonella* outbreak happening.

Johne's disease is a chronic type disease with an extremely slow onset. It is often contracted at birth when a calf sucks on a manure laden teat. The bacteria then colonize the gut of the calf, only to become apparent as a diarrhea condition when fresh for the first time (or even older). There may be familial genetic resistance or susceptibility to this disease. It is a fairly common disease in dairy intensive regions with lots of cattle movement between farms. In other words, it is a disease which you can buy into your herd. Typically, a Johne's cow will continue to eat and have lots of effortless dark green thick pea soup-like manure. Once on the ground, the little air bubbles will burst, leaving pock mocks in the pile (looking like a dark green pancake that needs flipping!). As the disease continues, the cow will become thinner and thinner, even though she is eating. This is due to malabsorption from the thickened intestinal lining as a reaction to the bacteria in place there. Eventually, with impaired protein retention in the vasculature, there will be a fluid build-up underneath the jaw. This is called "bottle jaw." If I am called to see a cow with longstanding diarrhea, really skinny, she eats and has bottle jaw, she is diagnosed as a Johne's cow and usually shipped off for salvage. Once a Johne's cow, always a Johne's cow. There is no treatment for normal, commercial dairy cows. I have seen some cows regain normal manure consistency (using the treatments described below) while waiting for lab results if early on. The important step here is not so much for the individual sick animal unfortunately, but rather to assess the prevalence of the disease in the herd and to take management steps to prevent it form infecting youngstock. Feeding acidosis inducing diets (to any age animal) will likely create a worse situation to an infected animal.

Bovine Virus Diarrhea, or BVD, is a virus which can wreak havoc in a herd. It has a lot of faces, but in its classical form it will cause severe diarrhea, fever, and possibly death within a few days. It is distinguishable from *Salmonella* in the barn by checking for the telltale ulcers in the mouth that BVD causes.

When called out, I may find a normal temperature as the fever spike happened a day or two before when the cow was just hinting at something being wrong. It is by inspecting the mouth for ulcers which then becomes important. Depending on the cow's immune status, the BVD may only cause mild to moderate diarrhea and sickness or perhaps kill her. BVD vaccination should be considered as there are other more insidious ways for it to ruin a herd (abortion, reproductive infertility, etc.)

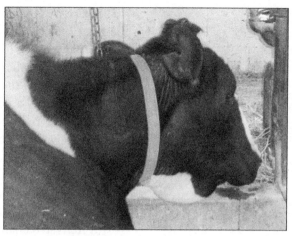

Bottle jaw due to Johnes diarrhea and hypoproteinemia.

Clostridial diarrhea is not common, although it is seen occasionally in a just fresh animal. They will look very sick, not want to get up, have a below normal temperature with an increased heart rate with lots of watery diarrhea. Clostridia bacteria normally live in the gut and do no harm when the other entire micro flora are in balance. However, if for some reason the bacterial equilibrium of the gut is thrown out of kilter, the deadly clostridia can become overwhelming and cause death, usually within a day.

Parasite infestations certainly can cause loose, diarrhea-like manure. However, in adult dairy cows, this is not usually a primary problem as the adult can live in equilibrium with a given load. Grazed animals will tend to have a higher potential for worm infestations, *especially first calf heifers*. Worms are not high on my rule-out list in older cows; however, I will often check manure for presence of eggs in first calf heifers. If there is an infestation, conventional worming will probably be needed and then following up with diatomaceous earth (DE) to keep a new equilibrium and prevent infestation (ivermectin is OK for organic). (See section on internal parasites.)

For most diarrhea conditions, try these treatments, but also be ready to call your vet in a couple of days if no improvement is noticed.

Check for fever To do so in a cow with diarrhea, it is wiser to place the thermometer into the vulva (exit of birth canal) in a clean fashion in order to get a true reading. This is because of a lot of air entering the rectum during bouts of diarrhea with straining which causes a potentially false low reading on the thermometer.

1) Feed a cow with diarrhea like you would normally feed a horse — grass hay — NO ensiled feeds and only a very little pasture, if any. Feed hay for 3 days. If no better, think Johnes.

2) Stomach tube 1 gallon mineral oil (counter-intuitive, but it helps to coat the intestinal lining and quell hyper-fermentation in the small bowel).

3) Vitro-tox, Pepto-Bismol, any type clay to help constipate.

4) Astringent (Anasorb) pills, give 2 for next 3 milkings.

5) Oats — "Quaker Oats": 2 cups twice daily. Oats are a great normalizer of the gut.

6) Offer her two buckets of water, one regular and one spiked with electrolytes. It is amazing but cows will often prefer the salts and sugars because they instinctively know they need them.

7) Hematinic, a colloidal mineral suspension of tannins, iron and trace minerals, 20cc mixed with equal parts water and syrup/molasses (60cc total per dose), given 1-2 times a day for a few days.

8) 5cc vitamin B12 at acupuncture point GV-1 a few times a day (see Acupuncture charts).

9) 2.5cc vitamin B12 mixed with 2.5cc atropine sulfate (large animal concentration) at BL-21 and BL-25 (bilaterally) (see Acupuncture charts).

10) Activated charcoal.

11) *Arsenicum alb* 4 x 3 for stinking, offensive diarrhea; bloody diarrhea; cow takes frequent small sips of water and she seems chilly.

12) *Merc corr* 4 x 3 for straining and blood-tinged diarrhea and ulcers on gums in mouth (may be valuable in a BVD outbreak).

13) *Podophyllum* 4 x 3 for "pipe stream" diarrhea (as seen in Johne's); watery, brown, no straining, lots of it.

14) *China* 3 x 5 for periodic, intermittent diarrhea (every AM only, or every other day, etc.). Also for washed out cows.

Conventional: Astringent boluses, mineral oil, probiotics.

(Dadd, 1897)

Powdered slippery elm	1 oz.
Powdered charcoal	1 tbsp.
Boiling water	2 qts.

Common starch, or flour, may be substituted for slippery elm. The mixture should be given in pint doses, at intervals of two hours. When the fecal discharges appear more natural and less frequent, a tea of raspberry leaves or bayberry bark will complete the cure.

When the disease assumes a chronic form (Johne's — author) and the animal loses flesh, the following tonic, stimulating, astringent drink is recommended:

Infusion of chamomile	1 qt.
Powdered caraway seeds	1 oz.
Bayberry, powdered	½ oz.
Mix for one dose (p.136)	

Severely dehydrated calf — needs IV fluids.

Dehydration gives sunken eyes — needs IV fluids.

CALF DIARRHEA

It is extremely important to keep calves hydrated as they have fewer reserves than adults. Do not be too forceful in drenching calves as they might get aspiration pneumonia. It is best to keep calves on milk while with diarrhea, as milk provides nutrition and not only fluid replacement as electrolytes do. A few years ago, the idea was to remove milk all together while sick. It is of utmost importance to get colostrum to the calf as soon as it is born. The cause of diarrhea in calves usually can be determined by what age of life it is, although this is definitely not always accurate. But generally, calves 0-10 days old will break with either coliform, Rota or Corona caused diarrhea, whereas calves 2-3 weeks old will break with a coccidia or cryptosporidia (protozoa induced) diarrhea. Calves near weaning, at about 2 months of age, will either have protozoa induced or a gastro-intestinal worm (nematode) induced diarrhea. *Salmonella* induced diarrhea can occur at anytime. The cause of diarrhea should be investigated by your veterinarian to make sure the proper measures are taken to prevent it in the future. Taking a few manure samples and sending them out to a diagnostic lab is valuable when a few calves have become ill.

Coliform induced diarrhea will make a calf septic, in other words, high fever, bloodshot whites of the eyes and severely dehydrated. To have a calf survive this, antibiotic treatment may be the only help. If let go, convulsions may occur due to the bacteria causing a meningitis. Rota/Corona virus can also cause a fever and diarrhea in the first week of life and is potentially life-threatening but with good fluid support a calf will usually survive. Coccidia or cryptosporidia are protozoan organisms that parasitize the calf's intestines, irritate it and cause the animal to have a very dark, almost black, diarrhea. The calves have a rough hair coat, eat fairly well, but not gain in weight and size as do others. It is a debilitating condition, which, if neglected, can lead to a slow, long-drawn out death. Most times this condition is seen when calf pens are continually wet and not bedded enough regularly. In other words, poor sanitation.

Worms are often a cause of loose manure and diarrhea in youngstock. The worms inhabit the stomach and intestines and are attached to the lining of each. They are true parasites in that they live off the calf's blood and will eventually cause death due to anemia. They generally have a three week life cycle and thus are not seen any earlier in calves (usually 2-3 months is when first a problem). Calves can ingest worm eggs and larvae by nibbling at bedding with manure on it or splattered nearby. This is a good reason to offer hay to a calf in order to satisfy its natural urge to eat fiber. Often calves will have a pot-bellied appearance, rough hair coat, stunted growth and a lighter hair color (black becomes brownish black). Diarrhea or loose manure is concurrent. Life threatening anemia occurs in the springtime when hibernating worms emerge from the gastro-intestinal lining and begin actively feeding again. It is easy for your veterinarian to diagnose eggs in the manure and this should be done routinely with calves, especially if you are not using conventional wormers on any animals. (See Parasites.)

For calves still on milk:
At first sign of not finishing bottle or loose manure: Check for fever
1) *Arsenicum* 4 times for one day.
2) Passive antibodies (Bo-Bac-2x/Borisera/Quatracon) specific for coliform and *Salmonella*.
3) Colostrum bolus or yogurt.
4) Chamomile tea (body temperature) for 1-2 feedings.
5) Consider giving Biocel CBT or other commercially prepared probiotics with passive antibody transfer properties.
6) Lime water — add lime to cold water and make a supersaturated solution and give carefully as a drench, 1 pint 3-4 times a day. This will help neutralize acidosis as well as hydrate the calf.
7) Immunoboost, 1-2 cc. It is best to give at very first signs or a few days before, if a pattern is noticed.
8) Homemade electrolyte mix: 8 tbsp. honey, 2 tsp. baking soda, 2 tsp. salt in 1 gallon water.
9) Vit E and selenium (Bose™).
Conventional: Sulfamethazine (antibiotic) pills.

Calves near, or already, weaned:
1) If calf is already eating solid feed (hay, grain), only offer grassy hay for a few days as well as water and electrolytes.
2) If dark or blood tinged manure, use *Merc corr* 4 x 3 with *Ipecac* 4 x 3 or *Phos* 4 x 3.

3) Hematinic (colloidal iron, minerals and tannins), 5cc mixed with 5 cc water and 5 cc syrup/molasses once daily for 3-4 days. If worms are suspected, then dose once daily for 7 days in a row.

(Winslow 1919)
Calcium carbonate (chalk)   2 oz.
Catechu tincture   2 oz.
Ginger fluidextract   2 oz.
Opium (paregoric can substitute)   ½ dram
Shake; give 2 tablespoonfuls in half pint of flour gruel 3 times daily

(Steffen, 1914)
Creosote (Beechwood)   15 minims
Capsicum tincture   1 oz.
Cajeput oil   1 dram
Castor oil   1 oz.
Give as one dose for scours, repeat in one hour if necessary (p.75)

BLOAT IN CALVES

Calves will sometimes bloat for no apparent reason. This bloat can kill them if left unnoticed by the farmer. The rumen quickly fills with gas, and for obscure reasons the calf cannot burp the gas off or pass it down stream. The symptoms are obvious. This usually happens when calves are on milk replacer of poor quality. Sometimes it occurs when the calf is already weaned. I have not seen this on milk-fed calves or those dairies that have nurse cows that calves can suck on. A good prevention is to feed hay during the milk feeding time (already discussed in a previous chapter). This allows the rumen to have the feed it will normally have, instead of just grain and milk. Additionally, having calves drink from bottles at nose height will help as their sucking produces saliva (which has bicarb) and will send the milk to the abomasum to form a curd due to the rennen. Drinking milk from a bucket does not create this action. Occasionally a weaned animal will bloat. This is a rare occurrence and not easily explained. It could theoretically be due to problems with the nerves that run to the digestive tract, some of which originate on the brain stem and could have been damaged by a bout of calfhood pneumonia, especially if caused by mycoplasma (which sometimes can be traced back to an umbilical cord infection).

In any event, treatment generally consists of passing a stomach tube to immediately relieve the dangerous rumen gas build-up (which will choke the calf internally), then followed by a little mineral oil, about 1 pint. I have had to re-treat these calves at times but no longer need to when dispensing the Digestive Ø as a follow-up (about 5cc orally as needed). I also like to put peppermint

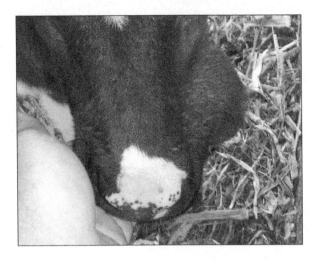
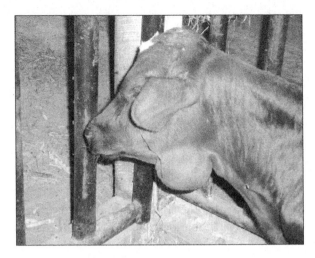

Two examples of jaw abscess. The one on the right is quite larege.

oil in the mineral oil as an initial carminative. Mineral oil, if given as a follow up, should be given a flavor so the animal just doesn't inhale the tasteless material that mineral oil is.

FLEXED TENDONS

This problem is rare but occasionally affects a calf. It is not necessarily due to a hard calving. A calf will be seen with its front hooves curled under and walking on its "wrists." This problem is correctable, but measures must be taken *immediately*. Too often I've had a phone call about this and upon questioning the farmer says it's been going on for a few weeks, not thinking much of it because the calf was eating fine. Too late! The tendons will be frozen in place in short order — even within a few days. A splint type device should be affixed to the area to keep it straight as soon as possible. Use of *Calc fluor,* 5 pellets 3 times daily, may also help to correct the problem more quickly once the splint has been affixed.

Conventional: Splint (+/- IV oxytetracycline).

JAW ABSCESS

Occasionally a calf is noticed to have a small lump on its jaw. These are almost universally a small abscess, commonly due to a tiny puncture wound from whatever the calf may have chewed. Sometimes they can become a bit large, but they inevitably open and drain. To help them along, *Hepar sulph* in low potency (10X), 5 pellets 3-4 times daily will open them up more quickly than if simply letting time takes its course.

PARASITES
(See Calves and Parasites, discussed earlier)

CALF BLINDNESS (PEM)

This condition is due to a lack of healthy rumen flora and subsequent lack of vitamin B1 (thiamine). This is caused by a lack of fiber in the calf's diet or in rare instances by too much sulfur in the drinking water. By far the more common cause is not enough effective fiber in the young animal's ration. Unfortunately there is a fad to feed only milk and grain to young calves until weaning at 6 weeks of age. This disease could occur near weaning but more commonly a month or two after weaning especially if the effective fiber needs are not being met. The signs are nearly pathognomonic — the animal leans as it

Hernia or abscess?

walks, is blind and depressed. The treatment is a *slow* IV injection of thiamine with follow-up by the farmer with vitamin B complex once daily for the next 3-4 days. So it doesn't occur in the others, a serious discussion about getting enough effective fiber into the calves is a must. Conventionally, a short acting steroid may be given (dexamethasone) in order to reduce swelling and inflammation within the cranium. However, I have never used that for this condition and all animals treated with the thiamine have responded well (if indeed it is a thiamine deficiency and not some other neurologic disease).

WHITE MUSCLE DISEASE

Occasionally a calf or group of calves will be born weak and cannot stand. Although many possibilities exists for the cause of this, geographical areas which are known to have selenium deficient soils can be the reason. Therefore, any weak calf, can rationally be given an injection of selenium (combination of vitamin E and selenium as BoSe). A necropsy on a calf which dies and shows whitened muscle, especially the heart, is highly suspicious of white muscle disease. As prevention, cows can be given a MuSe (adult formulation) 2-3 weeks before calving (this also helps prevent retained placenta). During the early weeks of life, a farmer can give any calf a BoSe injection or give homeopathic selenium in low potency (3X) every day for a few weeks in the drinking water or milk bottle.

UMBILICAL HERNIA/ABSCESS

Sometimes a calf will have a swelling under its belly right in the area of its navel. There are basically two conditions which give rise to this — hernia or abscess. An umbilical hernia is an inherited disorder and if the calf grows up to be a reproducing female cow, her offspring have a greater likelihood of having the same condition. An abscess is due to an environmental contamination of the navel in the first few days of life when the umbilical cord has not yet

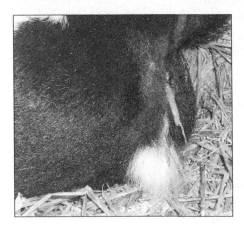

Congential oddities (above, left and right) and cogenital oddity: Atresia coli — only mucus from rectum.

dried and shriveled up. To know the difference between the two is critical to proper resolution. A hernia, which is basically the intestines dropping out through a too large umbilical ring, will be able to be reduced (pushed back up in through the ring). An abscess will not. The abscess will be more painful to the calf, especially when being felt and pushed upon. To definitely know if an abscess exists, I usually pour some alcohol on the lowest spot and plunge a needle in to see if pus drains out. If it does, it is an abscess. These can go away with time, but if large, they are better to be opened and flushed with 3% hydrogen peroxide. To do this I clean shave a small area at the lowest spot, clean the area with alcohol and make a small "X" with a scalpel blade. (Have someone "tail jack" the calf.) Then I drain out the pus contents and then flush liberally with hydrogen peroxide. The farmer can follow-up with the hydrogen peroxide flush twice daily for about 2 days, then use tincture of iodine to help dry it up. Using *Hepar sulph* (10X) or *Silica* (30C) twice daily for about a week will help dry it up a little quicker. A hernia is treated differently. If not too large of an umbilical ring (1-2 fingers wide at most), a belly wrap with Elasticon will usually work, if left in place for two weeks and a new wrap is used for another two weeks. Use of *Calc carb* is said to help this condition if using a belly wrap. Use of hernia clamp or surgery should be done for larger hernias. Do not let a hernia continue un-addressed. It will enlarge as the animal grows and at some point the intestines may become incarcerated and twist when they are hanging down in the sac. The animal will then die. This is entirely preventable.

ULCERS IN CALVES

Baby calves can get stomach ulcers, usually due to withholding hay from their diets. Much to the consternation of nutritionists which advocate with-

 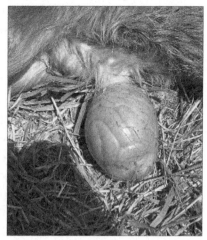

holding hay to calves until weaning, the all grain and milk replacer diets drop the acidity of the little calf's system to levels which are precarious. Anything which then tilts the system into a problem, will do so much more easily in these hay starved calves. In addition, some bacteria are thought to be associated with stomach ulcers. In addition, normal clostridia that inhabit the gut can raise havoc in these calves if given a chance. Ulcers in calves are unfortunately only found post-mortem. Signs may be grinding the teeth, if not sudden death due to a perforated ulcer. To put it simply — feed hay to calves.

RESPIRATORY PROBLEMS

Cattle that develop lung problems may develop chronic, recurrent problems later on. For example, a calf that has respiratory difficulty early in life is prone to show problems in the sum-

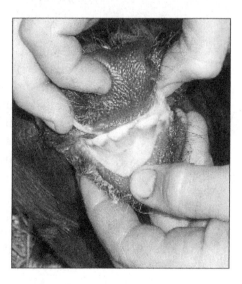

Calf with intestines prolapsed through umbilicus (above, close up, left) and ring below teeth showing toxic state (bottom).

mer heat as a bred heifer, if it is not treated in a timely manner. To help calves get a good start, raise them in a sunny spot outside the barn but in an area protected from strong, direct winds. Individual hutches minimize pneumonia in calves, in my experience.

With calves, entire groups will usually display respiratory symptoms to some degree. In weaned calves, the calves with the best body condition will generally withstand infection and those that are poor condition, especially if parasitized by stomach worms, will show the worst symptoms. Obviously, a cough will tip most farmers off. But is the cough dry or wet? Is it deep or a mild hacking? Is it only when rustled up or even when resting comfortably? How do the ears look? Are they perky or drooping? Generally speaking, calves that are still eating, have a hacking dry cough, perky ears and low grade fever (not above 103+/-) will do OK with treatments shown below. However, calves that have droopy ears, a deep wet cough at rest, don't eat and fever of 104+ will need antibiotics. Normally it is the one really sick calf that grabs the farmer's attention, while all the rest in the group can do fine with natural treatments. Antibiotics are not evil (as some individuals seem to firmly believe), they just should be used extremely sparingly and only when absolutely needed. Vaccination for some of the viral respiratory pathogens may be warranted in young-stock or animals approaching a stress period (shipping, calving) since it is the viral respiratory pathogens which set the animal up for deadly secondary bacterial pneumonia.

Another age of animals that seem to get pneumonia is fresh 1st calf heifers. These animals can get it anytime of year and it is probably due to joining the milking herd and being in a stuffy air situation and a new feed ration (along with the new stress of lactation for the first time in their life). If the farmer is truly alert and takes fresh cow temperatures daily, these animals can respond well to the Nasalgen or TSV-2 intranasal vaccine, Immunoboost, vitamin A, D, E and the herbal antibiotic tincture along with 1-2 bottles of vitamin C IV — if given early enough. I used to fear pneumonia in certified organic animals due to the ban on antibiotic use. However, the aforementioned treatment has really lessened my angst regarding this disease *if caught early enough.* This is probably the main disease that needs absolute attentiveness by the farmer for quick and aggressive intervention with natural treatments.

The animal's temperature should always be checked, at least twice daily at milking times, in order to stay on top of the animal's situation. Farmers that have kept respiratory symptoms in check generally have dosed the animal with one or more of the following remedies:

1) Tincture of garlic, echinacea, goldenseal, wild indigo and barberry, 10-20cc 3 times daily for adult cows (also IV in dextrose); use 5-10 cc 3 times daily for calves.

2) Aconite Ø 3-5cc in mouth for cow (or 2-4cc IV in dextrose) or 1cc in mouth for calf every hour or two until less restless and fever drops.

3) Passive antibodies (Bo-Bac-2X/ Quatracon/ BoviSera) 60cc/100 lbs. subcutaneously or intravenously. Giving 500cc subcutaneously is probably enough at one time. Repeat in a day or two if needed.

4) Immunoboost intramuscular or subcutaneously, 1-2cc for a calf, 5cc for a cow.

5) *Antimonium* for first sign of cough, especially in calves.

6) *Bryonia* worth a try whenever a cough is present.

7) *Spongia* for a wet cough.

8) *Ferrum phos* for restlessness, fever and red, ulcerated mouth.

9) *Pulsatilla* for yellowish-green mucoid nasal discharge.

10) *Phos* for nosebleeds and bright red, slow-clotting blood.

11) *Aconite* for harsh cough, restlessness and fever, especially from cold, dry winds.

Sudden onset of coughs. Aconite Ø in calves has worked well with Immunoboost and TSV-2.

12) "BUB" *(Belladonna/Urtica/Bryonia)* a good starter for early respiratory problems.

13) Vitamin C, 250 cc IV, or 5cc/100 lbs. in muscle once daily for 3 days (give no more than 30cc per injection site).

14) Vitamin E & selenium, or vitamin A, D & E, 10cc in muscle one time.

15) Colloidal silver IV has mild antibacterial qualities and is safe to administer if given only once or twice.

Conventional: Antibiotics.

Essential oils

- All essential oils are antiseptic, increase resistance to disease, promote elimination of toxins and parasites and revitalize by giving new energy.
- Using a diffuser at the rate of 2-5cc/hr for 2 hrs a day or 12-24 hrs a day as curative.
- Essential oils can be added to food at 5-10 drops a day.
- Revitalizing treatment can be carried out at 2-3 times a year for prevention.

(Grosjean, 1994)
Respiratory blend for diffuser or for internal use (makes 1 liter):

EO	Volume (cc)
Eucalyptus	450
Pine	150
Thyme	50
Rosemary	50
Terebinth	200
Tea Tree	100

5 drops/50 kg. animal twice daily in food for 3 weeks as preventative or 1 week in acute cases

Anti-Infectious Blend

EO	Volume (cc)
Thyme	100
Wild marjoram	150
Lavandin	300
Rosemary	300
Cinnamon bark	50
Basil	100

Anti-bacterial: Thyme 1% solution (p.93-94)

(Herbal, 1995)
(p.124) Usnea lichen "Old man's beard" (has usneic acid — strong respiratory and urinary antibiotic)
(p.127) Yerba Santa leaf
(p.103) Garlic
(p.111) Lomatium root
(p.116) Pleurisy root
(p.97) Chaparral leaf
(p.153) Compounded Echinacea/Goldenseal (antibiotic, antiviral, antibacterial): Fresh Echinacea spp., fresh goldenseal root (hydrastis), fresh Oregon grape root (Berberis aquifolium), fresh barberry root bark (Berberis vulgaris), fresh St. Johnswort flowering buds (Hypericum perforatum), Propolis (Bee harvested tree resin)
Initial: 2-3 tsp. in small amount of water, follow with 1-2 tsp. in warm water every 2 hrs. for a max of 5 days

(Nuzzi, 1992)
Coltsfoot (Tussilago farfara), Elecampane root (Inula helenium), Licorice root (Glycyrrhiza glabra), Peppermint leaf (Mentha piperita), Pleurisy root (Asclepias tuberosa), Stillingia root (Stillingia sylvatica), Wild Cherry bark (Prunus serotina), Yerba Santa leaf (Eriodictyon spp.) (p.37)

(Dadd, 1897) To produce a sweat:
Lobelia herb   2 oz.
Spearmint   1 oz.
Boiling water   2 qts.
Mix; add 2 tbsp honey; give ½ pint every hour. Also give fresh air (p.123)

(Steffen, 1914)
Sparteine sulfate (from Broom), such that one ounce represents 20 grains; give an ounce every three or four hours during the day. The attack probably will terminate in 4-5 days with the temperature approaching normal. (p.72)

(Waterman 1925)
Fluidextract aconite   1 ½ drams
Fluidextract belladonna   1 oz.
Fluidextract colchicum seed   1 oz.
Saltpeter   2 oz.
Water to make pint
Shake; give 2 oz. 3-5 times daily (p.367)

Mustard applied to chest (p.370)

(M.R.C.V.S., 1914)
Cough ball:
1 dram each — Camphor, Digitalis, Gum ammoniacum, Nitre, Hard soap, Licorice, Ipecac powder, powdered squill, chloral
3 drams — Ground aniseeds (p.195)
Chill drench (given in a warm gruel):

Camphor	2 drams
Pimento	1 oz.
Ginger	1 oz.
Nitre	1 oz.
Gentian	1 oz.
Salts	6 oz. (p.202)

(Grosjean, 1994)
Vit. C im muscle, 20 grams, every 2 hours
Vit. B12, 10cc, daily and oral/injectible Vit. A, D, E
Vit. E, orally, for 2 weeks

(de Bairicli Levy, 1984)
Two days honey and water only. Then fluid milk, molasses, honey, slippery elm powder; 2 tbsp. of each stirred into 1 qt. warm milk.

No solid feeds while high fever present.

Garlic — 4 roots shredded and brewed in 1 pt. water, a pint drench being given twice daily. Or, 6 4-grain garlic tablets given twice daily.

Pine tree shoots and spruce make a good internal drench.

If high fever continues, use strong brew of yarrow herb.

Steambag/nosebag with hot, wet bran and grated friar's balsam or oil of eucalyptus. 2 oz. of friar's balsam to the nosebag or 1 tsp. of eucalyptus oil stirred into the bran. Tie around nose and horns and well steam out the nostrils.

Make balls of 2 tsp. grated camphor, 3 tbsp. molasses and a little barley flour to bind them. Makes 8 balls. Give 2 balls twice daily.

Also use horse blanket and fresh air. (p.229)

Decision Tree for Coughing Animal

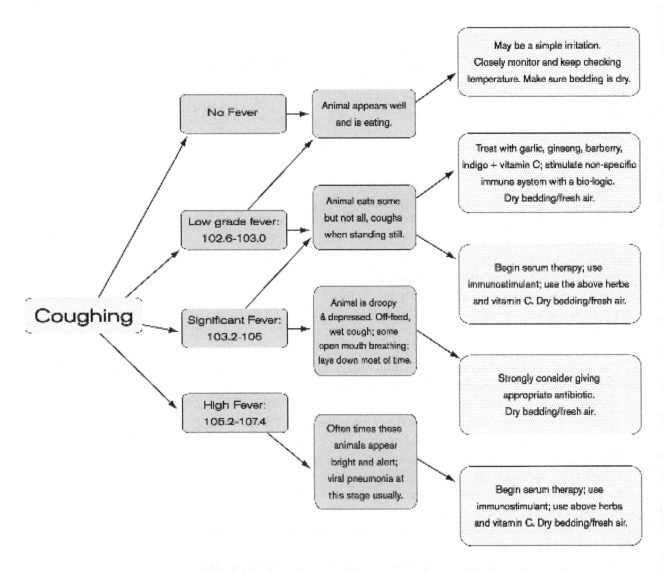

No Fever → Animal appears well and is eating. → May be a simple irritation. Closely monitor and keep checking temperature. Make sure bedding is dry.

Low grade fever: 102.6-103.0 → Animal eats some but not all, coughs when standing still. → Treat with garlic, ginseng, barberry, indigo + vitamin C; stimulate non-specific immune system with a bio-logic. Dry bedding/fresh air.

Significant Fever: 103.2-105 → Animal is droopy & depressed. Off-feed, wet cough; some open mouth breathing; lays down most of time. → Begin serum therapy; use immunostimulant; use the above herbs and vitamin C. Dry bedding/fresh air.

High Fever: 106.2-107.4 → Often times these animals appear bright and alert; viral pneumonia at this stage usually. → Strongly consider giving appropriate antibiotic. Dry bedding/fresh air.

Begin serum therapy; use immunostimulant; use above herbs and vitamin C. Dry bedding/fresh air.

Case Study: Severe Outbreak of Pneumonia in a Certified Organic Herd

This true case illustrates major points that I have made throughout earlier parts of this book.

I received a call from a farmer who said he had a bunch of coughing cows. They were mainly first calf heifers which had arrived on the farm many weeks previously. Some of the older cows were also coughing, but not too many. He said the affected animals were "slow" and not eating well. I said that I should perform physical examinations to determine the extent of illness, but he already had let all the cows out to pasture for the day. It sounded like "shipping fever" by the description of the symptoms and the history of new animals arriving on the farm. Shipping fever is dangerous and highly contagious; it starts as a viral pneumonia (usually IBR or PI3) and quickly gives way to a secondary bacterial infection (Pasteurella), which will kill cows if left untreated. Since this was a certified organic farm, I knew the first treatment choice would not be antibiotics. I'd always feared bacterial pneumonia on organic farms because of the potential for quick demise of animals. In any event, we discussed the options: either Naxcel (an antibiotic designed specifically for shipping fever) or do a whole host of labor intensive natural treatments, namely, using intranasal IBR/PI3 vaccination, intramuscular injections of the biologic, Immunoboost, intramuscular injections of a combination vitamin A, D, E and oral administrations of herbal antibiotic tincture. Why vaccinate sick animals? The intranasal IBR/PI3 vaccine is the only kind of vaccine which can be used in the face of an outbreak due to its stimulating non-specific immunity by increasing the gamma-interferon levels along the respiratory mucosa. After a little discussion the farmer asked for me to drop off the supply of natural treatments so he could use them (but not any Naxcel), which of course I did that day.

A few days later, Friday mid-afternoon, the farmer called and asked me to come out and check on his younger brother's herd, which were all just brought together as a new herd next door. They were all coughing, some worse than others. Indeed, the younger brother's cows were sick. Upon physical examination, many had muco-purulent nasal discharges, muco-purulent ocular discharges, temperatures ranging between 103.2-105° F, increased respiratory rates with shallow breathing and rough, wet lung sounds in the lower lung areas, were slow in eating and obviously depressed. About 25 cows were in this condition of the 40 total. Interestingly, they were all on the south side in a beautiful, brand new curtain barn — the side which faced the other brother's farm a hundred yards away. The younger brother's new herd was definitely ailing; the worse part was that he had just entered into the certified organic world, having recently sold his conventional herd at another location and now had this new herd in a brand new barn. All the cows were at least eating some, occasionally chewing their cuds, and none were dehydrated.

I felt bad for the younger brother as he basically had no idea how to go about treating a whole bunch of cows having obvious pneumonia without being able to use Naxcel/ceftiofur (otherwise they would all have to be sold due to the organic Rule). Just before coming to the younger brother's, I had stopped by at the first farm to see how those animals were doing. They seemed better and the older brother indicated that they were improving. This gave me some confidence. So after examining the sickest cows, I left the intranasal vaccine (TSV-2), the Immunoboost and the vitamin A, D, E for all 25 cows at the younger brother's farm. I also left the herbal antibiotic tincture (garlic, golden-seal, echinacea, baptisia) and aconite tincture (a bit hesitantly due to its power — but he insisted) for the worst 6 cows. He liked the idea of using the herbal antibiotic tincture since he knew of the ingredients as a remedy for pneumonia while growing up. We put one cow on Naxcel since I felt it was the absolute worst one clinically (from the lung sounds). Although she would have to leave the herd, I knew she would get better and live. I wasn't at all sure about the outcome of any of the other ones. I also left epinephrine, in case there was an allergic reaction to the vitamin A, D, E or Immunoboost (epinephrine is OK for organic use but only for anaphylactic reactions). I knew the younger brother would be a very busy person for the next few hours after milking time.

The next day, Saturday, the older brother called and said that he had talked to a phone consult veterinarian who hadn't examined any of the cows but prescribed homeopathic aconite and echinacea for use in the water bowls. Upon my asking, I found out that he had not told the phone vet what plan I had already put into place the previous day while I was on the brother's farm for two hours examining the cows and discussing options. And although we already instituted using aconite tincture and echinacea tincture the day before, I agreed to drop off the homeopathic versions of these medicinal plants to satisfy everyone that we were doing all that could be done. While there, I checked on all the cows I had examined the day before, and they were still all alive, although not too much better. I also found out that, despite my leaving written directions of exactly which cows were to get the herbal antibiotic, the younger brother had given all 25 cows two doses of it, but now none was remaining. So I left more of the herbal antibiotic tincture for follow-up on the 6 worst cows for a few days.

Over the rest of that weekend, I kept thinking if there was anything else that I could do, short of using antibiotics. Then I had remembered a biologic product, Bo-Bac-2X, which is a commercial source of passive antibodies for *Pasteurella multocida, Actinomyces pyogenes, Salmonella typhimurium,* and *E. coli.* (OK for organic). So I picked up 10 bottles at the local farm store on Monday and stopped by (un-announced) at the younger brother's in the afternoon after my other farm calls were finished. First thing I did was walk in front of all the cows to see how they were doing. None were worse, which was a relief, but they weren't too much better, either. The only problem was that now some of the cows in the north facing row were beginning to cough and show the same symptoms as the others from the previous week. When the farmer came in the barn he was happy to see me, for no cows had died yet and he liked my concern (believe me, it was genuine). So we decided to treat the north side the same as we had initially treated the south side (intranasal vaccine, immuno stimulant, and vitamin A, D, and E — as well as give each of the sickest cows (by his determination) a bottle of the Bo-Bac-2X passive antibodies. I left a renewed supply of the herbal antibiotic tincture. The cow treated with Naxcel three days earlier was doing fine, but now she had to leave the herd due to the federal organic law.

Over the course of the week, I stopped by there two more times to check on the cows, although the farmer wasn't there. The third time I stopped by, the farmer remarked how the cows seemed to be getting better — they were "starting to act more like regular cows." That was a relief — for when I had visited the two times previously, I thought the coughing had decreased, but couldn't be sure and chalked it up to wishful thinking. At this point, about 10 days had transpired since my initial examination of the cows — none had died yet and there was now visible improvement. A feeling of success was starting to take hold, which I hadn't let happen yet, since some six years previously, I had witnessed 8 of 40 cows die within a week's time from the same condition at a different farm (I had used only homeopathic remedies and intravenous vitamin C on those cows).

But there was a new twist to the situation — about 5 first calf heifers had just freshened and were now in the barn, and they were beginning to get sick. By this time we knew what we were going to do and so we set out to give each of the 5 a bottle of Bo-Bac-2X, Immunoboost, and vitamin A, D, and E. At least we were getting them earlier and these animals had been given the intranasal vaccine the week before (when the others were). One particular animal was very sick, so I decided to give her 2 bottles Bo-Bac-2X and 250cc 3% H_2O_2 IV after the Bo-Bac IV. During the H_2O_2, she started breathing very heavily, so I immediately stopped. Both the farmer and I stared at the cow for minutes on end wondering if she would begin breathing normally again. After about 5 minutes she was stabilizing and so we began treating the other 4 with the Bo-Bac (one bottle each). After finishing up giving a bottle of Bo-Bac to another cow I was walking away to wash out my IV she all of a sudden dropped over — anaphylactic shock. So I *ran* for the epinephrine which I had already left off for the farmer and gave her a shot — she started breathing again, regained consciousness, and got up in a few minutes. As I was leaving, the farmer thanked me again for all the help that afternoon. I said, "Sure thing!" as I usually do and drove off. Needless to say, I was quite worried about the two animals which showed severe stress from the treatments that afternoon. So I stopped back again the next day, wondering if the "H_2O_2" cow was still living. She was! And to make things even nicer, the farmer said how something seemed to really work since these 5 cows were right on track again. By this point all the cows had gotten back on track and none of the cows, except one, had to leave the farm because of antibiotic use. Unfortunately, the "H_2O_2" cow and a cow with a combination of pneumonia and toxic metritis died a few weeks later. However, from this case, we can see that a multi-prong approach was the correct course of action for nearly the entire herd. As far as costs go, the farmer had a very steep bill that month. However, ceftiofur for all the cows actually would have cost more than the natural treatments and he wouldn't have had any cows left for organic use. Best of all, he gained confidence in using natural treatments and still had cows producing certified organic milk.

BLOOD LOSS

Acute blood loss is seen usually when there is a cut, especially at the milk vein or an obvious vein on the udder. A cow can bleed out and die in a rather surprisingly short time. The most urgent thing to do is to stop the bleeding until the vet can get there to stitch the cut shut. Temporary measures include using a clothes line pin to clamp it shut or holding your hand up against the cut. I would recommend the clothes line pin so you can make the call to the vet. Minor cuts, such as seen sometimes along the tail or other areas (when the blood is only a steady drip rather then a stream of blood like from the milk vein) can benefit from a liberal amount of cobwebs. Also, homeopathic *Phos* or *Arnica* can help. Wonder Dust would also be good (all its ingredients are OK for organic use). Sometimes, I have been called to a cow that seems to leave blood puddles around her every day for a few days. This is usually due to a cut which needs stitching but gets scabbed over somehow (usually by dirt). Eventually this will open up, and a massive hemorrhage will take place. That is why I usually scuff up the area with my antiseptic (chlorhexidine soaked — OK for organic) gauze pads to fully expose the entire cut before stitching it. Usually only a few stitches are needed. Cows are very much different from horses in regards to cuts; cows don't get cut often probably due to their thicker hide, but when they do it usually involves a blood vessel.

If there is substantial blood loss, as evidenced by pale mucus membranes with weakness and rapid breathing, either many liters of lactated ringers solution will be needed to expand the volume of blood to retain normal circulation to all vital organs, or a blood transfusion will be needed. Blood transfusions in cows are relatively simple — and only a veterinarian should do this procedure! A blood transfusion is warranted if the cow is too weak to get up and is breathing with difficulty after significant blood loss. To do a blood transfusion, first select a cow, preferably a relation but this is not critical with cows, that is near the end of her lactation since this is a relatively stress free time. Then dump out 2-4 one liter size bottles and add in 10cc of sodium citrate (heparin is not allowed in organic animals). Then take a 12 or 14 gauge by 2 inch needle (an IV needle) and jab the milk vein of the donor cow while someone jacks her tail. Then allow a free flow stream of blood to fill each liter bottle, swirling them the entire time to keep the anticoagulant mixed in well. If there is enough labor available, the donor can be drained from each side at the same time in order to quicken the procedure (it usually takes 5-8 minutes to fill a liter). Then administer to the recipient cow through a standard IV line and 12 or 14 gauge needle.

This will brighten up the recipient almost immediately. Sometimes I have opted to give a few liters of lactated ringers solution after stitching (instead of a transfusion), only to do a blood transfusion a day or two later when the cow simply needed more red blood cells for normal activity to resume.

Cows which are bleeding internally are not good candidates for blood transfusion since it is not possible to stop the bleeding and it may simply continue on and on. This would be in the case of an abomasal ulcer as evidenced by black tarry manure (melena) or a cow with a bleed occurring into the udder. Emergency salvage of these cows is usually the least loss to the farmer.

ANEMIA

It is accepted by many that most lactating cows, especially high producers, are anemic. I can believe this. If I notice pale mucus membranes and a heart murmur in a sick cow that I am examining, I usually think of parasites (especially in young-stock), sub acute ulcer (in high producing cows), poor kidney function or old age. I have not given a blood transfusion in these instances. I mainly rely on Hematinic as a blood builder with its colloidal minerals and large amounts of iron and good nutrition. If parasites are the inciting problem, the environment must be addressed otherwise there will be recurrence in the same animal or more likely in future animals occupying the same place. In organic farming, systemic problems such as coccidia or worm infestations occurring every year on the same farm raise red flags with certifiers since it is obvious that the farmer is not addressing the root cause of the problem and only using band-aid measures.

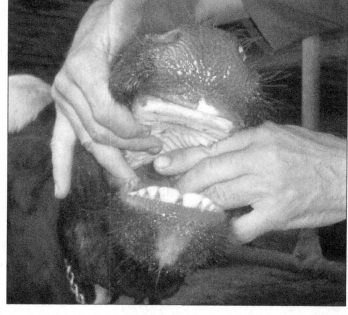

Very pale gums indicate anemia.

CANCER (Leukosis, Leukemia, Enzootic Bovine Leucosis, EBL)

Cattle can suffer from cancer just as any animal can. Unfortunately, there is a virus (bovine leukemia virus, "BLV"), that can be transmitted cow to cow by *a single drop of blood*. About 85% of the dairy cattle in the United States are thought to harbor antibodies to the virus. In other words, they have been exposed to it at some point in life. However, it is well known that it depends on familial genetics whether or not an animal will actually become cancerous. In other words, immunogenetic strength or weakness will determine the outcome

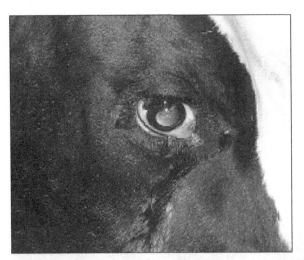

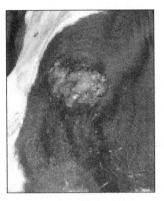

of an animal that is exposed to the virus. In general, only 1-3% of the animals exposed will get clinical cancer. New born calves can be exposed to it via colostrum and milk.

The symptoms can be quite plain or they can be very vague. The main symptoms are tumorous growths which are visible or palpable anywhere over the body, but mainly at lymph nodes. The supramammary lymph nodes are commonly visualized and palpable. The pre-femoral and pre-scapular lymph nodes also are occasionally enlarged. On rectal palpation, lymph nodes can be felt internally in the abdomen. Sometimes, it will affect an eye. Early on, the third eyelid (nictitans) will have a small growth on it which, over time, becomes larger and larger, eventually creating an infected, swollen eye. Sometimes, doing surgery for a displaced abomasum, I will see small tumors on the stomach wall and alert the farmer to their presence. These are the most common places. However, small tumors in the spinal cord are also a real possibility. And so, with older down cows, this must be considered as a possibility. But, unless I can palpate or see tumors, I really dislike diagnosing "cancer" in a cow, because of lack of physical confirmation.

Early cancer tumor of eye. Full blown cancer of the eye (and infected), right.

Obviously, cancerous tumors, especially by the time they are diagnosed out in the field, usually keep progressing and ultimately the cow dies from it. There can be, however, spontaneous remission, but don't ever count on it. I have observed this only one time when I palpated a large tumor at the internal iliac node on a fresh cow whose leg was bowed in (see picture). I thought the farmer would ship her right away, but he didn't. So at the next herd check I asked if I could reach in her to check on the tumor since she was standing better and it had shrunk to near normal. That was the only time.

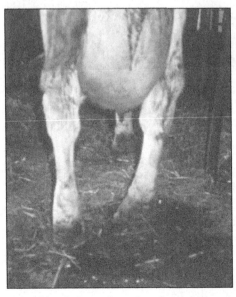

Since there is no cure for cancer, the best thing to do is try to prevent it from spreading. Since only one drop of blood transmitted to another cow is all it takes to infect a cow, individual sleeves should be used for rectal palpation, individual needles should be used for injections, and scrupulous detail to fly control is paramount. Flies can transmit the blood easily. They take a bite from an infected cow, the cow swats at the fly with her tail,

Bowed leg due to enlarged node resulting from bovine leukemia.

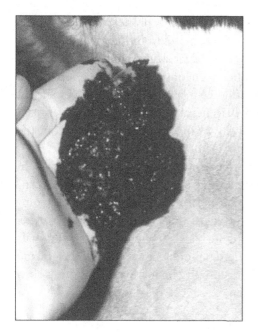

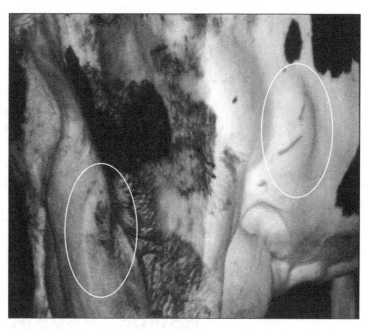

Black, tarry manure (melena) indicates an abomasal ulcer or a bleed in the digestive tract.

Circled areas indicate enlarged lymph nodes associated with bovine leukemia.

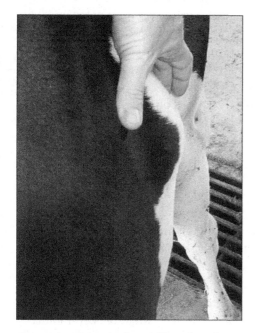

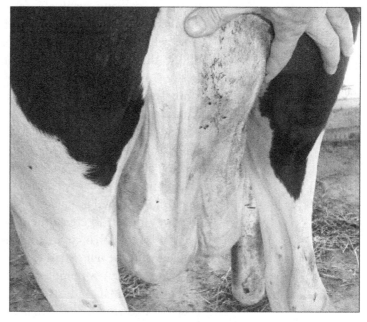

Cancer involving prefemoral lymph node.

Cancer involving supramammary lymph node.

the fly goes to the next cow with the wet blood on its proboscis and proceeds to bite her, thus infecting her with the still wet blood from the first cow. Unless all three measures are taken seriously (individual sleeves, individual needles, serious fly control), the chances of reducing enzootic bovine leukemia from a herd will be very difficult. It can be killed by pasteurization. Calves should either receive colostrum from known BLV negative animals or be given pasteurized milk on those few farms actively trying to eliminate BLV from the herd.

There are certainly as many cancer types as there are different cell types. In other words, cows can get bone cancer, kidney cancer, brain cancer, gum cancer, skin cancer, ovarian cancer — but all these are exceedingly rare as compared to the enzootic bovine leukemia virus, due to its contagious nature. Perhaps these other cancers are not seen much also due to the younger animals that are in production these days. As a herd has older cows (usually the grazing herds) that haven't been burned out but get to live longer lives, conditions such as cancer (all kinds) are seen more often.

BOVINE SPONGIOFORM ENCEPHALOPATHY (BSE/MAD COW DISEASE)

Bovine spongioform encephalopathy (BSE) is a disease which affects the brains of cows and eventually causes them to become uncoordinated and have odd behavior. Now that BSE has been diagnosed in the United States, down cows will likely become suspect of carrying the disease, especially if acting oddly. Although this is tragic for any country's livestock trade, organic farmers in the United States need not worry so terribly since the practice of feeding animal by-products to ruminants is strictly banned on USDA Certified Organic farms. The prevailing theory is that the disease is caused by the re-feeding of rendered diseased bovine nerve tissue to bovines. This practice has been for the most part banned in the United States since the late 1990's. However, the feeding of blood meal (a cheap protein source) and tallow (an energy source) has not been completely banned. It does not take a rocket scientist to entertain the idea that blood is a possible source of the disease. Why? Well, if the prions that are thought to create the disease are initially ingested into the intestinal tract, it is likely that circulating blood transports the prions to areas in direct contact with the nerve tissue. As stated in Chapter 1, cows were created to *eat grass* and occupy a unique agro-ecological niche in that respect (along with other ruminants). Industrialized, corporate agriculture cannot claim innocence in this realm, as the *modus operandi* is to feed whatever commodity is absolutely cheapest, no matter what the source. Hopefully, all the governments of the world will completely outlaw the feeding of any animal parts or by-products to herbivores. This will be a gigantic step, a step in the right direction as far as protecting the health of livestock and people. As Rudolph Steiner (the founder of the Biodynamic movement) correctly predicted back in the 1920's, creating cannibals out of herbivorous cows will cause them to become crazy. How right he was.

However, there is an alternative theory as to why BSE has surfaced. This has more to do with pollution and pesticides creating mineral imbalances in the soil and affecting nutritional qualities of native vegetation. Apparently, diseases similar to BSE (Scrapie in sheep and Chronic Wasting Disease in elk and deer) show geographical clustering. In these areas, the soils are high in manganese and low in copper. Normal prions in the brain require certain levels of copper and when the copper is not supplied, manganese can replace it, especially if it is in high concentration in the diet. Both are metals with known electrical conductivity, however, their conductivities are not identical — copper is an efficient conductor of electrical impulses while manganese holds on to electrical charges. In effect, they do opposite things. The theory then goes on to suggest that loud noises create shock waves of sound which energize the manganese, causing free radical damage and eventual brain damage.

Whichever theory may be correct (perhaps both), it is obvious for those of us in ecological agriculture to see that holistic management which focuses on proper balanced nutrition, whether it is nutrition of the soil or of the cow, is paramount to maintaining and creating a healthy way of life.

PARASITES

Parasitism can kill animals, if left unchecked. I believe parasitism is the "weakest link in the chain" as far as organic livestock farming systems go. If you are concerned, it is always best to have a manure sample analyzed to see which type of worms are present and the quantity. If a heavy infestation is present, it may best to use conventional therapy to eradicate the gross imbalance, but then use botanicals to prevent recurrence. In addition, many dairy farmers have successfully prevented internal parasite burdens on cattle that are intensively grazed by adding Diatomaceous Earth "DE" (sea sediment)

Filthy conditions — parasitism and ringworm.

to their grain mixes. This is done at the rate of 8-10 lbs. per ton throughout the grazing season, either continuously or "pulsed" every two weeks. It is very dry

Parasititzed blind calf (from pinkeye).

and should not be top-dressed (unless liquid molasses is mixed in) due to its potential to irritate the respiratory tract.

Hematinic is a colloidal mineral-tannin-iron solution and is an excellent product. It is mined in Mississippi and is very high in iron and tannins, as well as many minerals. The tannins act to cause gut spasm and expel parasites while the iron helps to build up the blood which was lost to the unfriendly "guests." Given at the rate of about 5cc per calf per day orally for 7 days, I have documented dramatic decreases in the worm egg load by observation under the microscope. The product is very nasty to taste and should be diluted with an equal part of water and molasses or syrup to make it palatable.

Conventional: Anthelmintics as needed (fenbendazole, ivermectin, eprinomectin, moxidectin). Ivermectin is OK for organic if other management and treatments acceptable to organics fail.

Calves and parasites

At about weaning age, and especially thereafter, calves generally are very susceptible to parasite (worm) infestations. The worms most harmful to the calf are called nematodes and generally live in the abomasum (true stomach) and intestines. They rob energy and blood from the calf and can lead to deadly anemia if not handled correctly. Although I am not certain of specific numbers, I would say that nearly all young animals are potential carriers of these parasites. Why? Because of the way calves are housed there is often a constant supply of worm eggs and or larvae available in the animals' bedding which they naturally nibble. If you casually observe calves for any length of time, you will see that their foraging instinct is already at work as they pick through and nibble at their bedding. As they ingest their desired fibrous material for their developing rumen, they also inadvertently ingest parasite eggs and larvae. Once ingested into the correct host, these larvae will develop into mature worms and repeat the cycle of life, at the calf's expense, until another generation of eggs and larvae are deposited once again into the environment with the calf's daily manure output.

So that the reader gets an idea of parasite burdens that an animal may harbor, a description of how a veterinarian diagnoses an infection is described here. First, about a teaspoonful of fresh manure is needed (taken directly from the animal or quickly recovered when the animal deposits it on the ground). The sample is then either refrigerated (so the parasite eggs won't hatch) or prepared for analysis. About 1 gram (teaspoonful) is placed in a small vial to which a sodium nitrate solution is added to fill the vial, and then a microscope cover glass is placed upon the liquid surface. The salt solution will force any parasite eggs to float. After 10-12 minutes the cover glass is gently lifted and placed upon a microscope slide glass. This is then placed under the microscope to examine. The slide is scanned in a systematic way, observing all areas for type of parasite eggs and total quantity of each (if any). Worm eggs (*Ostertagia, Haemonchus* and *Trichostrongylus*) and coccidia can be visualized. If from this one drop of liquid 10 eggs or less are observed, I then call it a mild infection. If 10-25 eggs are observed, I deem it to be a moderate infection

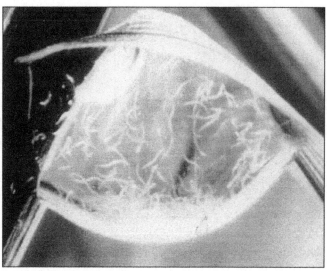

A dewdrop of grass with free living stomach worm larva.

and anything greater than that is a heavy infection. I have occasionally observed 70-80 eggs on one slide — that is severe! Remember that the eggs I am observing under the microscope *are derived from only 1 gram of manure.* Just think about the total parasite load there is in all the manure contained within the average 35-70 feet of gut length in the calf or cow!

It is nearly impossible to prevent exposure to parasites, but to help guard against parasite infestation it may be wise to keep calves on whole milk up to three to four months old in order to get really good body condition. This also can be accomplished by using nurse cows that have high somatic cell counts. Whatever you do, do not abruptly wean calves (especially from nurse cows) as their beautiful body condition can rapidly wither away, leaving the animals weakened and more susceptible to parasitism than may have otherwise been the case.

Unfortunately, calves don't have the same bodily reserves that adult cattle have in terms of being able to live in balance with a certain level of parasites. This is because the immunoglobulin which is specific for parasites, IgE, is not adequately developed until about 6-10 months of age (depending on nutritional status). The worms can overwhelm the young animals during the following spring, especially if carried over the winter (hibernated in the stomach lining). There will be an onslaught of highly active worms emerging from the stomach lining when they sense the correct conditions with the arrival of spring. At this point two things can happen: either the calf will suffer from a severe anemia and die, or an explosive amount of eggs and larvae will be deposited in the animal's environment. Thus, using a wormer at a strategic time will help prevent this. Generally, it would be wise to use a wormer as a calf begins its first autumn of life and then again twice, three weeks apart, in early spring. If potentially doing "calf work" in summer, it still is wise to worm the

Severe parasitism and blind from pinkeye.

calves at that time if it is their initial time. Calves with a heavy worm burden generally look pot-bellied (also called "hay belly"), have a rough hair coat and will have any black hair look a bit reddish-brown.

Homeopathic remedies may help, but use them in low potency in order to get material amounts of the remedies into the system. Remember that parasites can be present in large numbers (especially if the animal actually looks wormy) — therefore we need to use material amounts of remedies to "fight fire with fire" rather than just the energy essence of a homeopathic remedy. Botanicals, as used by the Eclectic and Regular schools, are probably the most rational to use as far as natural treatments go.

Coccidiosis

1) Hematinic, 5-10cc once daily for 7-10 days

2) Keep growing areas clean and dry. *Rotate areas.* Do not let moist manure build up. Put lime down (Barn-Grip, etc.) to alter the floor environment and help break parasite cycles.

3) *Ipecacuhuana* 30C with *Merc cor* 6 30C pellets 2 x 7.

4) See treatments for internal parasites below.

Conventional: Medicate feed with coccidiostats or use amprolium for actual treatment.

Nervous coccidiosis

This is more often seen in beef calves but could occur in dairy calves as well. Infestation with coccidia in severe winter weather may give signs of severe tremors, constant lateral movement of the eyes, "star gazing" (the head is stretched back and slowly but continuously is swaying from side to side) and not being able to rise. Thiamin deficiency (vitamin B1 deficiency) resembles this condition but it is usually caused by lack of dietary fiber or high sulfur content in the water. Nervous coccidiosis animals usually die if these signs are seen due to the nerve damage, although it is not known how the coccidia infestation leads to the nerve damage. Treatment with amprolium and thiamine may be of benefit (amprolium has not been approved for organic farms). Obviously, prevention is the best treatment.

INTERNAL PARASITES

When giving internal medications like parasiticides to ruminants it is probably a good idea to *give about 5cc weak copper sulfate solution before giving the medicine* in order to close the esophageal sphincter and deliver the medicine to the abomasum (the true stomach).

Hematinic — 10-15cc once daily for 7-10 days.

Eclectic Agents Acting upon Intestinal Parasites (Anthelmintics) (Elligwood, 1919)
Artemisia pauciflora (Levant wormseed)
Spigelia marilandica (Maryland Pink)
Chenopodium ambrosioides (American wormseed)
Aspidium felix-mas (Male fern)
Convolvulus scammonia (Scammony)
Mallotus philippinensis (Kamala)
Brayera anthelmintica (Kousso)
Punica granatum (Pomegranate)

Severely parasitized calf — had 80 stomach worm eggs per drop of manure.

(Nuzzi 1992)
Parasite Formula: Black Walnut Hulls, Butternut Bark, Chaparro, Echinacea Root, Garlic bulb/seed, Goldenseal Root, Marshmallow Root, Quassia, Wormwood Leaf: for 1-2 weeks (p.69)

(Herbal, 1995)
Fresh Wormwood, Quassia bark, Fresh Black Walnut Hulls, Neem leaves, Bilva bulb, Embelia ribes, Eclipta alba, Phyllanthus amarus, Gentian Root, Fresh Ginger Root
30-40 drops; 3-5x/day in water. 6 days on, 1 day off
Avoid sugar, refined carbohydrates

(Grainger and Moore, 1991)
Fresh Garlic, ground Cayenne, fresh/dried/powdered Eucalyptus/Rosemary/Rue/Wormwood, ground Pumpkin Seeds; if fresh 4-5 tbs of each or 3 tbs dried or powdered
$1^1/_2$ cup oat or wheat bran
$^1/_2$ cup water
2 tbs Slippery Elm bark powder

(Lust, 2001)
Pinkroot/Carolina pink, Worm grass, Starbloom, Indian pink; Use dried rootstock powder, 4-8 g., twice daily for several days, then a strong laxative to remove worms

Combination: Equal parts: pinkroot, aniseed, male fern, senna leaves and turtlebloom. Steep 1 tsp of mix in 1 cup boiling hot water. Take 1 cup at a time (p.309)

(de Bairicli Levy, 1984)

General wormer: 6-8 drams aloe (1 dram=1/8 oz.) — 1st giving 24 hrs cold bran mashes

Tapeworms: (1) 24 hour fast, (2) in evening, 2 oz. turpentine and 1pt. linseed oil or molasses, 1 hr later give large warm mash of bran and molasses. 4 days later, 1 pt. linseed oil without turpentine. Repeat in 2 weeks if needed.

Garlic AM & PM: 3-4 grated roots of garlic to bran/molasses mash, for several weeks. Or give in PM for 2 weeks: 6 4-grain garlic tabs.

48 hour fast, then PM and AM dose of 3-4 grated garlic roots bound with flour and honey; followed 1 hour later by a drench of 1 pint linseed oil (or 4 pints molasses) stirred into a mix of equal parts of warm water and milk — one pint, with 8 tbs. castor oil added.

Be feeding pumpkin, mustard, seeded parsley, nasturtium, melons, etc.

Powdered red cayenne pepper mixed with powdered wormwood (4 tsp. cayenne to 2 tsp. wormwood) made into balls with honey and flour. (p.254-255)

(Mowry 1986)

Garlic, Black walnut, Butternut (bark of root) (p.229)

(Dadd 1897)

Garlic root, 1 oz. boiled in 1 pt. milk given in AM 1 hour before feeding (p.300)

Powdered wormseed, skunk cabbage, ginger (equal parts); 1 tsp. AM & PM (p.233)

(Waterman 1925)

Tapeworms — ½ oz. Oil of Male Fern twice a day in a pt. of milk for 3 days, then on day 3 give 1½ lbs. of Epsom salts.

Roundworms — 2 drams sulphate of iron 3 times a day, mixed in moistened grain, then on day 3 give $1^1/_2$ lbs. of Epsom salts. Oil of Turpentine (1oz.) with pt. of milk or raw linseed oil, or santonin in 1 dram doses in feed for 3 days, then on day 3 give $1^1/_2$ lbs. of Epsom salts. Calves — $^1/_4$-$^1/_2$ dose of above. (p.397-8)

(Alexander, 1929)

Sheep: 1% solution of copper sulphate — $1^3/_4$ oz./lamb and $3^1/_2$ oz./sheep after fasting 16-18 hrs.

$^1/_2$ oz. Lugol's iodine + 1 qt. water 2-4 oz. /lamb

1 oz. Lugol's iodine + 1 qt. water 4 oz. / strong lamb or sheep

2 oz. Lugol's iodine + 1 qt. water 4-6 oz./badly infested sheep

Give 1 oz. Epsom salt 9 hrs. later

Tapeworms fresh powdered kamalaà 1 dram/lamb & 2 drams/sheep

Liver flukes Essential Oil of Male Fern (p.94)

Equine: Fast 36 hrs. then 1-2 drams each of dried sulphate of iron and tartar emetic, each AM for 6 mornings in 1 pt. dry feed. Give 1 qt. linseed oil at end of 6 days OR

2 parts of salt, 1 part each: dried iron sulphate, Tartar emetic, flowers of sulfur (yellow sulfur powder) and powdered fenugreek. 1-4 tsp. AM & PM in wet feed for 1 week. Wait 10 days, then repeat if needed. (p.54)

Pigs: Santonin, 3 times a day (PM, AM, noon), 5 grains. 12 hrs. later Epsom salt in tepid milk: repeat as needed.

Oil of Chenopodium: 20-30 drops/50 lbs. along with 1 oz. each of castor oil & raw linseed oil, after fasting 16-24 hours: feed a little just before treatment (p.108).

(Burkett, 1913)

Salt saturated with spirits of turpentine (1 gill to 4 qts. salt). Box with tobacco stems free choice, 1 inch lengths. Add wheat bran occasionally on top of stems. No young worms can stand the diet, except tape worms. Won't kill mature worms, just prevents multiplying. (p.251)

(M.R.C.V.S., 1914)

Roundworms — turpentine & santonine

Tapes — Areca (p.204)

Extract of Male Fern	30 drops
Venice Turpentine	2 drams
Santonine	40 grains
Aloes	4 drams

Give above mix after 24 hrs. fast (p.196)

Emetic tartar	1 dram
Santonine	40 grains
Iron sulphate	40 grains
Powdered chamomile flowers	1 dram
Salt	1 oz.

in 1 pt. bran mash after 20 hrs. fast (p.198)

Turpentine and linseed oil, santonine, thymol, tobacco, aloes and tartrate of antimony (p.38)

(Dun, 1910)

Start by fasting 12-15 hours.

Tapeworms — 2 dram asafetida, 1 dram savin, 30 drops male fern extract. OR areca nut, 3-5 drams male fern, kamala, kousso, pomegranate root bark, turpentine, chloroform.

Roundworms — same as above and also: santonin, oil of chenopodium, bitters, arsenic, lysol, creolin.

Strongyles — turpentine and essential oil of thymol, tannin, Lysol, carbolic acid, napthol, turpentine oil, with enemata of common salt, ferric chloride solution, lime water. (p.118)

Copper sulphate no more than 12 drams for cattle & horses and no more than 1 dram for sheep or swine. (p.244)

Iron sulphate ("copperas") Tonic: 1-2 drams FeSO4 and $^1/_2$ oz. each ginger and gentian in pint of ale or gruel. Give at feeding or shortly after. (p.257)

Iodide of sulfur: parasiticide and stimulant. (p.321)

Topical: 4 parts iodine & 1 part sublimed sulfur & heat until liquefies. This is soluble in 8-10 parts glycerin.

Topical: 2 parts sulfur & 1 part K2CO3 dissolved together with heat, in 10-12 parts lard or oil, give weekly if needed. First soak and wet existing scabs then apply. (p.322)

Borax $^1/_2$-1oz. dry (p.205)

Pigs: Santonin (Artemisia, Wormwood), give with aloe or jalap. $^1/_2$ oz. mixed with food is enough for 15 young pigs. (p.612-14)

Oil of Chenopodium in castor oil (p.614)

Topical: Stavesacre seeds (Staphisagria semina)

1 part bruised seeds boiled with 20-30 parts water, for 2 hrs. (p.614)

Kousso — Fast first, then 2 doses with interval of hrs, followed by purgative. Put in honey & peppermint water. (p.615)

Topical: Camphor in oil or Vaseline/oil of turpentine (p.633)

Topical: Sanitas (oxidized oil of turpentine/camphoraceous bodies/hydrogen peroxide) (p.635)

Thymol — strongylosis of foals. (dissolved in glycerin and alcohol, once daily for 4-5 days, followed by laxative). (p.637)

Balsam of Peru (Myroxylon Pereirae) (p.651)

(Udall, 1943)

1 oz. aloe or enema of soap and water or vinegar and water and quassia chips. (p.215)

Equine strongyles: Oil of chenopodium, spirits of turpentine, linseed oil, mineral oil. (p.213)

16cc oil of chenopodium, followed y 1qt. linseed oil after 36 hr. fast. OR 60cc turpentine followed by 1 qt. linseed oil, fast for 12-24 hrs. OR 8 gm. tartar emetic in feed for 5 days. (p.215)

(Winslow, 1919)

Copper sulphate (p.148)

Amebic dysentery (coccidia/ giardia/protozoa):

Ipecac with opium: 1 powder/bolus/pill every 4 hrs. (p.351)

(de Bairicli Levy, 1984)

Worm when moon is waxing and near full — worms are stirring, breeding and easier to dislodge.

Bot flies (topical): 8 lbs. tobacco powder + 4 lbs. fresh lime steeped for 2 days in 1 gal. water. Strain through course sack and apply with stiff brush. (p.226-227)

1-2 garlic plants, finely chopped and mixed with flour and treacle (molasses) into a size of walnut. If bad, use linseed turpentine: lambs — 10-50 drops turpentine + ½ oz.linseed oil + pinch ginger. Adult sheep — 80 drops turpentine + 2 oz. linseed oil. Max for sheep is $^1/_4$-$^1/_2$ oz./dose. Can soak oatmeal with turpentine + linseed oil. OR for 1 sheep use 2 or 3 4-grain tablets once daily.

Mustard seed (black or white): 2 oz./lamb in milk is a safe vermifuge. Can let mustard be grazed.

Feed additives: raw grated carrot, raw desiccated coconut and seeds of: pumpkins, nasturtium, papaya, grapes and melon. Raw grated common radish and horseradish and turnip with charcoal. (p.166-167)

Plant to eat: mustard, wild turnip, couch grass, brambles, green fern and broom tops. (p.253)

Equine: Garlic — AM & PM. 3-4 grated roots of garlic to a mash of bran & molasses for several weeks OR 6 4-grain tablets in PM for 2 weeks. Feed plenty of pulped carrots.

Topical: Rub garlic brew or oil of camphor or eucalyptus. (p.254)

Fast 2 days, then AM & PM 3-4 grated garlic roots bound together with honey or molasses, followed in 1 hour by a drench of 1 pt. linseed oil or substitute 4 parts cane molasses into equal parts of warm water & milk — then add 8 tablespoonfuls castor oil + seeds of pumpkin, nasturtium, papaya, grapes and melon; then graze. (p.255)

Aloe, 6-8 drams aloe juice after 24 hrs. of cold bran mash.

Equine tapeworms — 24 hr. fast (including water), then wineglassful (2 oz.) turpentine mixed in 1 pt. linseed oil or molasses substitute, 1 hr. later give large warm mash of bran & molasses. 4 days later, give 1 pt. linseed oil with no turpentine. Repeat in 2 weeks. Can put turpentine on sugar cubes.

Equine tapeworms — Male fern, give as extract in capsules. 3 x human dose. (p.255)

4 tsp. Powdered cayenne with 2 tsp. powdered into honey-flour balls. (p.255-256)

Calves — 1 qt. buttermilk (good and mild worm expellant) OR wormwood herb/purging (mountain) flax/nettle foliage boiled in whey.

Lice — Buttermilk + vinegar in equal parts, well rubbed in. Some brew of quassia chips can be added in to mix. (p.249)

Common plant derivatives for internal parasites:
Areca, garlic, pumpkin seeds, black walnut hulls, quassia, wormwood, ginger, thymol, chenopodium, male fern, kuosso, aloe, santonin, turpentine

(Burkett, 1913)
Equine: Largeworms —

Oil of turpentine	2oz.
Oil of male fern	½ oz.
Castor oil	4 oz.
Pure raw linseed oil	8 oz.
Milk	1 pt.

Give above mix after 14 hour fast, repeat in 1 week, then follow with "Worm powder":

Common smoking tobacco	8oz.
Powdered wormseed	6oz.
Powdered sulphate of iron	4oz.
Salt	½ lb.
Sugar	½ lb.

Give above mix in AM, before feeding, give a heaping teaspoon of powder in 4 qts. wet wheat bran for 10 days (p.201-202)

(Grosjean, 1994)
Worms — worm every two months:
Aromatic worming blend:

Essential Oil	cc
Bergamot	15
Caraway	15
Wild marjoram	10
Sassafras	20

Give 5 drops of mix 3 times a day to food for 3 days along with fresh garlic

(Burkett, 1913)

Lungworms in calves/lambs: Common spirits of turpentine with salt and sulfur (gill of turpentine to 4 qts. of common salt) + ½ pt. sublimed; place in a covered box from which animals can lick

Intratracheal injection 1tsp. turpentine, chloroform, olive oil mix (p.214)

(Udall, 1943)

Coccidia — astringents and protectants:

tannic acid   1 oz.

bismuth   4 oz. in warm milk

Catechu powder 2 tbsp. in 1 pt. milk, 2-3 times per day/400 lbs.

Mineral oil   1 quart daily

Severely sick ½-1 liter blood transfusion; 1-2 liters 0.9% NaCl with 5% dextrose; epidural lidocaine if needed (p.168-169)

EXTERNAL PARASITES/ SKIN PROBLEMS

There are not too many common problems of the skin in dairy cows. The most common are lice, mange, ringworm and warts. Care must be taken with ringworm as it is contagious to people. Ringworm is a fungus and lives on the most outer layer of skin, which is shed continuously. Therefore, with or without treatment, ringworm will usually disappear on its own in about a month or two. The ultraviolet rays of the sun help turn over the skin more quickly and that is why "Dr. Green" (green grass) is often the treatment since it needs to be sunny for the grass to be growing. Usually heifers in

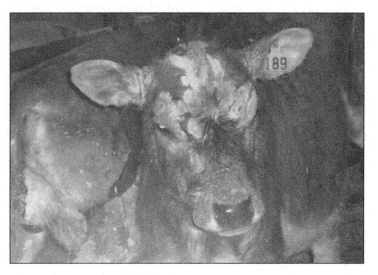

Calf with severe ringworm and mange.

a box pen in the winter time will show signs of ringworm, which will disappear over the course of the spring. However, if adult milking cows start showing ringworm (or if a solitary animal in a group shows a skin problem like the heifer with mange in the picture on the following page), I usually consider poor immune system capabilities and try to stimulate the immune system either by using biologics or improved nutrition.

Ringworm:

1) Scrub with Betadine™ solution. Wear rubber gloves!

2) *Bacilinum* 200C one time; can repeat in a month.

Sever parasitism and blind from pinkeye.

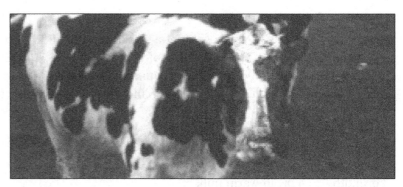

Generalized mange in bred heifer possibly reflecting a poor immune system. None of the other animals in this group displayed similar symptoms.

3) "Dr. Green" — sunshine and green grass as much as possible.
4) Tea Tree Oil rubbed or sprayed on to the area. Tea Tree Oil has strong antiseptic qualities and is ideal as a natural treatment for skin problems. *(Warning: do not give Tee Tree Oil internally)*
5) Powdered sulfur (+/- tobacco dust) rubbed into area.

Warts are usually due to the papilloma virus and generally appear near the neck and ears, although on a bull they sometimes will be on the tip of the penis. They are contagious and can spread from animal to animal, especially if instruments are not sanitized between animals (see picture of warts in heifers' ears — all at the tattoo and metal tag sites). If they are large and have a stalk, they can quickly be pulled off. If they bleed a lot, they may need a stitch to stop them. Warts located on a bull's penis need to be surgically excised. If not, they can become large and also get infected. Sometimes cows have warts on teats which impede normal milking. These either need to be cauterized or somehow removed. However, I have had successful results using homeopathic *Thuja* for the 'cauliflower' type warts on teats and *Sabina* for flat type warts. I would recommend giving these 3-4 times daily for about 5-7 days if electing to use this route. There are wart vaccines, however, I do not have any experience with them.

(Waterman, 1925)

Teat warts — Olive oil applied thickly after milking; acetic acid (vinegar) once every 3-4 days, if many do only a few at a time (p. 439)

(Grainger and Moore, 1991)
Lice
Pyrethrum powders
Essential oils — Anise, camphor, eucalyptus, pennyroyal, pine rosemary & sassafras: 1 part of each with 2-3 parts olive or other oil; rub in well
Place cellophane wrap on for 2 hours, then betadine scrub shampoo
Repeat 1-2 times in following 10 days to get survivors (p.108-111)

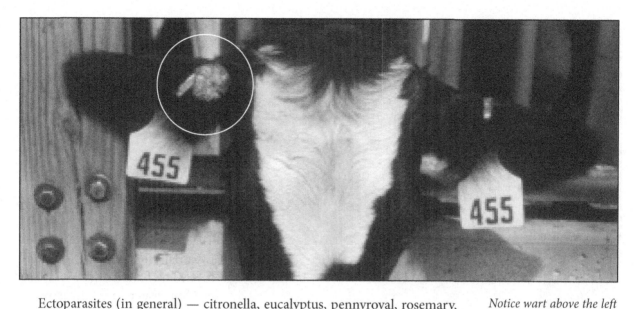

Ectoparasites (in general) — citronella, eucalyptus, pennyroyal, rosemary, rue and wormwood (p.11)

Mange — "Lemon lotion": lemon rinds + 1 qt. boiling water, discard pulp & sponge areas needed

Ringworm — Tea tree oil dabbed onto ringworm patches repeatedly for at least a week

Chronic skin problems — Comfrey root + aloe vera gel, rubbed into balding spots or dry, itchy, scaly patches (p.33)

Notice wart above the left metal ear tag.

(Dadd, 1897)

Lice — Wash AM & PM with powdered lobelia seeds (2 oz. in 1 qt. boiling water); let stand a few hours and apply with sponge (p.196)

Mange — give internally:

Powdered sassafras	2oz.
Powdered charcoal	1 handful
Sulfur	1oz.

Mix, then divide into 6 equal parts, give AM & PM until finished.

(Waterman, 1925)

Internally for ectoparasites:

Common salt	10 lb.
Sulfur	2 lb.
Saltpeter	$^1/_2$ lb.

Mix & put in box free choice (p.714)

(Alexander, 1929)

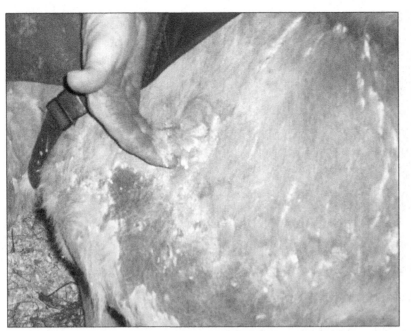
Haircoat lifting off the skin.

Powdered pyrethrum (in cold water). Blanket after treatment, and brush a few hours later (preferably outdoors) OR sulfur (burning) fumigation of entire body. (1929 USDA No. 1493 "Eradication of lice, ticks and mange"). (p.48)

Ringworm — soften areas by saturation daily with sweet oil, cottonseed oil or castor oil — then remove them without bleeding. Then rub strong iodine into area twice daily for 3-4 days OR if many spots, 4 oz. bluestone (CuSO4) in 1 pt. hot water. 1-2 times/week. (p.79)

Lice — found mainly in dark, dirty, damp, hot and badly ventilated stables. Dip with coal-tar dip OR powdered sabadilla OR powdered sabadilla, powdered pyrethrum, powdered tobacco leaves or snuff and finely sifted flowers of sulfur (elemental yellow sulfur). Blanket the animal and groom outdoors a few hours later OR groom with a hard brush dipped in raw linseed oil. (p.74)

(Burkett, 1913)

Ringworm — (1) ash with warm solution of 1oz. potassium carbonate + 1 qt. water to remove crusts. (2) 2 tsp. iodine + 4 tsp. Vaseline mix rubbed on.

Mange is not noticed when on grass. Destroyed by direct sunlight almost immediately. Live 7-10 days in protected places. Wash premises with 5% carbolic acid (phenol).

External: "salted" with 1 lb. flowers of sulfur + 10 lbs. common salt.

Lice (can be seen with eye): Spray premises with kerosene.

Boil stavesacre seeds, 1 part to 20 parts water, for 1 hr. & let simmer for another hr., then add water to make it up to the original volume. Repeat application in 7-10 days.

To make dip — 3 lbs. flowers of sulfur, 2 ½ lbs. unslaked lime, 15 gallons water. In making the unslaked lime into a thick paste, sift in the sulfur and stir well. Put mix in a kettle with 5 gallons water and boil for better than ½ hr. When chocolate looking mass settles, the clear liquid is drawn off and water enough is added to make 15 gallons. More effective if warm (just above body temperature). Apply with scrubbing brush, clot or sponges and all scabs and crusts should be thoroughly saturated.

(Udall, 1943)

Lice — Lime sulfur liquid brushed on; retreat in 2 weeks.

Raw linseed oil, 1 pt/4 cows, applied with stiff brush.

Spray: 5cc Blackleaf 40 tobacco to 1 gallon water (nicotine sulphate — kills on contact). (p.321)

Mange — "Sulfur ointment":

Sulfur	1 part by weight
Soft soap	1 part
Lard	5-6 parts

Above mix is an effective dressing after 1st soaking in soft soap with plenty of warm water.

Lime-sulfur dip:

Sulfur flowers, 5oz. + Freshly slaked limestone, 1 oz in 2 gallons water, let boil 1 hour. Use the orange colored fluid (smells strongly of H2S). Will kill mange but repeat in 5-8 days to get eggs which hatched. (p.150)

(Winslow, 1919)

Ringworm:

Tincture of iodine; glycerite of phenol; creoline; creosote; chrysarobin ointment; cantharidies; croton oil; formalin; salicylic acid; boric acid; thymol; sulfurous acid

Mange:

Sulfur; lime-sulfur dips; SO_2; Tar; Crude petroleum; Peruvian balsam; styrax; phenol; salicylic acid; cantharides

Lice:

Staphisagria; oil of tar; Peruvian balsam; styrax; oil of anise; creolin; tobacco; pyrethrum; coal-tar preps; creosote preps. (p.57)

Staphisagria ointment (1-2%) (powdered in Vaseline). (p.44)

Tincture of larkspur (1 part larkspur seeds to 16 parts alcohol).

Fleas: Pyrethrum (dried root of Anacyclus pyrethrum). Tincture 1-4% in 10 parts water.

Mange or scab in sheep (preventative and curative)

Manufactured tobacco	1 lb.
Flowers of sulfur	1 lb.
Water	5 gallons

The tobacco is soaked in cold or tepid water for 24 hours and, on the night before dipping, the solution is brought to a boil for a minute and the tobacco allowed to remain in it overnight. Mix the sulfur in a pail to the consistency of a gruel with water. Strain and press the liquid out of the tobacco and add it to the sulfur and enough water to make 5 gallons. After dipping, the sheep must be turned into a clean yard or barn to drain. (p.314)

(de Bairicli Levy, 1984)

Lice, Mange, Ringworm (air excluding glazes are best):

Raw lemon juice (not bottled or canned) or light paste of egg white and liquid lime.

Sheep (cows are 3 x sheep):

Tobacco powder/dust	1 lb.
Elder leaves	4 handfuls
Geranium leaves	2 handfuls
Quassia wood chips	1 handful

Bring to boil & simmer for 2 minutes, then allow to brew for 6 hrs. do not drain, rub well into body.

Afterwards use dry tobacco

Derris powder (has rotenone) for mild cases either dry or as a dip.

Also be giving abundant garlic for internal consumption.

Lice and ticks — Strong solution of derris, preferably powdered OR eucalyptus oil.

Into sores — Apply strong solution of garlic leaves and roots. Garlic internally and eucalyptus externally. Dust with derris root and quassia chips.

Tobacco dust	1 lb.
Elder leaves brew	2 qts.
Eucalyptus oil	1 tsp. (p.174)

Conventional: Ivermectin, eprinomectin, moxidectin; be wary of organophosphates.

FLIES

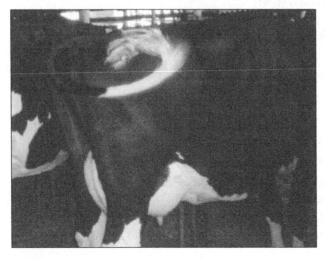

A tail at work on a nice clean cow.

The main principle behind fly control is to reduce moisture. It is this simple: moisture + manure = flies. In humid geographical areas when the summer heat gets into high gear, flies can be torture to livestock. Flies can best be controlled around your barn area by keeping the manure more dry than moist. It is known that keeping moisture in manure to less than 40% will significantly decrease fly populations. This can be done by adding straw or other bedding to it (this will also increase your carbon to nitrogen ratio which helps soil microorganisms decompose manure applied to the land); adding rock phosphate to it (which will help soil fertility); hauling out manure from the gutter more than the typical once a week in the summer; composting; and using the tiny parasitic wasps that specifically feed off emergent fly pupae. Keeping youngstock well-bedded will also help. Pens with "soup" for the animals to stand in are the most offensive. Try also to keep the manure area covered so rainfall doesn't keep things moist. During those hot and humid summer

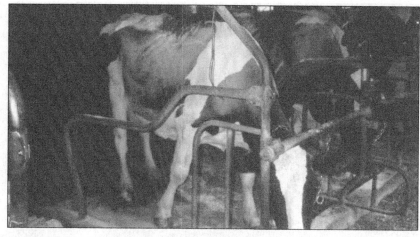

A cow liberally dusted with field lime — keeps the animal looking dry (flies are attracted to moisture) and it is cheap.

weeks, especially after some thundershower activity, fly populations can explode. Intensively grazed cows can help keep fly populations down around the barn due to the manure not building up in the barn and barnyard in the first place. In my experience, liquid manure pits act as fly magnets. It is not the pit itself, but the slow moving slurry of manure leaving the barn via gravity that attracts flies.

The best and most effective alternative to commercial sprays that I've ever seen is the use of pulverized limestone. This is liberally dusted onto the backs and sides of the cows, enough to make a black cow look almost white. It helps keep the cow from showing moisture on the skin, and thus she won't attract flies. I couldn't believe how few flies were on cows in a particular certified organic herd on a hot and humid summer day, especially when I had just been in a barn dense with flies. If the cows then go out in the rain, it will wash off onto the pasture which is usually helpful. It needs to be applied as often as needed; however, field lime is very cheap in general. Also good are the sticky lines that get reeled in — they can collect enough flies to get filled up in half a day sometimes! Simple crates that the cows walk through (without electricity) can also be made so that flies essentially get trapped behind screens as the cows brush by the screens. A pheromone attractant behind the screens certainly would be good as well. Window sill box traps with pheromones are another weapon to attract flies and trap them. Using spray mixtures of essential oils seem to work nearly as well as pyrethrums. But no matter what, the key word in fly control in your barn environment is to keep things as DRY as possible. Ventilation will also go along way in keeping flies at bay — tunnel ventilation is probably the best single method of non-chemical fly control within a barn. Usually, however, graziers don't invest in such equipment since the cows are to be outdoors. The strategy for fly control best typifies what I mean by a multi-prong approach to problems if not using quick-fix chemical solutions.

Keeping your cows as clean as possible will attract fewer flies than damp, filthy cows. In addition, keeping tails on cows will go a long way towards battling

Walk through box with screening and pheromones to brush off flies, attract them to the screens, and then trap them.

flies, at least as far as the cows are concerned. As for helping keep your animals remain comfortable during especially heavy fly times, you can always use commercial sprays. Certainly the essential oil sprays are the most appropriate for organic farms. As a herdsman, I came up with a mixture of essential oils that appeared to work fairly well when sprayed directly onto the cows, and it makes the barn smell pleasantly aromatic compared to the normal odors. Although it appeared to work, before using it a lot I would advise anyone to check to see if it gives off any aroma or taste to the milk. There are some formulas which you can make up on your own, however, there are some good commercial essential oil sprays now available on the market which certainly seem effective (as part of

an overall multi-prong approach). The best results I have seen are from using Ecto-Phyte from Agri-Dynamics — this formulation is allowable for organic usage. Whichever product you may choose, make sure that the carrier oil is allowable for organics. These can be sprayed directly onto the cows during milking time and also "self-administered" by weanling heifers out in their pasture. This is done by using a plastic 50 gallon barrel hanging from a frame with its own solar panel to power a small sprayer that the animal triggers when it puts its head into the barrel to get at salt or other free choice "bait."

If you are certified organic, you are limited in regular commercial products to the pyrethrum type sprays which give a knock down effect, with very little residual effectiveness. Flies will build-up resistance to any spray used repeatedly because of their quick

Fly trap box with a pheromone attractant underneath the dome-shaped screen.

reproducing abilities. Pyrethrums are allowed *if used infrequently and with other management corrections*. These cannot be "extended" (diluted) with diesel fuel and also *should not combined with petroleum products* (like piperonyl butoxide) on organic farms. I am amazed that someone hasn't yet come up with a pyrethrin spray that does not include piperonyl butoxide. Chrysanthemum powder itself could be used on the cows, but only in small quantities.

While researching old texts regarding common animal problems, not once did I come across any specific treatments for flies. I was rather amazed. Either the farmers simply lived with them during those times or the lower animal densities (and subsequently less manure) did not attract as many flies. Free-ranging chickens also probably helped by pecking through manure and eating insects or their pupa offspring. The only compound which seemed to be used would be pyrethrum powder ("Dalmatian insect powder").

Conventional: Insecticide fly sprays, pour-ons, fly-bait.

FLY STRIKE & MAGGOTS

This is an unfortunate situation usually found in calves that have either been dehorned with a gouging/chopping instrument or a calf which has experienced a rather prolonged bout of diarrhea. In any event, the area where the maggots are burrowing into must be thoroughly cleaned. Put on some latex gloves or dishwasher gloves and scrub the area well with soapy water until no further maggots are seen. Then use liberal amounts of hydrogen peroxide. Then use a pyrethrin fly spray in case you missed a few (not allowed in organic due to the piperonyl butoxide synergist) or an essential oil as a germicide (perhaps citronella). Clean again the next day and keep a close eye on it. These cases often pull through OK with daily cleaning for a few days. I would recommend the herbal antibiotic mix to be given orally a few times a day for a few days in a row (while cleansing the area daily).

Barrel feeder with solar powered sprayer (above). Close up of solar powered sprayer.

Conventional: Scrub and apply ivermectin (OK for organic).

PINKEYE
(and other eye irritations)

Pinkeye is usually the result of a classic case of a disorder of: stress, environment and nutrition. The following list of homeopathic remedies are given in

Early pinkeye.

order of increasing severity of the disease pinkeye; from a slight haziness to the eye with increased tear production all the way to an ulcer on the surface with a central area of pus ringed by red. In addition, frequent spraying (4-6 times per day) of the eyes with a non-alcoholic calendula spray mixed with *Euphrasia* 2X, *Hypericum* 30C liquid and boric acid (3%). Placing an eye patch may help decrease transmission of the causative organism (of pinkeye) by flies. This will also keep irritating light out and retain needed moisture for healing. But then again, I have seen eye patches actually trap flies and create a much worse situation than was initially seen. Try keeping affected cows in at day and allow out to graze at night. For any eye irritation, give one of the following and change as symptoms change:

1) *Euphrasia* 30C, 4 x 3 for eyes watering all the time, has small blisters on eye, hazy cornea and swollen eyelids.

2) *Hypericum* 30C, 4 x 3 for excessive pain, as shown by squinting of eye.

3) *Aconite* for red, inflamed eyes with swollen eyelids.

4) *Hepar sulph* 200C, 3 x 5 for eye with ulcer on cornea, pus in eye and profuse discharge with great sensitivity to touch and air.

5) *Conium* 200C, 4 x 3 for sensitivity to light and excessive watering of eye; intense squinting and pain.

6) *Kali bichromium* 30C, 4 x 3 for thickened discharge, ulcer on eye but not painful or squinting anymore.

7) *Silica* 30C, 2 x 14 to help remove or lessen "scar spot" after disease has resolved.

8) MaxiGuard pinkeye vaccine (for prevention and control).

9) Inject eye with hyper-immume plasma.

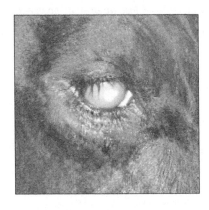
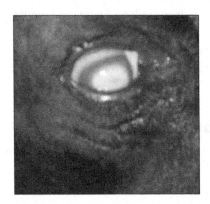
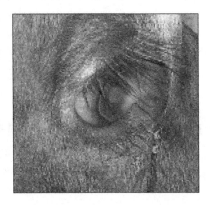

Conventional: Antibiotics in the muscle and/or inject under first layer of cornea with an antibiotic-steroid-mydriatic combination.

EAR INFECTIONS

Ear infections are occasionally observed, mainly in youngstock and rarely in adult cattle. The signs are a drooping ear with a fetid odor and usually a fever. It is important to treat these early as the infection can go to the inner ear which is in direct communication with the brain. If the calf is stumbling or off-balance, the inner ear is probably involved. Calves can get ear infections by having their milk or milk replacer work its way backwards into the eustacian tube. Calf milk infected with mycoplasma is the probable reason. Adult cows probably get it if they are lying flat out or thumping their head into a dirty puddle of water or a muddy area when they are weak with milk fever. Cleansing solutions with antiseptics (chlorhexidine) can be helpful if applied a few times a day. Squirt the solution into the ear, close off the ear and work it into the ear further, then let the calf or cow shake its head. Homeopathic pyrogen would be indicated for a fetid smell and fever. Infusing 15-20 cc of "silver water," twice daily for 3-4 days, has given nice results. The conventional treatment is antibiotics, usually an antibiotic tube into the ear as the topical and perhaps a systemic antibiotic if feverish, dull and depressed.

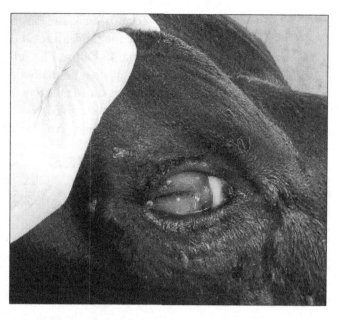

Pinkeye treated with injection of plasma.

PUNCTURE WOUNDS

1) Try to flush out the puncture with hydrogen peroxide right away.

2) *Ledum* orally and ointment at site.

3) *Hypericum* and *Ledum* early, if tetanus suspected. (Also get tetanus anti-toxin and consider penicillin before serious lockjaw sets in.)

4) If classic lockjaw occurs in a bovine, it is difficult if not impossible to get the animal over it. However, the author has seen a case of classic lockjaw in a yearling heifer be treated successfully in the following manner: a rumen trocar is screwed in the rumen to relieve the bloat and frequent doses of oligomeric proanthocyanidins "OPCs" are funneled into the rumen. This along with *Hypericum* 30C 4 x 5 brought the cure of a 500 pound heifer within a few days. The yearling went on to be a successful dairy cow.

(Steffen 1914, p.60)

Tetanus/Lockjaw: *Passiflora incarnate* as a sedative to abate nerve irritability. Use the fluidextract at one ounce every three hours, day and night, during the first 48 hours. The full physiologic effect should then be obtained. The effect can then be maintained indefinitely by giving an ounce every three or four hours during the day. This should be kept up until only an occasional dose is needed to be given.

Conventional: Antibiotics, local irrigation, promazine and anti-toxin.

ALLERGIES & INSECT STINGS

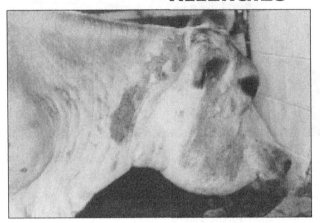

Swollen face and hives on cow with anaphylactic reaction.

Occasionally, cattle get an allergy, but never as frequently or chronically as horses. Usually an allergy is due to an insect sting or an injection reaction (most commonly vitamin E/selenium, B complex or vaccine). The symptoms are readily apparent: swelling and hives in the case of an insect sting or severe respiratory distress with the cow dropping to the ground due to anaphylaxis from an injection. In the case of anaphylaxis, epinephrine must be administered or the animal will die. (Epinephrine is a naturally occuring substance and is OK for organics). With insect stings and hives (urticaria), the homeopathic remedies *Apis mellifica* and *Urtica urens* are truly homeopathic. *Apis* is made from honey bee venom and is indicated when the animal likes cool or cold application to the swellings. *Urtica* is made from stinging nettle and is indicated when the animal likes warm applications to the swellings. Non-life threatening allergies tend to subside with time if the inciting cause is taken away.

Conventional: Antihistamines (and steroids if not pregnant).

LUMPY JAW & WOODEN TONGUE

These two conditions are rare (especially wooden tongue). The cause is usually a traumatic puncture injury to the mouth, either near the jaw or the tongue. Cows, being creatures that can eat ligneous forages, may sometimes chew upon a long splinter or something of the like. A cut due to a sliver of wood will inoculate the tissue with bacteria, giving either actinomycosis (lumpy jaw) or actinobacillosis (wooden tongue). Lumpy jaw is rather self-describing and can come on slowly, only ulcerating the skin of the jaw when it is advanced. Wooden tongue, on the other hand, comes on quickly. With an alert herdsman, the initial swelling of the lymph nodes below the jaw will be observed well before the infection has taken hold. It is at this time that sodium iodide given IV, once weekly for 3 weeks, will prove beneficial. Dosing with

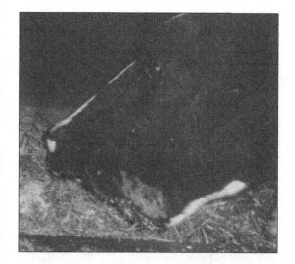
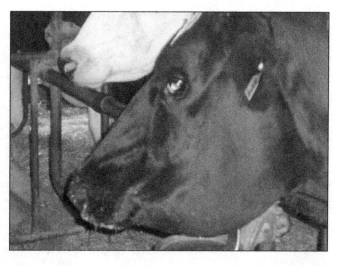

Lumpy jaw (left).
Oral infection — self-limit-ing or fatal? (right).

phytolacca (poke weed) either in very low potency or in herbal tincture form may help the lymph node swellings. Animals may recover from lumpy jaw with a permanent swollen jaw but do fine for many lactations.

Conventional: Systemic antibiotics +/- sodium iodide intravenously.

FORELIMB LAMENESS

Lameness in a forelimb which is *not* due to a hoof problem or some obvious inflammation is uncommon in cattle. A forelimb that appears normal but won't bear weight may well be due to nerve injury in the shoulder/elbow area or in a nerve coming off the spinal cord. Oddly, forelimb lameness is occasion-ally seen in a fresh cow that has been paralyzed in the back end due to calving paralysis. In any event, this kind of lameness does not have a good prognosis, if only because the inciting cause is usually unknown. I have had success using electro-acupuncture in these cases. Also, homeopathic *Curare* would be indi-cated. Putting a cow in a sling may be beneficial, or even an Aqua-Float if one is available. Sometimes, an inflammation will appear and this should be dealt with either systemically with IV colloidal silver, *Hepar sulph,* or antibiotics (penicillin) since the eventual outcome may be a large abscess.

Electro-acupuncture can be very helpful. Use GV-12, SI-14, BL-39, TH 10-02, LI-11, TH-4.

Conventional: Anti-inflammatories +/- antibiotics.

LEG ABSCESSES

Not infrequently, I see an abscess upon the back leg of a cow. These are usually the result of an injection through a dirty area of skin (manure). These abscesses generally are not painful, but they can become enormously big. Sometimes they will spontaneously open, however, more often than not, the farmer does not have the patience to wait and asks me to work on it. The first

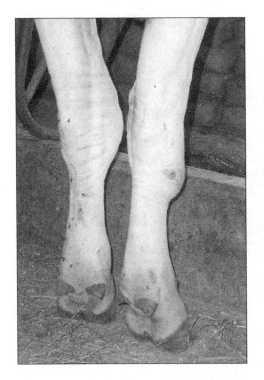

Massive quantity of pus being expressed from a leg.

thing that needs to be done is to aseptically determine whether or not the swelling is indeed an abscess. This is done by cleaning the lowest area of the abscess (where lancing may potentially take place) and thrusting in a 14 gauge needle to see if pus drains out. If it is pus, then it can be lanced. If it is not, it should *not* be opened. I usually shave the area first, douse it with alcohol and then use a scalpel blade and make an "X" at the lowest spot so gravity can help draw down the pus. Usually the flow starts immediately and then I help it by massaging it out. Then a few infusions of 3% hydrogen peroxide (60cc) are squirted into the opened abscess. The farmer should continue to do this for 3 milkings, then no more since hydrogen peroxide will hinder healing.

These abscesses usually dry up over time and leave a thickened, roughened appearance to the skin. Some farmers who like to use homeopathic remedies a lot usually opt to use either high potency *Hepar sulph* or *Silica* to help clear up the drainage. It is best to leave the lanced area open to drain out and not cover it over with a bandage. However, one farmer seems to help his cows by applying a bandage with goldenseal on it.

Conventional: Antibiotics.

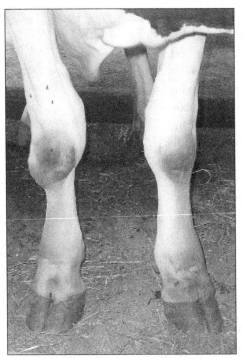

Healed leg infection — notice old puncture at back of knee (top) and slightly enlarged knee area (bottom).

SWOLLEN JOINTS/ARTHRITIS
(degenerative and septic)

Arthritis means inflammation of the joint. It can be mild or it can be severe and it can be either a sterile inflammation or a septic, purulent inflammation. Older cows will get arthritis and you can hear sometimes a clicking sound, especially of the hook bone area when they get up or with every step they take when moving the affected limb. There really isn't much to do for this except palliation, so pat yourself on the back for having a cow long enough in the herd that she gets "old-age" arthritis. One possible treatment would be MSM or chondroitin added to the animal's feed for an extended period of time. However, this may be cost prohibitive. Swollen joints can also be the result of hoof problems (see Hoof problems). Always consider it. Cows kept on concrete for longer periods will tend to have more swollen hocks than those exercised outside on real ground. Generalized swollen joints of all limbs can be indicative of a bacterial infection, especially in young calves. It is usually due to an initial umbilical cord infection giving rise to a systemic infection which settles into the joints. The prognosis is poor for

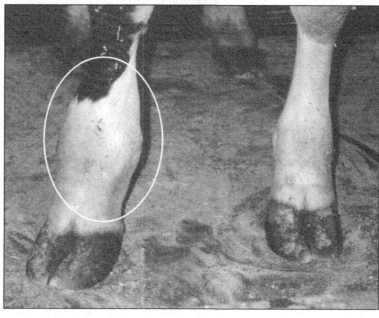

Swollen joint due to a twisted ankle out in pasture.

calves with this condition. Septic arthritis joint infections are usually due to trauma (puncture, tear, rip, pressure sore, etc.). Lyme disease is not recognized in cows but may occur. A swelling that is purple-blue and invading the fetlock area should be tended to quickly as it usually indicates either a deep seated abscess of the hoof deeply invading into the upper joint or possibly a small broken bone. A severely swollen long bone that is painful to the touch may indicate a cellulitis (sub-surface infection). This can be a disaster and usually will require an antibiotic like penicillin for the animal to stave off a generalized septicemia from the condition. If an animal is non-weight bearing, certainly call your vet. Herds that have many swollen hocks are usually kept on concrete stalls with inadequate amounts of bedding. With the advent of stall mattresses, these kinds of swollen hocks are a rarity (at least in my practice area).

Check for fever

1) Colloidal silver seems to help farmers well with infections of the skin and just below its surface. If there is a swollen leg with no obvious focal abscess to open, consider giving colloidal silver IV or "silver water" orally or intrave-

nously. It is not legal for a veterinarian to use colloidal silver, but creative farmers seem to be able to know how to on their own. The reason colloidal silver cannot be used is that *chronic* administration of it can lead to a graying of the mucous membranes. In using it in an acute case of cellulitis for a few days at most, such graying does not seem to occur. Colloidal silver has been historically used as a mild antibacterial solution.

2) *Hepar sulph* low potency (<12X) 3 x 5 to help open a painful abscess.

3) *Hepar sulph* high potency (>30C) 3 x 5 to keep a site open that is already draining.

4) *Gunpowder* (30C) 3 x 7 for extremely infected, swollen, painful joints.

5) *Myristica* (30C or 200C) 3 x 7 follows *Gunpowder* well.

6) Apply Strong Iodine (7%) to the swelling for a few days. This should help to dry out the skin surface so it cracks open to allow drainage.

7) Apply 20% icthammol twice daily for a week, especially for swellings at the hock.

8) If there is a swelling at the hock and it opens, flushing the inside with 3% hydrogen peroxide is of benefit early on. Use only for 2-3 milkings as it will hinder healing if used for longer.

9) If not infected and a soft swelling due to trauma, try *Apis* 4 x 4 and apply *Arnica* ointment locally as well as orally.

Conventional: Antibiotics.

BLACKLEG/MALIGNANT EDEMA/MYOSITIS

Blackleg is a highly fatal condition caused by a clostridia bug. A limb will be severely swollen, the animal off-feed and having a fever. There is no treatment for this once it becomes clinically obvious. However, penicillin at high doses may help cure it if given very early. It is much better to prevent this infection. This is usually accomplished by vaccination; the blackleg vaccine is one of the oldest and very effective. In my practice area, blackleg is rare and is usually due to a puncture wound, if the cause can be verified. Blackleg is endemic in the U.S. Midwest and West, much like anthrax, as the alkaline soils and flood plains help to germinate the clostridial spores. A more likely problem in my area is that of a severe cellulitis which really swells the leg, but doesn't make the cow too systemically ill and off-feed. Either IV colloidal silver or penicillin is needed in these cases in my experience. Often, the eventual outcome is either a large focal abscess or a chronically enlarged leg which regains almost total mobility.

Conventional: Antibiotics.

HOOF PROBLEMS

It is important to have the hoof lifted, cleaned and trimmed before deciding on the course of action. In my opinion, veterinarians who refuse to lift cows' hooves to assess the problem (and simply toss you some medicine instead) do not have the best interest of your animals in mind. Cows should have their hooves trimmed by a professional hoof trimmer at least twice a year. Those confined to concrete should be done every 3-4 months. Even intensively grazed cows will need periodic trimming. A safe way to lift a hoof is to use a come-along thrown over a beam; then get six loops of baling twine and create a slip knot just above the cow's hock; attach the come-along to the baling twine and jack the cow's hock up to just below the level of the pin bone. It's good to have somebody give the cow balance on the other side or have the cow's head haltered facing the opposite side and tied. Use some gritty substance for good footing for her

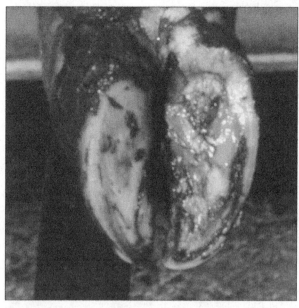

Solar ulcer, typical location. Due to rumen acidosis.

other three hooves. If the cow still loses her footing and slips, you can simply cut down the baling twine to allow her to fall fully so to not have her hanging by the leg. For a front leg, use a cow collar set to be jacked up as it is under the "armpit" of the cow. Front legs don't need to be jacked up more than about 6-10 inches off the ground to be worked on. Usually, lameness is seen in the rear leg. And in rear legs, it is the outer toe that usually has the problem. In front legs, the inner toe has the problem. This is because of the ways cows walk and how they bear weight as they walk.

Whenever cleansing a hoof after paring it to see what is the problem, tea tree oil is a great antiseptic and can be applied before wrapping the hoof with whatever ointment or powder you will be using. Tea tree oil may be valuable in spraying the heels of cows for strawberry/hairy heel wart during milking times. However, tea tree oil is extremely aromatic and care should be taken to not have milk take on any aromas that it comes near. Many farmers use copper sulphate as a dressing. I find this works only in early and mild cases of foot rot. As stated earlier, never use tea tree oil internally.

Do not be afraid to pare and nip away lots of hoof. I don't mind if some bleeding takes place — blood will bring new nutrients to the area and whisk away bad things. Also, always check to see if you can slip a finger between the sole and the inner hoof. If you can, keep paring away. There should be a tight junction between the sole and the inner hoof when normal. If there is space between the two, dirt and other matter can get in. In cattle, you must really open up the area because cows too easily can wall off an infected area quickly.

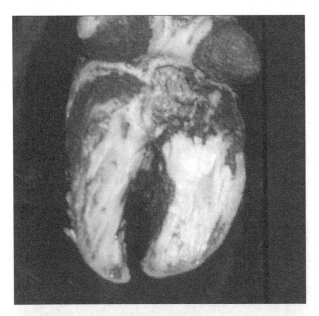

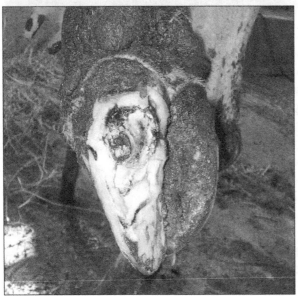

Strawberry heel, hairy heel wart (interdigital dermatitis), above, and hoof abscess opened wide.

As a rule, the first attempt at working on a problem hoof will yield your best results. I am always leery about being the third person to get a chance at correcting a hoof problem. Your first shot is definitely your best shot when it comes to hoof abscesses.

A dry foot bath consisting of a deep box of dry hydrated lime is an excellent antiseptic and preventative for hoof problems (OK for organic as long as the box contents are not eventually applied to fields). It is not a sloppy mess like copper sulphate water foot baths and doesn't need changing as often — only when the level gets too low to "push up" in between the hooves as the cows walk through the box. One farmer added boron to the feed (illegal for a feed company to do) and clinical hairy heel warts cleared up quickly and did not return.

1) Cleanse with 3% hydrogen peroxide and 7% iodine.

2) Thick paste made of sugar and iodine (Betadine), packed into the foot rot area and wrapped. Repeat one to two times. Remove necrotic/dead materials when changing bandage.

3) *Pyrogen* is truly homeopathic to foot rot cases, since the remedy is made from rotting meat originally and foot rot itself has decomposing flesh between the toes as its main symptom.

4) Hoofmate™ (available from IBA only) is an herbal ointment that has shown excellent results for strawberry heel. If used on foot rot cases, re-wrap will be needed.

5) Icthammol wrap for ulcer on sole and a block on the good claw to allow cow to walk. This condition usually comes and goes once established. If seen in a first calf heifer, it is usually due to a "hot" ration (high in energy). I don't see many of these in my practice.

6) *Hepar sulph* (<12X) to encourage hoof abscess to drain. *Hepar sulph* (200C) to keep abscess draining once opened. Have hooves trimmed and a block put on the good claw and wrap bad claw in icthammol.

7) *Silica* 30C can help a long-standing, deep abscess continue to drain and dry up — but only after the actual problem has been directly addressed by hoof lifting to inspect and pare whatever areas need to be opened up.

8) Fibromas — these are also called "quittor" or interdigital corns. These are no problem for the most part. However, if the animal ulcerates a fibroma, it gets infected very quickly with foot rot *(Fuso-bacterium necrophorum)*. The main treatment is to manually cut out the corn, cleanse with hydrogen peroxide, iodine and/or tea tree oil, and then wrap with sugar & Betadine paste. This should be re-wrapped in about four days.

Conventional treatment for infectious problems (foot rot and strawberry wart) would be to apply oxytetracycline or formaldehyde (formaldehyde is carcinogenic and *illegal* to use even on conventional farms) to effected hoof before wrapping and then run all the cows through a foot bath of lincomycin (follow label instructions) or copper sulfate (1 lb. per 2 gallons hot water). An alternative for prevention is to run the cows through a shallow container of dry hydrated lime powder.

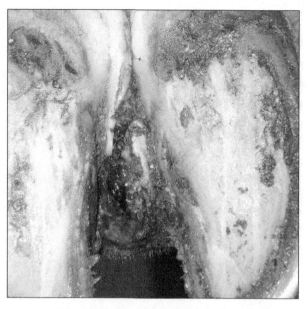

9) White line disease — This is when the junction between the hoof wall and the hoof sole becomes infected. Ideally, if the hoof is in good health, this junction is very tight and non-porous. However, if for some reason the hoof health is not good, usually due to poor nutrition (acidosis), various materials from the environment that the cows walk in will become lodged in the white line area and cause an infection. Upon lifting the hoof to examine it and finding an infected spot along the white line, it is usually obvious that the rest of the extent of the white line will have darkened areas (not good). This should be noted and then correct the actual bad spot as you would for any abscess.

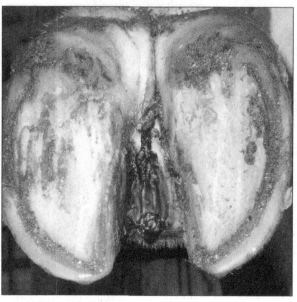

Infected interdigital growth — corn — (above) and Infected interdigital growth removed (below).

10) Fractures — These are uncommon except some herds seem to have a few calves with fractures within a short time. This is usually due to calves freely walking around among the cows and being stepped on by a cow when the calf is lying down. If the bone is showing through the skin, the prognosis is poor due to probable bone infection occurring (bone infections usually require IV antibiotic treatment for humans in hospitals); however, if the bone is not showing, having the vet place a cast on the calf's leg usually provides the cure. Calves, being young animals, heal bone fractures quickly. The cast should

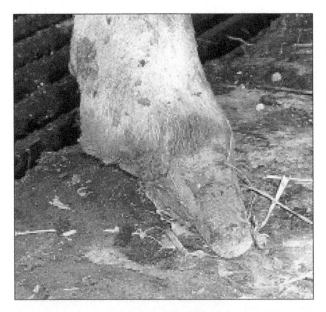

Successful amputation of toe (above), and ruptured gastrocnemius tendon in right rear leg — no cure.

be left on for about 4-5 weeks. Giving *Symphytum* twice daily for 2-3 weeks seems to help the bone union if the leg was set correctly.

11) If an abscess is found when paring a hoof, and then opened up only to expose a roughened bone, the prognosis is very poor for the cow to return to normal — unless the toe is amputated. Bone infections are extremely serious. Removing the offending digit is usually the best solution unless the farmer opts to send the cow for salvage. Amputating a toe is a rather simple procedure for a veterinarian to accomplish. The farmer must be prepared, however, to re-wrap the remaining stump twice weekly for a month to promote a healthy outcome. Animals can remain in the herd for more than one lactation, if the amputation is done before major swelling occurs in the fetlock above the hoof. My general success rate is about 70%, all depending on how long the condition has been festering, how infected the fetlock is, and how well the farmer cares for the hoof afterwards.

(Dadd, 1897)
Poultice for Foul Foot:

Roots of marshmallows, bruised	½ lb.
Powdered charcoal	a handful
Powdered lobelia	a few ounces
Meal	a tea-cupful
Boiling water sufficient to soften the mass	

Another:
Powdered lobelia
Slippery elm
Pond lily, bruised
Take equal parts, and mix with boiling water; put the ingredients into a bag and secure it above the fetlock

Give the animal the following at a dose:

Flowers of sulfur	¹/₂ oz.
Powdered sassafras bark	1 oz.
Burdock, (any part of the plant)	2 oz.

The above to be steeped in 1 qt. of boiling water; when cool, strain and give as one dose

Whenever any growth or ulcerated growth makes its appearance between the claws, apply powdered bloodroot and burnt alum.

If a fetid smell remains, wet the cleft, morning and evening, with

Chloride of soda	1 oz.
Water	6 oz.
Mix	

Another:

Common salt	1 tbsp.
Vinegar	a wine-glass
Water	1 qt. (p.154)

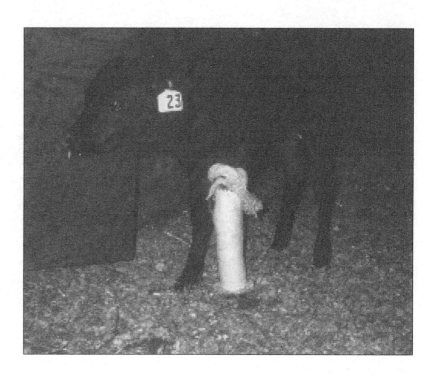

Leg sustained a simple fracture when stepped on by an adult cow. This type of break heals quickly in calves.

Time to head out to the pasture for the evening.

Treatments, Part 2

Reproduction & Calving

Note: Please also refer to chapter 8 where data is presented. All the organic farms shown in those tables, either individually or as a group, use the following treatments as their only reproductive treatments. I know this to be the case as I am the herd veterinarian.

Heat Cycles

If using natural treatments for reproductive purposes, the animal care-giver must be looking for the *slightest* signs of a cow being in heat to breed. The usual standing heat is wonderful to see, but often one must really be on their toes and detect *any* slight changes in milk letdown or not finishing feed quite right on potential heat day. Good dairy farmers can sometimes "just see" that a cow is in heat by the way the cow looks at you when milking her. In conventional terms, this is not a reliable heat indicator, but with natural therapies this is not discounted and should be considered as a potential sign to breed the cow.

Most natural minded dairy people could probably help their cows by using kelp at the rate of 2 oz. per head per day. Kelp has some 54 different vitamins and minerals in it, including iodine, which is an important player in the thy-

Cow Clock (±21 days)

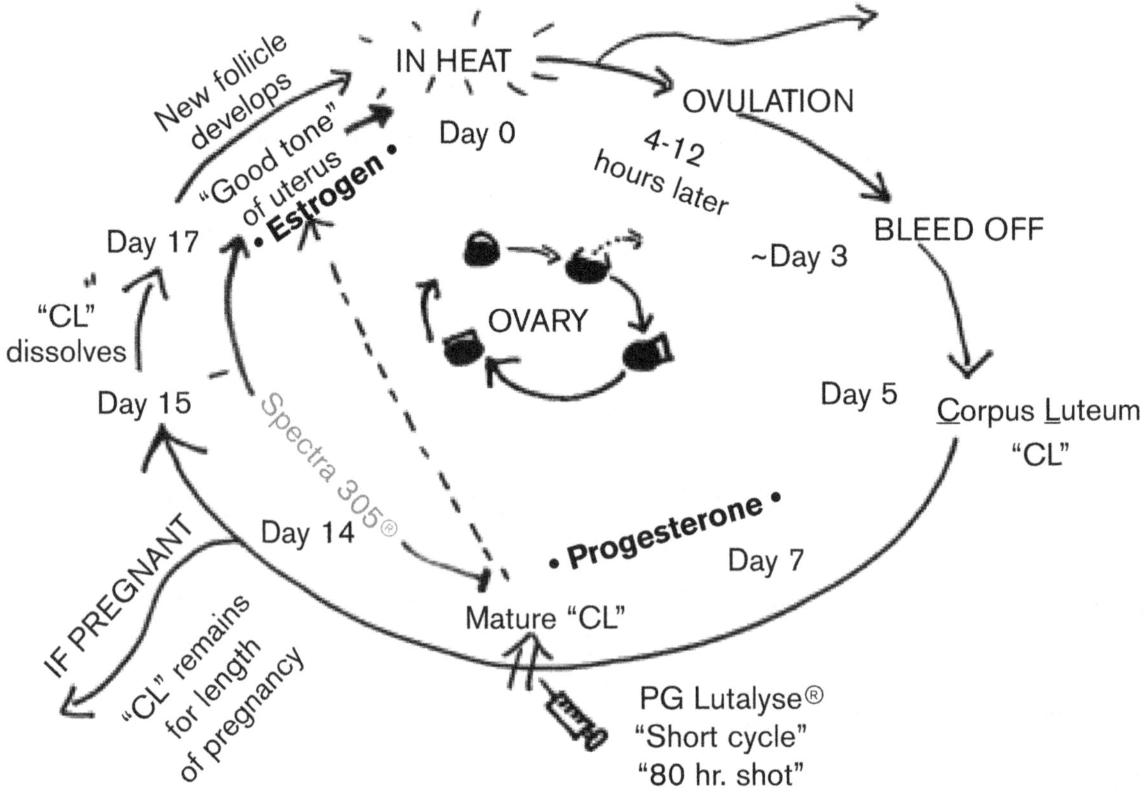
IN HEAT
Day 0
OVULATION
4-12 hours later
BLEED OFF
~Day 3
Day 5
Corpus Luteum "CL"
Day 7
• Progesterone •
Mature "CL"
PG Lutalyse®
"Short cycle"
"80 hr. shot"
Day 14
Spectra 305®
IF PREGNANT
"CL" remains for length of pregnancy
Day 15
"CL" dissolves
Day 17
New follicle develops
"Good tone" of uterus
• Estrogen •
OVARY

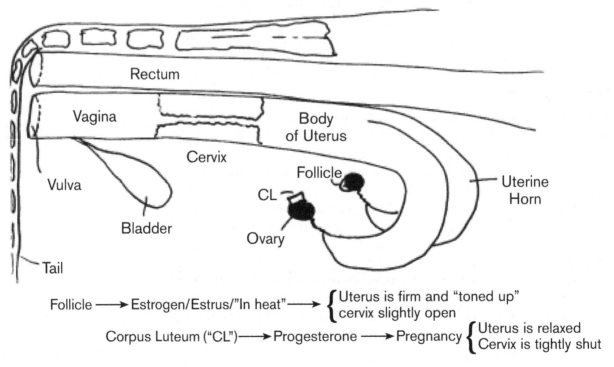

Bovine Reproductive Tract

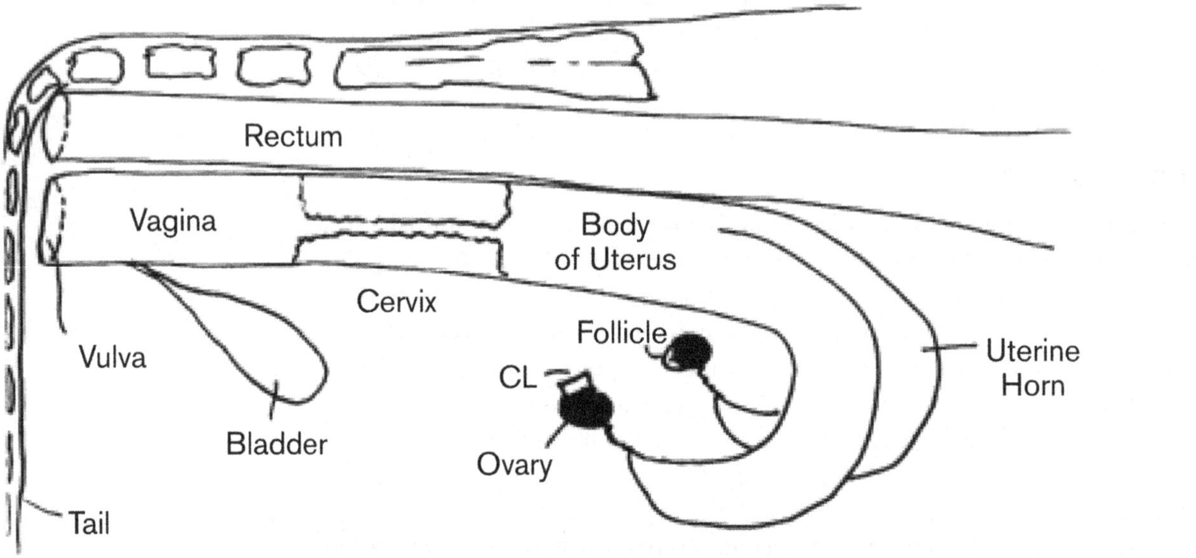
Rectum
Vagina
Body of Uterus
Cervix
Follicle
CL
Ovary
Uterine Horn
Vulva
Bladder
Tail

Follicle ⟶ Estrogen/Estrus/"In heat" ⟶ { Uterus is firm and "toned up"
cervix slightly open

Corpus Luteum ("CL") ⟶ Progesterone ⟶ Pregnancy { Uterus is relaxed
Cervix is tightly shut

roid's normal function in metabolic rates. Perhaps the most popular brand among farmers I work with is Thorvin kelp from Iceland, although there are more brands available now.

For cows cycling normally, being bred and not settling, try breeding the cow earlier than you normally would. There is good evidence (from studies at Virginia Tech) that breeding a cow between 0-8 hours after seeing a standing heat actually may be best. This is partly due to the egg being "fresher" early on and that it ages during the hours later. In addition, you may be seeing the last mount of the cow, and there really is no need to wait a whole 12 hours since you don't know when she first started to show heat. Bottom line is: don't wait as long as the traditional wisdom in AI has previously maintained. And anyway, does a bull wait 12 hours to breed a cow? If after 3 or 4 breedings she still shows standing heat, try a different breed *i.e.,* a Jersey or a Dutch belt on a Holstein. For some reason, often times a cow will then settle. I prefer a Jersey usually as you may get a nice milking animal and there is good calving ease.

Before I describe reproductive treatments for cows, it would be wise to give an overview of the normal cycle of a cow. The normal cycle is roughly 21 days long, give or take a day or two either way. If we consider "standing heat" to be day zero, the follicle usually drops an egg (ovulation) into the fallopian tube about 12 hours later. This egg is then able to be fertilized if there is sperm present (bull or AI). If the egg does not get fertilized, the follicle which released the egg becomes a transient *corpus hemorrhagicum* (CH), which is associated with the bleed off sometimes noticed by the farmer two or three days after standing heat. The CH then develops into the *corpus luteum* (CL) by about the fifth day after ovulation. The CL remains present until about day fifteen when it either recognizes pregnancy or not. If not pregnant, the CL begins to dissolve. At about the same time, another follicle that's "been waiting in the wings," seizes its opportunity to mature in a few days. By about day 21 (19-23 days), the cow will show signs of heat again due to the presence of a mature follicle with its egg. And so the cycle repeats itself over and over (if normal) until conception takes place. Then the CL remains in place for the full course of pregnancy until shortly before calving time when it dissolves. It should be mentioned that during the 21 day cycle, there will be small follicles that will appear on the ovaries which can give the uterus "tone" (firmness, rigidity) and will fake out a vet when palpating the cow. Even though a CL is present and the vet says the cow is mid-cycle, she may come in heat in a day or two because the small follicle is somehow triggered to all of a sudden become the dominant follicle upon which a heat is based. But, on the other hand (and this is usually the case), a uterus that is toned up due to a small mid-cycle follicle will fake the vet out to say "watch for heat soon," when in actuality the CL's progesterone still rules the environment and a dominant follicle isn't ready yet to bring about "heat."

To view the anatomy of the female bovine and to see a pictorial representation of the above explanation, see the diagrams (drawn by author)

There certainly are many reasons for the normal cycle of the cow not to proceed correctly. There may be problems right in the uterus itself, such as an unresolved low grade infection from a calving problem or previous retained placenta. Also, a follicle may not have properly released its egg and has become cystic (fluid accumulation). Or, if the cow is producing a lot of milk and not being fed enough to compensate for her work, she may be in a negative energy balance. Also, if being fed high amounts of grain, there can be too much protein in her system to metabolize it correctly and a subsequent excess of nitrogen in the form of urea which may be a cause of cysts.

Nutrition may affect fertility by acting on the hypothalamus and anterior pituitary gland or directly on the ovaries, influencing both endocrine function and oogenesis (egg formation). Keep in mind that the main problem in determining the influence of nutrition is the time interval between the start of a dietary change and its apparent effect. In cows, although protein, vitamins and mineral deficiencies are capable of poor fertility, the main cause is that of deficient energy intake. In heifers, sexual maturity is closely related to body weight, and the better fed and grown animals will need less services per conception.

There is a relationship between reduced blood glucose levels, excessive weight loss at time of mating and depressed pregnancy rates. When blood glucose is less than 30mg/dl, there is reduced fertility. Signs of energy deficiency are usually first shown in 1st calf heifers, followed by second calf heifers, with mature cows least effected. However, if a herd of mature cows is constantly too skinny, nutritional status should be looked at closely. Also, watching milk protein % may be of value. Milk protein % is affected primarily by energy intake, not the amount of protein intake. A 2.6% protein was related to a 105-day calving to conception interval compared with a 3.4% protein for a 94-day calving to conception interval. One study took 3 groups of cows on isocaloric diets (same energy level) but fed protein levels of 12.7%, 16.3% and 19.3%. The 19.3% protein intake gave the shortest interval to first estrus but the 12.7% protein intake gave the best services per conception and calving interval (Arthur, et. al., 1983).

Infertility in the Cow: Specific Nutrients

Phosphorus — supplemental P to grazing cattle increased breeding performance. There is a deficiency when there is less than 4mg/dl P in blood, usually when there is less than .20-.26% P in feed.

Copper — deficient either directly or indirectly by excess Mo or Fe.

Signs of copper deficiency are anemia, poor growth, bleached coat, diarrhea.

Cobalt — deficient usually when copper is low. Can diagnose by taking liver vitamin B12 levels.

Manganese — deficiency signs: anestrous, poor conception and deformed calves. Need about 80 ppm in the feed.

Iodine — need 0.8 ppm in feed. Deficiency signs include goiter as the thyroid gland desperately seeks out any available iodine in the body and enlarges. Also abortion, stillbirth, weak calves, fertilization failure and early embryonic death.

Phytoestrogens — excess can give anestrous, large ovarian cysts, vulvar and cervical enlargement and poor conception. Certain strain of red and white clover and alfalfa. Phytoestrogens can be the cause of unbred heifters developing udders (from which milk is sometimes produced).

Selenium — deficiency diagnosed by glutathione peroxidase levels. Supplemental selenium and vitamin E together decrease metritis and cysts postpartum (after calving) and increase pregnancy rates post treatment.

Vitamin A — 90mg/kg/day. It's been shown that if only 30mg/kg/day then there will be no pregnancies.

Beta-carotene — corn silage very low in this, only 2-4 mg/kg DM. Grass silage gives about 7 mg/kg DM. Fresh growing plants are generally high in beta-carotenes. Beta-carotene deficiency with adequate vitamin A gives increased follicular phases, delayed ovulation, silent estrus and follicular cysts. Bovine luteal tissue has some of the highest beta-carotene content of any tissue and it may be involved in ovarian steroid production or corpus luteum formation (Arthur, et. al., 1983).

In most cases the most important factor responsible for poor breeding is unknowingly *underfeeding*. This may be due to over evaluation of the feeding value of forages, over estimation of feed intake under self-feed conditions (esp. true of silage), failure to realize the reduction of forage intake caused by high concentrate intake, and/or contributions of mineral licks and free access sources being difficult to quantify.

Poor Heat Detection

Reasons:

1) Not knowing "true signs" of estrus.

2) As herd size increases, slight behavioral changes in individual cows are lost.

3) Short duration of estrus (10-15 hours). And of that, 20% of cows are in heat for only 6 hours. And, there is more activity in the hours of darkness and less farming activity.

4) Crowded areas, confined areas, poor footing.

5) Poor dry matter intake. Cows must have enough energy in their rations to show heats!

Methods of Improving Heat Detection

1) Improved cow identification for easy observation.

2) Good lighting.

3) Increased regular observation of cows. Best heat detection rates are made by 4 observations of 30 minutes each. Don't try to watch for heats during milking, when feeding or when mucking out.

4) Use of heat detection aids — Kamar™, tail paint.

5) Use of dogs. They can be trained to detect odors associated with estrus in cows.

6) Use of milk progesterone assays: can help anticipate next heat.

7) In sows and ewes, having a boar or ram present stimulates estrus cycles — this should be the same with a bull. And, if you have a bull, let him run with the cows. Don't only bring the cows to him, you might as well use AI and get better genetic advancement. But having a bull will normally yield high rates of pregnancy. Just keep him under control (use a bull ring with a 30-36 inch chain permanently attached).

Breeding Cows Following Hot Weather

Remember that ova (eggs) start their development toward maturation some 60-90 days before their moment of estrus.

During this time, any insult to the cow's basic physiology can lead to reduced fertility. Possible insults include excessive heat, lack of sufficient energy, toxins in the feed, illness, etc., will hamper normal cycling.

Hot weather itself can ruin embryo quality and decrease pregnancy rates. Also, it can ruin the sperm of bulls when body temperature is above 105 degrees for a few days on end. Cows sick with fever often abort or the calf is born dead if the fever is not reduced quickly enough.

Moreover, cows simply don't have as much drive or show the typical signs of estrus when it is hot and humid weather, especially if you breed cows on secondary signs of heat (don't letdown milk quite right, she gives you "that look," or only on clear discharge). Once the autumn cool weather arrives, more aggressive heat behavior is seen again.

Hot weather (above about 80 degrees) also reduces dry matter intake. If dry weather is present as well, then pasture quality is further reduced of common cool season plants. Energy is needed for fertility, so a drop in feed eaten will likely mean a drop in fertility.

A drop in dry matter intake may make a cow go into negative energy balance. Once they pick up in feed again in autumn, you may see an increase in cystic ovaries and twinning, due to the positive energy balance.

Fertility Treatments

For regular herd checks, try the following remedies depending on what your vet finds on rectal palpation.

1) *Sepia* 2 x 3, for a cow fresh about 3 weeks and needing some support to become fully involuted.

2) *Ovarian* (6X or 5C) 2 x 5, for cycling cow that hasn't shown visible heat; cow with corpus luteum (CL) and good tone; watch for any sign of heat, no matter how minor, over next 3 days.

3) *Folliculinum* (6X or 5C) 2 x 3, for cow with healthy follicle or if having shown previous heats and bred but not settled; give prior to heat or on day of standing heat.

4) *Pulsatilla* homeochord 2 x 5, for non-odorous pus discharges; followed by Sepia 1 x 3.

5) *Apis* 2 x 5, for right-sided cyst; or cysts on both ovaries; then follow with *Nat mur* 2 x 3. Have vet try to rupture the cyst with gentle pressure.

6) *Lachesis* 2 x 5, for left-sided cyst; followed by *Nat mur* 2 x 3. Try to have vet rupture the cyst with gentle pressure.

7) *Lillium tigrinum* for hyperactive, cystic, nymphomaniac cow; 2 x 7.

8) *Iodium* (6C) 2 x 10, for shriveled ovaries and no heat; cow is lean and has good appetite.

9) *Pulsatilla* homeochord 1 x 21, for no heat, poor uterine tone, no ovarian activity, but normal ovarian size; followed by *Sepia* (30C) 1 x 3 or *Sepia* (200C) one time.

10) *Palladium* 2 x 5, for cow with painful ovary when palpated; then *Sepia* 2 x 3.

11) *Ustilago* homeochord 2 x 7, then *Sepia* 2 x 3, for thin, watery, fluid in uterus without heat, especially when later than 1-2 months in lactation.

12) *Natrum mur* 2 x 5, for fluid in uterine horns, if present well into the lactation, followed by *Calc phos* 1 x 3.

13) *Causticum* 2 x 7, for older cow with sagging uterus due to lax ligaments.

14) 10cc vitamin E/selenium (commercially available) at 3 weeks after calving, especially in 1st calf heifers milking really well and losing body condition, or Selplex fed continuously (source of organically bound selenium).

15) Powdered wild yam, cramp bark, black cohosh, flax seed, evening primrose, dong quai and vit B6 every other day for 6 doses or until heat, whichever is first. For cows with normal CL, normal uterus, normal body condition and simply not showing heats. This will often bring a cow into an easy to observe heat. Breed her as you would any cow. Sometimes, however, three weeks later the same cow will then show a flaming heat — breed her again. Cows usually settle equally as well with Heat Seek as they do in natural heats. If there is just one medicine in this entire book that I would like to do com-

parative clinical trials against conventional drugs, Heat Seek would be it. I would say estrus (heats) are shown about 80-90% of the time when this is used as I directed above. Often times, it will not even take an entire full course before the cow shows heat. In these cases, it is also competitive on a cost basis with its conventional counterpart, PGF_{2aplha} to which I would compare it.

Acupuncture (see Acupuncture Charts for locations of points):

16) Pyometra (pus in the uterus): Moxa or traditional Chinese needles at BL-20, 22 and 28; inject 5cc vitamin B12 at SP-6.

17) General infertility: inject 5cc vitamin B12 at BL-22,23,26, 31, 40 and SP-6. Also any sensitive points — often BL18 or BL19.

18) Repeat breeders with normal heats: inject 5cc vitamin B12 at BL-27, 28 and BL-31 bilaterally.

19) Abnormally flaccid uterus: traditional Chinese needles at BL-27, 28 and 31, keep in place 15 minutes.

20) Abnormally contractile uterus: traditional Chinese needles at GV-2, 3 and 4, keep in place 15 minutes.

Pyometra

This is the condition of having pus in the uterus and a small bit of pus or cloudy discharge is seen from the vulva (leucorrhea). Usually it is a result of a retained placenta and subsequent uterine infection at calving time. (see also Retained Placenta.) Occasionally, a pyometra will occur independently of calving time — for example a pyometra is diagnosed during a herd check in an animal that previously checked OK and now has a slight muco-purulent discharge. The usual case of a pyometra following uterine infection from calving needs to be addressed as soon as possible and as aggressively as possible, with any available natural therapies made use of since these can become chronic if not cleared up by 3-4 weeks fresh. The first thing to consider is to give a vitamin E/selenium shot *2-3 weeks before calving.* This would be especially true for herds in geographic areas where soils are deficient in selenium and when unassisted calvings still produce cows having uterine infections. (Any assisted calving, twins, near-term abortion, or otherwise early calving almost invariably ends up with a retained placenta and subsequent uterine infection.) In any event, the general idea is to irrigate the uterus once a day maximum or every other day. Uterine infusions should be done in as clean a manner as possible and those individuals knowing how to inseminate cows will be the best ones to do it, especially if the cow is long fresh since the cervix will be open barely enough to pass an infusion rod. In any event, mild irrigation solutions include using a mixture of 50cc colostrum whey with 10cc (or more) calendula tincture; 30cc Clean Cow with 30cc herbal antibiotic tincture; 40cc of hypertonic saline with a 20cc 7% iodine (iodine is a two edged sword if used internally

— only use *early* in lactation as it takes a couple months to clear out of the uterine environment); aloe with saline; and tincture of pulsatilla and hydrastis (15cc each) in 30-60cc saline. In cows that are more recently fresh (within 2 weeks), use of 60-120cc 3% hydrogen peroxide or 12 chlorhexidine tablets can be of definite benefit. However, as both these agents foam when in contact with organic matter, the cervix *must* be open enough to allow escape of the foam. Do not use these two agents with a nearly closed cervix (one that you need to "thread" the rod through). If a bull is breeding the herd, sometimes the bull's semen can be mistaken for drainage. However, more important to note, is that bulls can pass infections from cow to cow, so a slow disaster can be brewing before the farmer is confronted with a lot of not pregnant cows or cows with mild uterine infections. Conventional farmers can clear up these mild infections with PGF_{2aplha}. Without doubt, I have repeatedly seen that grazing cows will clear an early lactation uterine infection quite effectively, as long as it is not a severe metritis, with minimal effort on the part of the farmer or veterinarian. This may be due to improved lymphatic drainage as the animal moves around freely while also eating fresh green forages. The slight, chronic pyometra that occurs in a later lactation cow is definitely more of a problem (grazing or not). A treatment which seems to work more often than not is to infuse a mixture of 40cc propylene glycol (consider as a "carrier") and 20cc 7% tincture of iodine *one time*. Some cows will strain for a day or two after infusing but the majority do not. Bringing them into heat with Heat Seek may also be a good solution and then infusing with the farmer's choice of solutions while in heat. (Also see the *materia medica* for compounds mentioning "leucorrhea.") These suggestions are the usual means that I take for this situation and perhaps the reader should review the reproduction data presented in Chapter 8 in order to confirm for him/herself that these methods are worth pursuing.

Ovarian Cysts

Cysts are not a major problem, as far as treatments go. It seems that there are certain times of the year when cysts are more prevalent. This may be in the autumn when the weather is cooler and cows begin to cycle again after the summer heat. Also, whenever cows have been in negative energy balance (milking well and becoming very lean) they may experience cysts when they begin to put more flesh on again as they rebound. Generally speaking, I will try to gently (very gently) rupture a cyst. If this is not possible, I usually prescribe either homeopathic *Apis* (for the right side) or homeopathic *Lachesis* (for the left side) twice daily for 5 days, with either one being immediately followed by homeopathic *Natrum mur,* twice daily for three days. This usually works. If, on a follow-up exam, a cyst is still present it will often then rupture spontaneously when I go to feel it. Otherwise, I may suggest using Heat Seek,

10 tablets orally every other day for *twelve* doses (24 day treatment). Normally I use Heat Seek to help show heats in a normally cycling cow with a CL that just won't "show a heat." I double the length of administration for cystic cows and it does seem to work quite well. Because the cost is higher than the homeopathic remedies, most farmers will wait to see if the homeopathic remedies work first (which they very often do). If for some reason there is a very stubborn cyst, I will then use the acupuncture points (BL-22, 23, 24, 25) and inject 5 cc vitamin B12 into each of these on the side where the cyst resides. The response I have seen with this is pleasing. Sometimes when doing the next herd check and I call a certain cow pregnant, the farmer will say, "Doc, that's the one you stuck the needles into when she was cystic and she came in heat the next week." Of course that is gratifying to hear. But sometimes the cystic ovary is still there but much more fragile and ruptures spontaneously when I go to feeling around for it. I have yet to see a cystic cow be culled from a herd (with the cyst being the primary reason for culling) when treated in these ways. Nearly all herds will experience cysts at one time or another, and I believe the data shown in Chapter 8 bear out that calving intervals, etc., are very acceptable in herds where I do the fertility examinations.

Eclectic Agents Acting upon the Female Reproductive Organs (Ellingwood)
Viburnum prunifolium (Black Haw)
Viburnum opulus (Cramp bark, High bush cranberry)
Senecio aureus (Life root)
Helonias dioica (Unicorn root)
Mitchella repens (Squaw vine, Partridgeberry)
Aletris farinosa (Star grass, Starwort, False unicorn root)
Caulophyllum thalictroides (Blue Cohosh)
Fraxinus americana (Black Ash, Elder leaved Ash, White Ash)
Polygonum punctatum (Smart-weed)
Leonurus cardiaca (Motherwort)
Gossypium herbaceum (Cotton)
Lilium tigrinum (Tiger lily)
Cypripedium pubescens (Yellow Ladies' Slipper)
Aralia racemosa (American Spikenard)
Cimicifuga racemosa (Black Snakeroot, Black Cohosh)

Uterine Sedatives — in cases of threatened abortion (Milks):
Viburnum prunifolium, Viburnum opulus

TABLE 13: Summary of Eclectic Remedies Acting on the Female Reproductive Organs (Ellingwood)

Action	Viburnum prunifolium	Helonias dioica	Senecio aureus	Cimicifuga racemosa	Aletris farinosa
General	Soothing to uterus; overcomes sterility	Pronounced tonic influence; specific in threatened prolapse	In atonic disorders of uterine function; increases ovarian activity	Relieves muscular soreness; acts directly on reproductive functions	Indicated in extreme weakness of the reproductive organs
Action on menstrual function	Dysmenorrhea with cramp-like pains	Is auxiliary; corrects reflex irritation	Restores tonicity; promotes regular flow; use for 2-3 cycles	Menstrual disorders accompanied by muscular soreness	Acts directly in females that are feeble and with weariness
Prevent abortion	The best of remedies for this purpose, either in emergencies or habitual abortion	Not to be relied on	Not to be relied on alone, facilitates action of Viburnum	Is not to be depended on; acts more like ergot	Not for immediate results; corrects tendency to habitual abortion
Prepare for birth	Decreases nerve irritation and erratic pains	Use with Mitchella and Cimicifuga to decrease pain	Not active, except for extreme loss of tone	Conduces to a normal, easy, short labor; less reliable than Mitchella	Important for females who seem too feeble to bring forth young
In labor	Promotes normal contractions; prevents hemorrhage	Not used in labor	Not used in labor	A most reliable oxytocic; prevents post-partum hemorrhage	Has no direct influence during labor
After labor	Restores normal tone; prevents subinvolution	Preserves tone, preventing subinvolution	Restores suppressed lochia; promotes normal contraction	Promotes normal involution; relieves severe aching; cures persistent leucorrhea	Acts in conjunction with iron tonics in overcoming general weakness
Other	Valuable in fevers with impending uterine inflammation and sepsis	Corrects liver and stomach disorders during pregnancy	Used persistently, its influence is plainly apparent and beneficial	Aching and muscular soreness are its plain indications	Strengthens uterine structures; prevents subinvolution

I am probably one of only a few dairy practitioners that do not have the hormonal synchronization programs memorized — I have no real need to do so, and only rarely do I suggest using any (OvSynch, if any). It should be pointed out that these programs are in widespread use on conventional dairy farms across the United States. None of these are allowed for certified organic livestock; but I am including it anyway in order to show all options available for fertility treatments. With all the shots needed, many needles will be used. I doubt that farmers change needles between every cow; thus the potential spread of contagious disease could increase.

Cow Synchronization Protocols

Ovsynch: GnRH day 0, PGF2a day 7, GnRH day 9, AI 10-24 hrs. after 2nd GnRH is given.

Co-Synch: GnRH day 0, PGF2a day 7, GnRH and AI Day 9.

Presynch: PGF2a Day 0 and day 14, GnRH day 28, PGF2a day 35, GnRH day 37, AI 10-24 hrs. after 2nd GnRH.

Heat Synch: PGF2a day 0 and day 14, GnRH day 28, PGF2a day 35, ECP day 36, Heat detect and breed day 37, AI day 38.

Select Synch: GnRH day 0, start heat detection on day 6, PGF2a all non-inseminated cattle and heat detect on day 7.

Heifer Synchronization Protocols

Oral MGA: Feed oral MGA at day 0 until day 14, GnRH on day 26, PGF2a on day 31, heat detect and AI on day 32 and on.

2 PGF2a: PGF2a on day 0, heat detect on day 2, PGF2a all non-inseminated cattle and heat detect on day 14.

Heat detection and PGF2a: Intense heat detection on day 0, PGF2a all non-inseminated cattle and continue to heat detect until day 6.

Select Synch: same as for adult cow (above).

CIDR — a vaginal implant that tricks the cow's system into thinking it is pregnant, thus a corpus luteum will be present. When it is taken out about 7-8 days later, a shot of PGF2 is given to dissolve the CL, and the cow will then come into heat/estrus.

Reproduction Problems

Uterine subinvolution (Winslow, 1919)

Give fluidextract of ergot in full dose thrice daily for two weeks, and it may well be combined with a moderate dose of quinine sulphate. In subinvolution or hypertrophy of the womb following labor, the use of hot vaginal injections also aids the action of ergot in restoring normal condition (p. 615).

Urine pooling

This is a condition where the urine, instead of advancing out, remains in the vaginal vault and can cause irritation and interfere with conception. This is an anatomical defect. However, I usually find this to occur just after the parturient period, and it usually seems to resolve within a few months. Perhaps the increased estrogens around calving time relax the musculature of the vaginal area so much in some cows that it simply takes a while for the normalizing of hormones to correct the lax muscle tone. Sometimes using *Causticum* 2 x 10 seems to help, but it probably would correct itself with time.

Conventional: "Tincture of time."

Wind suckers/Pneumovagina

This is a condition where air too easily enters the vaginal vault and is usually seen when cows rise to get up. This definitely will interfere with breeding — the cow comes in heat normally, but just won't settle. Again, *Causticum* may help for the lax ligaments. However, I usually have found it doesn't, and the condition only resolves if a Caslick's operation is performed. This entails anesthetizing the area with an epidural of lidocaine, then numbing the labia with lidocaine locally and then resecting the labia slightly and stitching it shut, except leaving enough room to allow the cow to urinate. Lidocaine is OK for organic use. However, the condition can be hereditary and by doing the Caslick's in order to get the cow bred may pass on the trait to the calf, if it is a heifer. After a Caslick's operation, the cow will usually settle easily. The scarred vulva needs to be incised (open up) before calving.

Conventional: As described.

Hydrops

This is a very rare condition in which the uterine fluids of pregnancy keep accumulating to an extreme degree. The pregnant cow will slowly become rounder and fuller looking over the course of a few weeks, usually in mid- to later pregnancy. They keep eating until they are simply too filled up with uterine fluid and in an unstable systemic hydration state. This is when I've been called in to check on these cows. There is not much to do except induce parturition (usually accomplished by hormones and steroids combined) or ship the cow for salvage. By inducing the cow, the fluids will burst forth through the vagina as she goes into labor. She will "decompress," however, she will also need intravenous fluid re-hydration to not go into shock. This is preferable to attempting an emergency C-section with rapid loss of gallons upon gallons of intrauterine fluids and sending the cow right into hypovolemic shock. If the cow is still standing when examined by the veterinarian, induction of labor may be a good option. If the cow is down and cannot rise, the prognosis is very poor if not grave. With excessive intrauterine fluid build-up, it might be thought that the calf may be somewhat smaller due to abnormalities existing. This may or may not be the case — one down hydrops cow that I induced had a huge calf which needed forced traction to extract. The combination of the hydrops condition (resolved by going into labor) and a hard calving resulted in the cow expiring, even though I administered lots of IV fluids and analgesics.

Conventional: Induce calving with steroidal hormones and treat with IV fluids or salvage.

Cystic Bartholin's gland

This is seldom seen, but when it is, it causes alarm to the farmer (while not really bothering the cow). Bartholin's glands are in the vaginal walls and occasionally get blocked for unknown reasons. It will at some point protrude from the vulva and be seen as a small round object the size of a golf ball or tennis ball with an opaque white filling. It will be seen when the cow is either standing up of laying down. These are benign and will resolve on their own in general. If worried, they could be surgically resected by the veterinarian by first applying a tourniquet at their base as they are sometime fairly vascular.

Conventional: Surgical removal or "Tincture of Time."

Prolapsed vagina

In dairy cows nearing parturition, the estrogenic effects of internal hormones will sometimes cause lax muscle tone in the vagina. This will be shown occasionally by a "ball" looking mass (about the size of a soccer ball) protruding from the vulva when the cow is lying down, but it will recede when she rises. In dairy cows, this is nothing to be much worried about. However, in beef cattle this can be a major problem in certain breeds (Herefords). Lidocaine epidurals and intra-vaginal devices are normally used to keep the mass from drying out in these beef cows. Unfortunately they usually keep straining and straining, and therefore an alcohol epidural might be used for longer action. This is problematic in that a permanent block may occur and the animal will lose use of its tail. It will thus be limp and manure will accumulate on it.

Conventional: Surgical resection or placement of anti-prolapse device.

Haemophilus

Although this pathogen primarily affects beef cattle in the form of pneumonia, it is sometimes implicated in reproductive problems in dairy cows. The main symptoms would be irregular heat cycles and difficulty in getting cows settled. Sometimes clear stringy saliva or mucous dangling from the nose is considered a sign of *Haemophilus* by some vets, while other vets consider the exact same signs to be due to chlamydia. Vaccinating for this is what most farmers that I know prefer, if they do anything at all. But for those that do vaccinate, they are very satisfied and would not do without the vaccine (given once yearly).

Abortions

In a 50 cow herd, there really should not be more than 1 spontaneous abortion a year (perhaps two, if they are not in a close time-frame). Abortion storms are the cause for investigation by blood work on the dam and the aborted fetus (if it is found in excellent condition). Take two blood samples from the dam, 3 weeks apart, and look for a fourfold rise or decline in any of the antigens tested for. In Pennsylvania, the standard

Possible BVD heifer (1 year older than mates).

"Reproductive Panel" test includes BVD, IBR, *Neospora* and the 5 strains of Leptospirosis. By taking the samples 3 weeks apart, you get to see the difference in antibody levels for the antigens within the cow. A full post-mortem necropsy of the calf can be valuable as well when combined with the blood work of the dam. However, sometimes there are no clear cut answers, much to the consternation of both the farmer and the veterinarian. But, usually there is evidence of some sort of infection through laboratory diagnosis. It sure beats shooting in the dark by clinical signs or stage of gestation.

In my practice area, *Neospora* has become a major player in abortions. This typically will cause a cow to abort at about 6-8 months of pregnancy; she will remain well, keep on milking, and usually conceive again in about a month's time. There is a vaccine out now, but its efficacy has yet to be firmly established. Usually there will be an abortion storm over a couple months and then it will subside as quickly as it arose. Dogs are now known to be the definitive carriers of the protozoa, shedding it in their stools. Make sure cow feed is not contaminated with this!

BVD is probably the most insidious cause of reproductive failure, early embryonic death and abortions. It is now known that a Persistently Infected (PI) carrier state exists and animals that are PI shed the virus in large amounts to their environment via bodily discharges (respiratory, urinary, etc.) Vaccination will not overcome this situation. Common herd symptoms include irregular heat periods, breeding cows that then come back into heat on irregular heat

Same BVD heifer 1 year older than Holstein.

cycles (due to conceiving and then early embryonic death), abortions usually between 3 and 6 months of gestation and weak calves being born. The PI animal itself becomes infected when it is still inside its dam during development — specifically between about 50-120 days pregnancy when the calf's immune system is developing. If the dam is exposed to BVD during that time period of pregnancy, the calf's immune system sees the BVD as "itself" and does not mount any immune reaction to it. Instead, the BVD slips into the calf's system and is not rejected. This will either abort the calf, bear a weak calf that dies soon after birth, give a calf that lives but is a poor doer through its early life and gets culled, or the calf can appear normal and make its way into the milking string —shedding the BVD virus to its herd mates everyday that it is alive. Testing for BVD is not complicated, but it involves a whole herd test, either by taking blood samples and checking for the virus itself (PI status) or taking ear notches and having a pathologist look at it using immunohistochemistry under the microscope. If a PI animal is identified, it should immediately be culled to eliminate the virus from being constantly shed in the farm environment. A PI animal will *never* cure since its immune system does not recognize the virus as anything but itself. If a PI animal gives birth, its offspring is definitely Persistently infected as well. In one herd a granddam, dam and yearling heifer were all PI. Never keep a PI animal in a herd!

Leptospirosis (Lepto) can be a cause of abortion. These bacteria live in stagnant water or slow moving water and can also be found in the urine of rodents. Abortions usually occur in mid-gestation. Sometimes a cow will be sick when aborting with Lepto. Classically, Lepto lives in the cow's kidneys and is shed in its urine. If bloody urine is a symptom when a cow aborts in mid-gestation, lepto should be considered as a cause and investigated. Using dark field microscopy has been the gold standard for Lepto detection (use of furosemide to help flush the kidneys helps recover any leptospires that may be there). However, titers can be of value as well. *Lepto hardjo bovis* is now thought to also be a persistent infection, much like BVD. Calves can be born with it and shed it into their environment. Research is being done on this and farmers will hear more about it from the Extension Service in the future. Vaccination for Lepto is useless if a persistently infected Lepto carrier is on the premises. The traditional Lepto vaccination in general gives only a short length of immunity, roughly 4-6 months and needs to be boosted then. Some farms which give a 9-way modified live vaccine will boost with Lepto alone at pregnancy confirmation if Lepto has been identified as a possible cause of reproductive failure. The new *Lepto hardjo bovis* vaccine is very effective and gives protection for 12-18 months.

Arcanobacter pyogenes (formerly known as *Actinomyces pyogenes)* are bacteria that are often found incidentally on fetuses submitted for investigation. It can be a primary cause of fetal pneumonia and death, but it seems to be more of a secondary condition (though still fatal). *A. pyogenes* infections in general are very purulent (pus). Abscesses and pyometra (pus filled uterus) are conditions where *A. pyogenes* is easily recovered. If a dry cow gets mastitis and it is *A. pyogenes,* the calf will often be aborted secondary to the cow's high fever or be born dead due to systemic spread of *A. pyogenes* to the fetus. Dry cows that experience coliform mastitis will have a pregnancy that ends the same way as described for *A. pyogenes,* but mainly due to the systemic endotoxemia the cow suffers due to the coliforms.

Immunization (vaccines or nosodes?)

Although some farms have truly "closed" herds (no additions whatsoever, no bulls brought on to the farm, visitors wash boots with a sanitizer, no animals are sent temporarily elsewhere to be raised), most farms do have traffic in the way of cows or bulls added in over time. I believe that it is paramount to have your animals protected. There are simply too many unknowns out there when bringing an animal onto your farm. Oftentimes, I'll hear a farmer tell me he just bought a cow that was vaccinated before shipping, but with what — the vaccine appropriate to the previous farm's environment? Perhaps a calfhood vaccination for brucellosis? And what if the animal comes in from an auction house where lots of animals have been congregated and could exchange germs? With the stress of shipping, internal steroid production is occurring which suppresses the immune system and weakens the animal thereby predisposing it to breaking with disease. All this is prior to having set hoof on your farm. Vaccines are proven within USDA guidelines, although clinical trials may not involve many animals (sometimes as few as 32 — just enough for obtaining valid statistical numbers). At least there is some statistical evidence to prove the vaccine's efficacy before releasing it for marketing. With homeopathic nosodes, there have been some studies in other species (canine kennel cough nosode) but not yet to any degree with dairy cattle. In addition, nosodes are intended for times of actual illness, rather than as a preventative. Therefore, if asked, and after discussing pertinent environmental realities of a specific farm, I may advise a farmer to vaccinate conventionally and follow with the nosode to blunt any potential untoward effects the vaccine. Vaccination is cheap insurance compared to a herd outbreak of BVD induced abortions and/ or pneumonia. Organic dairy producers should keep themselves open to vaccination as a preventative tool rather than be faced with hard to fight diseases (like bacterial pneumonia a.k.a. "shipping fever"). Three real life experiences with non-vaccinated animals have convinced me firmly that vaccinating *at least to prevent shipping fever* is worth every penny. Whereas normal monthly vet bills may be $100 for a herd of about 50 dairy cows, during an outbreak of shipping fever I've invoiced $1800 a couple times for the medicines and labor to combat pneumonia. And then, the cows only slowly come back to production. One of the herds was supposedly protected by the homeopathic nosode. The other assembled cows from many different locations without using the standard TSV-2 intranasal prior to shipping. However, I have also seen many cows die in a week's time as the ultimate penalty for not having an adequately vaccinated herd. Surprisingly enough, many producers I've talked with about organic certification are surprised to learn that they are allowed to use vaccines. For some reason they think they cannot because it's not philosophically "organic." Although there is a minimal amount of one type or another of antibiotic used as a preservative in commercial vaccines, the benefits outweigh by far this potentially "non-organic" component. The USDA NOSB has debated the issue of the antibiotic preservative in vaccines, and since it is not the active ingredient and reason for use, it is allowed at a fractional level as a preservative. Vaccines are categorically allowed by the USDA NOP, no matter how they are made.

No matter how you decide to immunize your animals, remember that any immunization program can appear to be effective if given at a time when there is no stress or pressure upon your herd. It is when there is a hot challenge that a farmer will see whether or not the animals are well immunized. The biggest negative aspect of conventional vaccination from the homeopathic view is that *continual, frequent* vaccinations will set up a low-grade chronic disease/condition that could deplete the animal's innate vitality due to a vaccine's ability to stimulate specific reactions in an animal's immune system. Although this is debatable and hard to prove, it certainly deserves attention and thought before vaccinating.

Immunization is a controversial topic and there are no ideal programs that work in all herds everywhere. The best program is one that is tailored to an individual herd and is based on factors such as movement of animals into the herd, environmental considerations and owner compliance. My opinions are based on seeing some non-vaccinated herds having terrible contagious disease outbreaks. And although conventional vaccines can effectively immunize a herd if given properly, they can be overcome if the environment or management is sloppy — such as no ventilation in the milking barn, calves tied right near cows, heifers in mucky pens, Persistently Infected (PI) BVD animals, flies everywhere, etc. If trying to use no antibiotics whatsoever, preventing disease sure beats trying to cure a "hot," contagious disease outbreak with only natural remedies. Keeping livestock well is definitely an art, but also depends on such real factors as genetic strength, environmental conditions and daily observation by the animal care provider. The animals' living environment and simple observations are two factors that the farmer has direct control over. In my opinion and experience, to rely solely on nosodes to prevent contagious disease is risky. However, if one is to use nosode made from actual disease discharges in a low potency like12X, I can definitely agree with that (over a high potency like 30C or 200C). One farmer I work with administered nosodes (30C) regularly to his herd over a few years, and then his herd came down with a terrible contagious pneumonia when he brought in just one cow from the neighboring farm. Therefore I tend to side with those who advocate vaccinating, albeit *sparingly*, rather than those who condemn vaccination. However, many of my farmers do not vaccinate and make out just fine (and we are in an area extremely dense with dairy cattle).

If you do decide to vaccinate, here is the standard one practiced in southeastern PA (consisting of all or part of the following):

1) Give a modified-live viral vaccine (BVD, IBR, PI3, and BRSV) at 5-6 months of age. This is an age when an animal is the least stressed in its life and their immune system should, at least theoretically, process the vaccine in the most effective way.

2) Repeat step (1) about 5-6 weeks before breeding.

3) Booster milking cows twice a year with a killed version of the above, OR,

4) Give the same modified-live combination to cows at about twenty to thirty days fresh, and, Optional:

> a) Give the "J-5" or "Endovac Bovi" and/or Rota-Corona vaccine three weeks before freshening, depending on your farm's history with neonatal calf diarrhea and fresh cow coliform mastitis. This is optional; depending on the amount of "hot quarter" watery mastitis is experienced early in lactation.

Note: If I give vaccine shots, I like to give them at the GV-1 acupuncture point, the cows don't kick, and they don't develop abscesses (see Acupuncture Charts).

If your veterinarian has diagnosed a common neonatal calf disease in your herd (especially diarrhea), it is possible to protect a future calf by vaccinating the expecting mother. This is usually done about 3 weeks before calving in order to let her udder accumulate the protective antibodies to be part of the colostrum. In general, 1st calf heifers are usually the source of a Rota/Corona virus problem. Vaccinating them prior to calving should help. Coliform can be vaccinated for, usually with the now common J-5 vaccine given to cows to reduce incidence and severity of coliform mastitis. There exist *Salmonella* and clostridium vaccines as well. It is important to communicate with your veterinarian about calfhood losses due to diarrhea. The age of the calf helps in diagnosing which pathogen (germ) is the culprit.

Vaccination cannot overcome sloppy housing and environmental conditions. It is the caretaker's responsibility to give the expectant cow clean and dry bedding in which to calve. This should ideally be changed after every calving, although it rarely is. Calf housing is critical in growing healthy calves as well. Calves should be housed individually in hutches to prevent transmission of infectious disease. They can be near each other for company, but in the first couple months they should be separated physically. There should always be a spare hutch not being used, being given a rest and turned up to the sun for beneficial UV rays to kill bacteria. When cleaning out a calf hutch or stall, let it air out for a couple hours and then put down a liberal amount of lime dust. Doing this diminishes the moisture and alters the pH of the floor area, thus creating havoc for the harmful environmental organisms.

Dry Cow Management

When cows go dry, consider using the biologic compound Immunoboost (an immuno-stimulant) to help their immune systems be better stimulated against infection in the early dry period. Referring back to chapter 8, it can be seen that the SCC of cows 0-41 days fresh on the organic herds was statistically equivalent to conventional herds — even though the organic herds do not use antibiotics for dry-off. A new product, Orbeseal, a combination of kaolin and bismuth subnitrite is showing very good results as a teat plug to prevent infections. It is not yet certain that it will be allowed for organic use. As cows approach freshening, they must start to prepare for a major change in their bodies. To do this it is good to give them some apple cider vinegar (2 oz. twice daily) for at least 2 weeks. The idea behind this is to have their bones be ready to release calcium with the onset of lactation. If too much calcium is in the dry cow ration, the bones will "stay on vacation" at calving, not releasing the amount drained out of the udder at the first few milkings. It is also important to limit the potassium in the dry cow ration, as well as calcium. It is being found more and more that high potassium levels in dry cow rations often will be the real cause of milk fever and weak cows — especially if many animals are experiencing the problem. Therefore, grassy hays and silage are a safe way to feed a dry cow. However, in certain micro-regions, grass hays may be high in potassium due to years over over-fertilizing with chemicals and manure applications. Slowly increasing the grain towards the day of calving is also good. Unless feeding anionic salts, don't feed alfalfa hay as this crop is known to take up potassium endlessly from the soil. Know your soil — if high in potassium even grassy hays can be questionable. Vitamin E and selenium can also be given in the dry period, once at dry off and again 3 weeks before calving. This is to help normal development of the placental "buttons" so detachment of the placenta occurs soon after calving.

Calving — An Exciting Time!

The cow is bred, you've fed her correctly in the dry period and now she is ready to calve. We all hope for a normal calving, but in cattle it is more common than in other animals to have difficulties at birth. This is due to the way the uterus is attached to the body wall within the cow.

If the cow is on or near her due date, is bagged up, relaxed and swollen yet loose at the vulva and also beginning to push, she is beginning to calve.

Here are some tips to help you out before you need to call the vet:

1) Have the cow in a separate stall, well bedded and dry for the calf and for the cleanliness of the cow's udder.

2) Have the cow haltered in a corner so she doesn't dance around when you reach in. Allow her enough rope length to fall easily when the calf is passing through the pelvis.

3) Tie the tail out of the way — but remember to release it later!

4) Have a bucket of warm, soapy water to wash off the cow's vulva thoroughly.

5) Put on a plastic sleeve, lube it well. Always have lots of lube and don't be afraid to pump a few pints into the uterus if needed to get things good and slick.

6) Reach in — What do you feel?

a) Two legs and the head? Good, then normal presentation.

b) Two legs and a tail? This is OK, but the umbilical cord will snap sooner in this position and therefore you must help it quickly, if you choose to help at all.

c) One leg and the head? **Do not** pull the calf in this position, you will break the leg. Try to feel for the other front leg, cover the hoof with your palm, bend it the way it will naturally bend and slowly correct the position. Often, you will need to push the head and one leg back before doing this.

d) One leg and a tail? Difficult to do correctly without tearing the uterus, call vet.

e) Only the tail, 2 tails, 3 legs, 4 legs, or head and tail — call vet.

7) When pulling a calf, make sure you twist the calf about a quarter to half turn once the head and front legs have come out — this helps the calf's hook bones to pass through the cow's pelvis more easily and helps prevent a hip-lock as well as helping to prevent nerve damage to the cow. To do this, perform a "half-nelson" by weaving your hand and arm underneath the calf's armpit and then curving your hand around the calf's neck. Then torque the calf. It is not usually necessary to torque the calf if it is coming backwards. To put it simply,

Normal Parturition

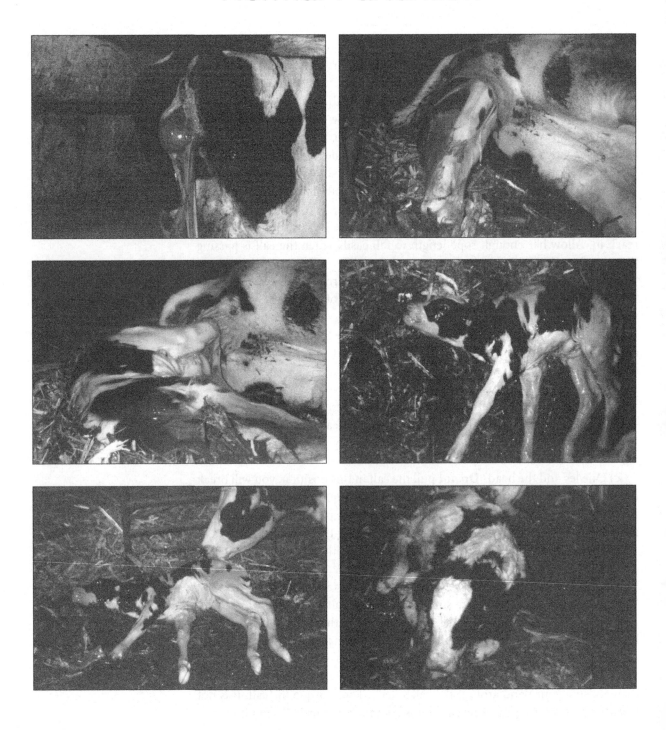

never have the calf's backbone 12 o'clock to the cow's backbone when delivering the calf.

8) If the cow is due, has been straining, but not progressing, this is usually a serious problem which should not be let go for more than 12 hours. If you reach in, which you should, you may feel the birth canal twisting like an auger and only way in can you feel a calf body part. This is most likely a uterine torsion and requires veterinary assistance. I've untwisted many uterine torsions giving a live calf, if the owner doesn't let it go too long! Uterine torsions are extremely common.

While assisting a cow in calving, make sure to cross the legs to turn the calf slightly as it is drawn out. Never have the backbone of the calf "12 o'clock" to the cow's backbone.

9) If you see the calves legs crossed with relatively big hooves when first appearing from the cow, be ready for a big calf. You may need to call the vet. You will probably need traction as well as definitely "torquing" the calf to prevent hip-lock.

Always check for a second calf! It may be in deep so go up to your armpit to check. Especially check if the cow is freshening 2-3 weeks ahead of time. Twins often cause an earlier than normal calving date. While you are checking to see if there is a second calf, also feel along the vaginal walls (birth canal) for any rips and tears. In first calf heifers, there often are rips of some sort or other. Inserting both arms into the birth canal and pushing outwards for many minutes on end can prevent these. This will help expand the walls before the actual calf is expelled (or extracted). If deep rips are felt, these will swell terribly in a few days time and cause great discomfort to the new heifer cow. Get some lard, freeze it, and cut it into

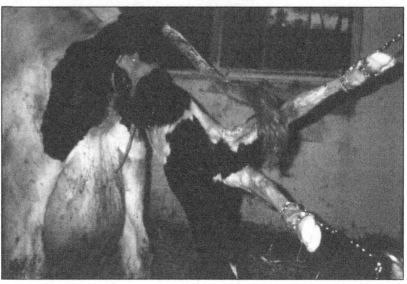

Crossing the legs early on during the delivery makes for a much easier passing of the calf's hips.

"butter sticks" and insert one "stick" into the vagina at each milking for about 5-6 milkings (tip from Dr. Joseph McCahon).

Lard has some mild anti-septic qualities as well as lubricating properties. Arnica given every 1-2 hours after a hard calving is good to do as would giving 3 aspirin at each milking for the first 3-4 milkings. If a steady dripping of bright red blood occurs, a vaginal artery may have been cut during a hard delivery. Give Phos or Arnica every 10-15 minutes, 5cc oxytocin, or tincture of ergot or caulophyllum. If still bright red blood after an hour, definitely call the vet immediately to prevent the animal from bleeding to death (it certainly can happen). A good tip is to *never* rush a first calf heifer. Instead allow the baby calf to expand the birth canal by letting it stay in there, even if this ends up with a dead calf. A live cow is almost worth more than the calf.

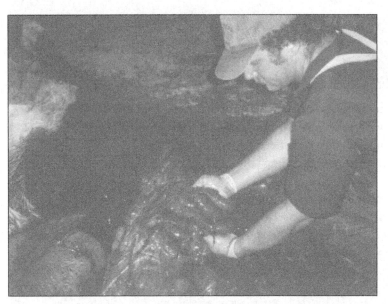

Prolpased uterus with incarcerated intestines — fatal.

Prolapsed Uterus ("casting the withers", "pushed out calf bed")

Apart from a bleeding milk vein, a prolapsed uterus is probably the most serious emergency a dairy practitioner is called out for. It is basically an obstetrics problem and is not usually an easy or clean job from which to emerge. One thing for sure, it takes major strength to replace the uterus into the cow. The condition of prolapsed uterus generally occurs due to two reasons: milk fever (hypocalcemia) in older cows or continual pushing after the calf is delivered, especially in first calf heifers. It is always a good thing to make sure a cow can get up a few minutes after calving — this will alert you to any possible problems to be dealt with and once the cow is up, the uterus tends to stay suspended within the abdomen due to anatomy.

These emergencies are usually observed by the farmer right before milking time, often times early in the morning. The uterus that is outside the body can be quite large, is fragile and is prone to becoming dirty with bedding, manure and other slop. This is the only time that I like to actually see a placenta still attached to the uterus because it will act as a barrier between the dirty environment and the sensitive uterine walls. A prolapsed uterus is unmistakable once you've seen one. The caruncles (buttons) on its walls look like really large mulberries or enlarged pepperoni upon the surface. If this is seen, do not move the cow. Instead, keep her tied so she will not move around on her own. If she is laying down, to keep it as clean as possible for the vet, gently and carefully either lay a plastic trash bag beneath it or place it into the plastic bag. If she is standing, tie her in place and get the vet. If she is out in the field and not near a fence post, have a digging iron ready to hammer into the ground so the vet can put come-alongs onto it. These will be used to bring the cows legs behind her in order to raise the rump slightly in order to facilitate replacing the uterus. The vet must have available help, either one but preferably two persons in order to keep the cow from lying flat out when replacement is being undertaken. I have heard one small animal veterinarian say to me that homeopathic sepia is what is needed for a prolapsed uterus. I am not sure if she really had ever experienced a prolapsed uterus,

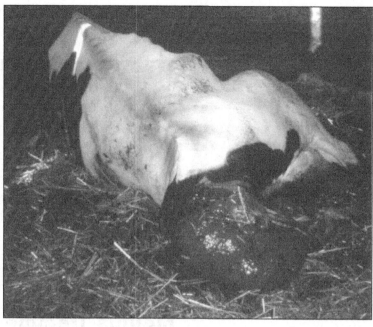

Cow with prolapsed uterus shortly after calving.

for there is major labor involved, and it won't go back in on its own with a simple dose of sepia! Sepia may be of use after the battle, but not alone for the battle! In any event, the cow can die at any time due to compromised circulation and possible blood clots released into circulation upon uterine replacement. However, once the crisis is resolved, most cows go on to become productive members of the herd and conceive again like other cows do. If it is an older cow, a bottle of calcium will be needed. If it is a first calf heifer, no calcium is needed. An epidural of lidocaine will probably be given by the vet (OK for organic), but not all vets do this. Most vets will want to give oxytocin immediately after the replacement is completed in order to shrink the uterus down and to give it very firm tone so it does not fall out of the body cavity again. (OK for organic by the USDA NOP, but some processors are

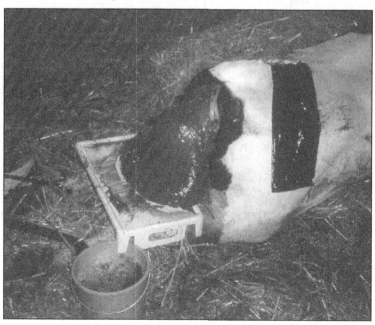

A cow in position to begin replacement of the uterus back into the body — usually a very difficult task.

adamant that their producers not use it since it is a hormone, although natural.) I doubt many veterinarians would know how to use ergot or even where to obtain it, but its physiologic actions and dosage are shown elsewhere in this book. Another compound, caulophyllum, is also described elsewhere in regards to how to use it and how much.

Uterine treatments.

Retained Placenta

After a difficult calving, there is often the potential for the placenta to not detach and pass out. The placenta should be passed within about eight hours after calving, if not, it is considered to be retained. The causes of a retained placenta are not completely understood, but, in the author's experience, they are strongly associated with hard calvings, twinning, and abortions — or low selenium. During the first few days after calving, always watch appetite, milk production and temperature. It is best to feed conservatively (best hay and some grain, but withhold silage). Call vet if any systemic illness is observed (fever). Be especially alert to problems in the heat of summer. Grazing cows weather this condition fairly well compared to those fed lots of corn silage and grain.

Metritis Treatments

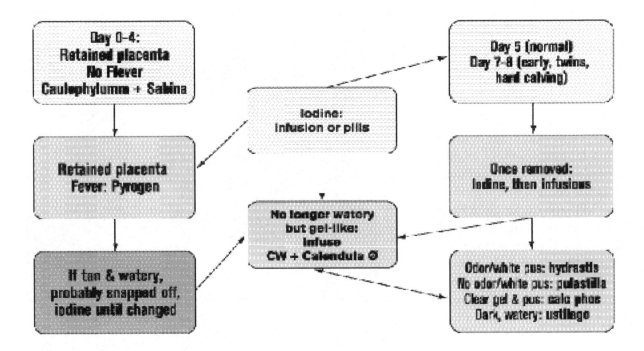

Clinical

1) Caulophyllum tincture alternating with Sabina homeochord, 3x4. Caulophyllum tends to help keep the cervix open to allow the placenta to (hopefully) pass. Sabina is for the bloody discharges usually seen after calving.

2) Can give 1-2cc oxytocin for 5-6 times in first two days fresh if begun soon after calving.

3) If temperature >103.5 give Pyrogen (high potency) and Sabina 3x5 until placenta passes, is removed or until temperature drops to within normal. Echinacea (herbal strength or homeopathic mother tincture) also would suit well if feverish.

4) Use 1 gram iodine (either as solid pills or liquid) every 24-48 hours as needed.

5) After 3-4 days, infuse 50cc Clean Cow with 10cc calendula tincture. This can be repeated every other day until the placenta will gently "tease" out completely (usually at day 5 or 6 after calving). One to two more infusions will finish up the problem in most cases.

6) Alternatively, you can infuse 3% (over the counter strength) hydrogen peroxide at the frequency mentioned above, but make sure the cervix is open to allow the bubbles which are generated to escape.

Conventional: Antibiotics and hormones.

Cows needing manual removal of the placenta will often have a dark, watery, putrid smelling discharge. If infused with Clean Cow and re-checked in a few days, the discharge will often have become more mucoid and pink or even creamy. This is a good sign as it indicates that the uterine glands are beginning to function again, trying to expel remaining debris. Homeopathic remedies can be given orally to enhance this process *i.e.,* Sepia or Pulsatilla.

All cows should always be given one dose of Sepia at day 21 post-calving, regardless if difficulties at calving or not. This may enhance general reproductive health.

Prevention

1) Increased levels of Vitamin E and Selenium in dry cow ration. Sel-plex™ is a source of readily bioavailable selenium which the gut easily absorbs.

2) Give 10cc Vitamin E/Se, at dry off and 3 weeks before projected calving date.

3) Caulophyllum at beginning of labor every 15 minutes until calving is completed.

4) Give Caulophyllum and/or Pulsatilla homeochords in bucket of luke warm water to cow immediately after expulsion of calf.

Vaginal Lacerations

Rips and tears of the vagina are not uncommon in calvings involving first calf heifers. It is unfortunate, but many calves these days are oversized for their heifer dam, either due to excessive feeding of the dam in the last third of gestation or due to the calf's inherent size. By cross-breeding, most calving problems disappear. In addition, impatience by the attending farmer or veterinarian to not allow adequate dilation of the vaginal birth canal causes such problems. Some of the rips can be long and deep, occasionally entering the peritoneal lining and causing a localized peritonitis. By using a lot of lubrication and delivering the calf in a somewhat turned way (as described previously) can minimize such rips. However, sometimes they are inevitable. In such a case, antiseptics and lubricants applied to the vaginal area for a few days after calving will help to relieve pain and swelling somewhat. My colleague, Dr. Joe McCahon, has recommended lard for such purposes, as it has both lubricant and mild antiseptic qualities. Freeze lard in butter shaped sticks and place intra-vaginally twice a day and allow it to melt in place. Also give such a cow arnica every couple hours as needed. Or use aspirin, if a fever is present. Cows with rough calvings usually don't pass their placenta, either. (see section on Retained Placenta). If there is severe swelling, do not try to place your hand into the vagina as it will re-open rips that may be beginning to heal through the natural inflammatory process. Some minor lacerations respond nicely to a few stitches (absorbable). Traumeel™ is a nice gel to reduce trauma & swelling.

Adhesions of the Uterus

Occasionally, an extension of a deep rip is an infection of the peritoneum in the area of the uterus. This causes inflammation and eventual scarring, which entails the uterus becoming stuck in place. An alternative way for this to happen is when an inexperienced person tries to infuse a cow's uterus by passing a pipette and poking through the vaginal lining instead of placing it properly through the cervix to deliver the infusion to the uterus itself. The severity of the ensuing infection (from either cause) will determine how much area is infected and for how long. A uterus that is stuck in place is usually only stuck at one point or area, not in its entirety. Tincture of time (the natural healing process as time passes) is what usually resolves this problem. I also recommend silica, 1-2 times daily for 3-4 weeks to help out. When I find a cow with this condition in a herd where I do routine periodic herd checks for fertility, I have found it usually takes about 5 months for the uterus to become fully moveable again, and thus be ready for breeding.

Conventional: Tincture of Time

Cervicitis

Inflammation of the cervix (os uteri) is a common sequel to a difficult calving. The cervix will feel enlarged upon rectal palpation during a routine herd check. I usually tell the farmer that the cow had a hard calving, much to the farmer's amazement. I then say that the cervix is enlarged and that is a tell tale sign. This is generally a benign condition early after freshening; however, a cervicitis that lasts well into lactation will be detrimental to the cow being bred back since the cervix is not tightly shut between heats and bacteria could enter the uterine environment. I usually call for silica to be used in these situations as recommended in the above section on adhesions of the uterus. Further re-checks on the cow will determine the resolution of the situation and future fertility.

Conventional: Tincture of Time

Cystitis
(bladder inflammation/infection)

This condition is infrequently seen; it is usually noticed when a cow urinates little amounts frequently and may have a reddish color. Rarely, dried urinary crystals are seen and adhered to the bottom of the vulva and will look like tan colored segments which will feel sandy or gritty if rubbed between the fingers. Often cystitis is a secondary infection due to pyelonephritis (infected kidney) or may give rise to pyelonephritis. If traced back in time it will not be surprising if the cow did not clean correctly and had a uterine infection early in lactation.

Berberis homeochord 2x7
Conventional: Antibiotics

(Dadd, 1897)

Powdered Black root	½ oz.
Warm water	1pt. (p. 126)

Thick urine, turbid deficient in quantity or voided with difficulty

Juniper berries	2 oz.
Boiling water	2 qts.

Strain; dose, 1 pint every four hours

Slippery elm	1 oz.
Poplar bark	2 oz.

Make a tea; sweeten with molasses, and give in pint doses every four hours (p.160)

Typical look of a pinched nerve cow (knuckled hoof).

Calving Paralysis

Sometimes there can be a mixed condition of milk fever and calving injury; if cow cannot rise after treating for milk fever, calving injury is likely. Nerve injuries can be mild or severe. The mild ones will be able to get up, but a hoof will knuckle and not be able to bear weight well. The prognosis on these cows is good. If cow can get up but is weak, staggers and falls, calving injury is still possible, especially if the symptoms persist despite milk fever therapy. First calf heifers will often be unable to rise immediately after the difficult calving if they have a pinched nerve (immediate onset). However, in older cows, I often will see a delayed onset of up to 24-48 hours before the animal is unable to rise due to a pinched nerve. Always remember to check the cow's milk as coliform mastitis can also make a fresh cow unable to rise. Diarrhea in a down cow should also make you think of coliform mastitis.

1) Tincture of hypericum or hypericum homeochord every two hours; can substitute or alternate with tincture of arnica or arnica homeochord.

2) If after 2 days she is not up, alternate hypericum homeochord with Conium mac.

3) Bellis perennis 3-4/day for hard-calf pulling.

4) Always have cow on good footing to help her get up. When she tries to get up, be holding the beginning of her tail to help balance her. This can be of tremendous value!

5) Place the knuckled hoof squarely on the ground once she is standing.

6) Rent a flotation device (Aqua-float) which fills with water; this rises the cow and lets her re-learn and gives confidence in footing.

7) Consider anti-inflammatories compounds prior to, or during, flotation (flunixin OK for organic).

8) Electro-acupuncture at GV-2, GV-3, BL-31, BL-32 (bilaterally) if obturator paralysis. If peroneal nerve paralysis, electro-stimulation acupuncture points BL-31, BL-32, BL-36, Bl-40, GB-31, GB-34, ST-41 and BL-67. It should be noted that paralyzed cows with no sensation at all at BL-67 (basically the fetlock area) have a bad prognosis. However, cows that eat hay or drink water while undergoing electro-acupuncture usually seem to have a good prognosis. It is especially interesting that while actually inserting the needles, some cows

begin to eat hay. My wife, Becky, calls this "the eating point."

Conventional: Anti-inflammatories (steroidal and nonsteroidal).

This condition can take up to two weeks to fully resolve. Be *patient* and persistent. If not resolved in about two weeks, she may have a broken pelvis and will never walk again. "Hip lifters" if used improperly can cause a broken pelvis. Never, ever lift the cow from the ground only by hip lifters. They should be placed on the animal and then she should be stimulated to get up on her own. Hip lifters should be engaged if she is almost fully up or is up, but unsure of her footing. The cow should be in the hoist for about a half an hour at a time every 3-4 hours. Putting a cow in a sling and "hanging her there" for a few days is dangerous in that the side poles of the sling can rub into the hook bone quickly. If this happens, the cow is damaged worse. Water flotation tanks are therefore the safest and give the best results. There are also belly slings which you can have the cow in to assist her to regain her strength in footing. These work well for mildly paralyzed animals. If an animal is eating well, have patience as she may be the animal that does take a week or so to get up and around. However, if an animal that started out bright and eating well but then begins to grind her teeth (evidence of pain) and has a diminishing appetite for feed and water, cut your losses and send her for salvage. Sometimes a cow that is down for a few days' time will experience muscle damage (usually due to not turning her from side to side every few hours), and her urine will be a brownish tea color (myoglobinuria). If this is seen, she will not rise again. This is because the amount of muscle injury is so great that the break down products from the damaged muscle enters the circulatory system and is eliminated in the urine.

Electro-acupuncture applied to down cow with calving paralysis. Prognosis is usually better if cow eats or drinks durning acupuncture session.

The Newborn Calf

First minutes

In nature, after the cow calves, a newborn will take its first breaths within a minute of being born. Normally the umbilical cord snaps as the calf is pushed out and the calf thus becomes responsible for its own airway breathing and circulation. The calf usually shakes its head a few times as it wakes up to its new environment. In about 5-10 minutes, when sitting up and breathing regularly, the calf will attempt to rise a few times, flop over, and give up temporarily. After the first 20 minutes or so, a calf will usually succeed in standing for a moment or two. Once they get a little bolder and hungrier, they rise and instinctively move towards the cow's udder to find the most mammalian of all nourishment — milk! All the while the calf is trying out its lungs and legs, the mother cow will be licking the calf, to keep stimulating it, and usually be dripping milk from her udder as well. This simultaneously cleans up the calving area so no predators will be attracted to the vulnerable newborn. This is also the only time in a cow's life when carnivorous (flesh ingesting) behavior is expressed. She will instinctively seek the placenta, which is normally expelled within 30-60 minutes after birth (sometimes up to 8 hours).

Some farmers like to sprinkle salt or minerals on the calf so the cow licking the calf will quickly get some mineral boost but more importantly this will make the cow thirsty. She will definitely become thirsty for a bucket or two of water, which all cows should be interested in after calving.

Colostrum

The first milk of the cow after calving (the colostrum) is vitally important for the calf to drink as calves are born without protective antibodies. This is due to the anatomy of the placental barrier between calf and cow. The calf does, however, have a somewhat functional immune system which can mount a defense. The calf's gut is ready and waiting for the colostrum's vital protective nourishment. The lining of the intestine remains fully "open" for colostral nourishment for about 12 hours and then rapidly begins closing the special pores and becomes a more mature intestine by 24-36 hours. This has practical implications. First, get colostrum into the calf within the first couple hours of life. If born overnight, make sure its nursing when you see it first thing. Don't wait until after all the morning chores are finished. If not sure if nursed, feel the teats, they will feel silky smooth if they've been nursed. If you are not sure, milk out about $1/2$-1 gallon of colostrum and bottle feed the calf. Make sure the teats are clean and dry. Failure of Passive Transport is a term which indicates that the calf did not receive enough colostrum or not enough colostrum of good quality. These calves will always be more susceptible to infections until their own immune system begins to fully function (at about 3-4 months old). If a farmer is certain the calf did not suck colostrum from the cow (cow died or other reason), the farmer should administer a source of passive antibodies as quickly as possible (not probiotics, but *antibodies*). There are many excellent commercial products now available.

The biology of the "open" intestine ready for protective colostrum is also unfortunately open to pathogenic "bad" bacteria and viruses. Although the newborn calf is fully competent to mount an immune response in the face of challenge, it doesn't have much in the way of reserves to weather a challenge. That's why colostrum is so important to get into the newborn. Colostrum, by its nature, is loaded with antibodies made in the mother cow. These antibodies are specific to the cow, any recent illnesses she's had and the environment of the farm itself. Older cows tend to have richer, higher quality colostrum than younger, 1st calf heifers. This is because they've endured more illnesses, whether the animal caretaker has noticed them or not. One way to ruin the nutritious nature of colostrum is to have a cow with manure on her teats. Manure can easily contain many of the bacteria and viruses that can kill a calf quickly. It is also the number one way to infect a calf with Johne's disease. (Johne's disease will initially show itself when a cow calves for the first time, although the cow was first infected as a young calf.) The cow's manure usually contains, among other things: coliforms, *Salmonella*, clostridia, parasites, Rota and Corona virus.

If the calf ingests these organisms by accident, even with a good dose of colostrum, it can contract the disease and show symptoms (usually life threatening diarrhea and dehydration). A client will sometimes ask how long should a calf stay with the cow. This really depends on each farm's unique situation. There's really no clear cut rule. If Johne's is a problem, then taking the calf away immediately may be best for the calf later in life. More rational is to not use milk from known Johne's positive cows.

Cow and calf together

In Biodynamic farming, the calf usually stays with the mother for a week, although the cow is milked regularly along with her herd mates. This allows a bonding to occur which fortifies the calf's vitality and natural strengths. Keeping a calf with a cow will also help the cow stay milked out, which can be beneficial if there is a mild mastitis problem. Generally, I like to see a calf with its mother for a whole day at least. The only problem with letting a calf stay with the mother is that you are not sure how much colostrum it is actually taking in. If committed to keeping calves with cows, then it may be good to check the antibody level with a colostrometer. I must admit, however, most calves lucky enough to stay with their mom longer than a day or two usually do perfectly well. Without doubt, the wealthiest most vigorous calves are those allowed to be with their moms until weaning. They grow fastest and rarely, if ever become sick.

Calves and hay

It should be stated clearly that a calf is born with only its true stomach being functional — much like a foal, piglet, puppy or kitten. The three pre-stomachs (reticulum, rumen and omasum) are stimulated to develop and begin functioning when a fibrous material is being consumed (grass or hay). I

Calf eating fodder (can be cause of parasites or choke).

believe calves which are fed hay as early in life as possible will be healthier due to the functioning of the rumen with a fiber mat. The rumen is one of the most important aspects of being a bovine (or any ruminant animal for that matter). With a functioning rumen, animals will automatically be strengthened. It is nearly criminal, in my opinion, to raise calves without hay and fibrous materials. Yet, incredibly, this is what many feed mill nutritionists now preach. The reason behind this is for faster growth with grains and pelleted protein feeds. I have gone to a nutritional meeting at a land-grant university where a top professor of nutrition showed small sections of rumen walls to look at under the microscope to magnify it. The walls were well developed, but perhaps displaying hyperparakeratosis (too thickened). When questioned, the graduate student of the professor maintained that no hay should be fed to calves, as this will negatively impact grain intake (and hence rumen development). Apparently the propionic and butyric acid generated by the grain diet stimulates faster rumen papillae growth than the acetic acid generated from the breakdown of fibrous hay fed. It was stated that a calf should be fed only milk and grain until weaning (with 5 pounds of grain being eaten daily at time of weaning for best rumen development) and then begin to introduce hay. As a vet, I have seen terrible digestive problems in calves fed this way —usually due to an overabundance of parasites of one sort or another. Or worse and more fatal are when clostridial bacteria which inhabit the gut take the upper hand and rapidly multiply when the animal's whole system becomes acidotic due to high starch feeding. The simple addition of hay, that the calves instinctually crave and gobble up, usually resolves or prevents bad situations from arising in the first place. Again, observe calves in a pen, you will see them nibbling at straw or hay that is in anywhere available, even partially submerged in the bedding pack. Doesn't it seem to make practical sense to have hay in a rack off the ground so they have the ability to eat clean fibrous material? In response to my question to the professor regarding watching calves eating fibrous material from the pen, he said that, yes, they would in any event, but by not feeding them hay in a rack we could keep the amount of fiber ingested to a minimum so they would eat maximal amounts of grain. I just had to wonder who had paid for that kind of research to be done (possibly the grain companies??).

Post-Weaning Disasters

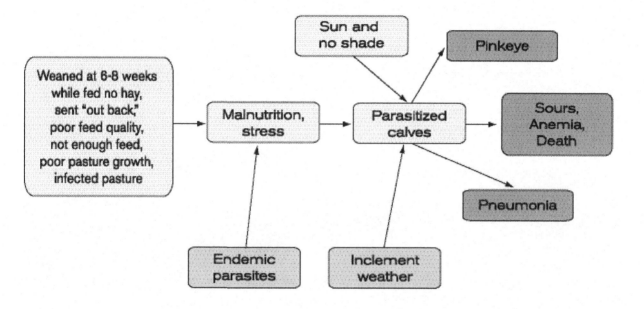

Treatments, Part 3

Udder Problems

Udder Edema (Udder swelling)

Udder edema is the swelling in the udder and in front of it that is commonly seen in first calf heifers but sometimes older cows as well. It is not totally known why it happens; however, excess salt intake in the dry period (especially in springing first calf heifers with access to free choice salt feeders) or the accumulation of colostral antibodies in the udder and subsequent drawing of bodily fluid to offset the concentration of antibodies in older cows are likely reasons. Udder edema also tends to remain longer after calving if there is a uterine infection — this is also usually in association with an elevated somatic cell count or even a few flakes of mastitis. If a cow that has a lot of edema develops a displaced abomasum, the edema goes away quickly, but then the cow has an arguably worse condition.

1) *Apis* 30C if cow likes cold applied to swelling (2-3/day).

2) *Urtica urens* 30C if cow likes warmth applied to the swelling.

3) Coffee/caffeine, dandelion, potassium nitrate (saltpeter), juniper.

4) First calf heifers will show this condition almost always; if severe and pressure is causing discoloration and the tissue is becoming devitalized, induction of calving may be the best treatment choice.

5) Apply a nice salve to any swollen udder or teats

Conventional: Diuretics (furosemide) and steroids (if not pregnant).

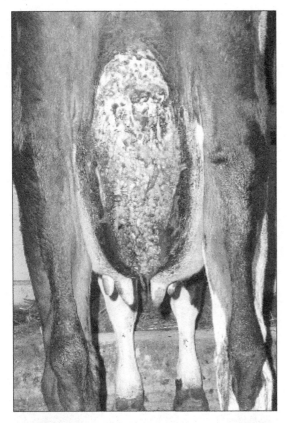

(Herbal, 1995)
Compound Juniper Berry: Fresh ingredients of the following — Juniper berry, Spring Horsetail, Corn silk, Goldenrod Flowering Tops, Cleavers, Marshmallow Root (p.51)

(Nuzzi, 1992)
Diuretic Formula: Burdock seed, Chickweed Herb, Corn silk, Cranberry, Dandelion Root/leaf, Goldenrod, Parsley leaf/root, Stone root, Uva ursi (p.47)

(Dadd, 1897)

Oil of Cedar	1oz.
Oil of marshmallows	2oz.
Soft soap	2 lbs.

Mix, give ½ lb. daily in feed (p.142)

Powdered mandrake	1oz.
Powdered lobelia	1oz.
Poplar bark	2oz.
Lemon balm	4oz.
Boiling water	3qts.

Let stand 1 hr; then strain and add a gill of honey;
    give ½ pt. every 3rd hr.

If surface extremities are cold,

Hyssop tea	2qts.
Powdered cayenne	1tsp.
Powdered licorice	1oz.

Mix and give; repeat if necessary

If inflammatory symptoms appear: omit cayenne and substitute with 1 tsp. cream of tartar.
    Also use laxative of:

Wormwood	2oz.
Boiling water	2qts.

    Boil a few moments, then add 2oz. castile soap, a gill of molasses or honey; give entire dose at one time (p.143)

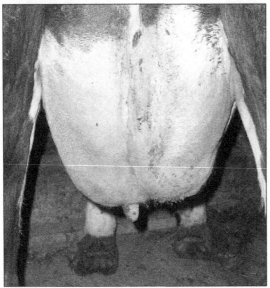

Lesion due to severe udder edema (above), severe udder lesion healed (below).

(Waterman, 1925)

Saltpeter (Potassium nitrate)

Acts on blood and kidneys, causing the kidneys to secrete a large amount of urine. Can help reduce fevers and dropsical swellings (udder edema, etc).

Cattle: ½-1oz.; Sheep ½-1 dram; Horse 1-4 drams (p.665)

(Burkett, 1913)

Saltpeter 4 tbsp/dose; 1time/day for 3-4 days (p.148)

(Dun, 1910)

Saltpeter (effects last 24 hours): 2 oz. orally for cattle/horses (p.183)

Juniper; 3-4oz. to cattle and horses

Oil of Juniper is best for diuresis: 1-2 drams to cattle and horses (p.593)

Turpentines (oil of turpentine): 1-3 oz. to cattle and horses, minimum, frequently repeated; mix with milk, oils, linseed gruel, mucilage, eggs or about 1/20th part of magnesia (p.600)

(Winslow, 1919)

Diuretic mechanisms: stimulate the renal cells or lessen absorption from the tubules, or both

Irritant diuretics (contain volatile oils, resins or aromatics): buchu, juniper, turpentine and cantharides

Irritant glucosides: scoparin and asparagin (p.45)

Caffeine, stimulates the secreting epithelium: $^1/_2$-2 drams to cattle and horses (p.293)

(Mowry, 1986)

Blue vervain, buchu leaf, burdock root, cayenne, celery seed, cleavers, corn silk, dandelion root, gotu kola, juniper berry, kelp, licorice root, parsley, pumpkin seed, queen-of-the-meadow, red clover, sarsaparilla root, uva-ursi (p.303)

Teat Blisters

This is a condition seen usually in first calf heifers, especially in the wintertime. It is associated with swollen udders and is usually caused by a herpes virus (herpes mammilitis). Either one or all four teats can be affected. Milk out is usually hindered and mastitis can occur. There will always be a sloughing of the original skin layer. Needless to say, teats are sore and painful, even before any potential mastitis sets in. There is no real good treatment, but calendula-echinacea ointment should be applied at least twice daily if not more frequently. Reduce or eliminate iodine based teat dips or other possible causes of excessive drying of the teat skin during the damp and cold times of the year. Caution! Teat blisters along with blisters on hooves and in mouth could be Foot and Mouth Disease.

Conventional: Antiseptic ointments (i.e., Nolvasan/chlorhexidine — OK for organics).

Pink Milk

This often occurs at calving and is generally hailed as a sign of an animal that really wants to milk. It is usually caused by hemorrhages of small blood vessels as the udder becomes activated. Therefore treatment is aimed at dealing with trying to help reduce the bleeding. Pink milk can also occur later in lactation and then is normally due to trauma to the udder, either by the cow banging it on something or a cow ramming her (especially a horned cow). These are treated the same way. Rarely, a cow will come down with classical leptospires with all four quarters giving pink milk. The cow's urine will also be red in this case. Call your vet, as Lepto is a serious disease and the cow is usually really sick. On even more rare occasions, if all four quarters have pink milk later in lactation, a tumor in the base of the udder may be the culprit.

1) Bottle of Calcium IV slowly and 250cc bottle of Vitamin C IV along with Vitamin K, 100 cc, SQ (divided into 5 sites).
2) Arnica homeochords 6x1. If no change, then
3) Phos 30C 4x2. If no change, then
4) Ipecac 30C 4x2day. If no change, then
5) Bufo rana 30C 4x3.
5) If from obvious injury, apply Arnica ointment on bruise.
6) Do not give aspirin, as this tends to inhibit clotting.
7) Keep the cow in a stall and do not let her move around much as the tiny blood vessels in the udder will keep bleeding by being jarred around.

Conventional: Same as step 1.

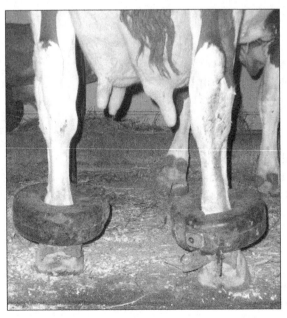

Donuts to protect long teats.

Stepped on Teat

Often damage is done only to the tip of the outside teat and/or the lower part of the milk canal. This hinders milk out. More rarely is when the milk canal itself is ripped open and milk comes out from many places at once. Do not wait a few days to see if it will get better, it won't (see Lacerations). Mastitis will soon set in. Corrective suturing (stitches) need to be done *the same day/night the milk canal is ripped open.*

For less severe problems:

Check for fever

1) Milk slightly impeded because tramped, use Thios ointment.

2) If a simple cut on outside of teat which hinders milking machine and cow kicks a lot, use Echinacea/Calendula/Hypericum ointment.

3) Give Arnica homeochord in mouth for pain.

4) Consider using a teat dilator between milkings. Use proper hygiene!

5) After teat has improved, try Silica 2/day for 1-2 weeks to reduce scarring.

6) Have your vet surgically open the damaged canal.

7) If mastitis occurs after dilators have been used and time has gone by, the usual germ is the *corynebacterium (arcanobacterium)*, especially if there is thick pudding mastitis with a foul smell. If this is the case, and the quarter is hard, try to kill the quarter (which see).

Blind Quarters

First calf heifers that have been sucked on as a calf will often come fresh with one quarter not functioning. Even though the quarter feels normal and the teat feels full, there is no canal and it will never milk. Sticking a bistoury up the canal to open it is stupid, as infection can be initiated. The way I check to see if a quarter and teat is truly functional in one of these cows is by rolling the teat between my fingers. If it is a truly blind quarter, there will be the feeling of a "string" or a thickening in the center of the teat as it is rolled between the fingers along with a limp feeling of the teat. Normal teats with normal streak canals will feel as though a hollow tube is there when it is rolled between the fingers and feel plump because they have milk in them. Obviously, there is no treatment for such a quarter and teat.

Lacerations (teat cuts)

Teats can and do get stepped on, either by the cow herself or by a neighbor cow next to her. I believe that most cows that cut their teat so that milk is gushing out from the streak canal (as opposed to its normal opening) do so with their dew claws. These cuts are often more vertical than horizontal, and it seems a likely cause. Some farmers elect to remove the inner dew claws from the rear legs of all their calves early in life (as when dehorning a calf). These herds have a very low (to no) incidence of tramped teats and teat cuts. Some folks who have had a particular cow tramp her teat in a previous lactation will buy a "donut" (soft rubber circle) that covers over the dew claws to prevent cuts. The problem is that you simply never truly know which cow will get a cut teat; thus prevention would be to remove the inner dew claw as a calf. I would only recommend this in herds experiencing a high prevalence of smashed and/ or cut teats. Neighbor cows could also be the cause (and you never really know) — if so, stall design needs to be evaluated. Stalls too small for the cows will give rise to smashed teats. As Holsteins have become bigger, older barns may not be able to accommodate the larger animals.

In any event, a cut teat with milk coming out of its side must be stitched within 6-8 hours before inflammation/swelling set in (making the tissue too plump for correct stitching and holding together). These cuts will never heal, no matter how small the cut is that gets into the teat canal. The reason is that when the milk letdown occurs, the milk will take the path of least resistance, which is where the cut is. Since milking occurs everyday twice a day, it will never heal. The exception may be if it is a very small pinpoint opening and the cow scars it over in the dry period. But as rule, they do not heal.

Cuts not involving the teat canal but only the outer skin of the teat need to be evaluated on a case by case basis. Some require stitching, especially if they are circular in nature and go more than a quarter ways around the teat. Cuts where there is a flap of skin which has been tended to by the farmer for several days with no real improvement can easily be snipped off and allowed to heal in by "second intention" (naturally). If the cut is not infected, using calendula-echinacea ointment on it and covering with a protective barrier (a hoof wrap) will often allow these to heal up over a few weeks time.

Teat Leakage

Cows that leak milk are prone to environmental mastitis (coliform or strep non-ag). They may leak due to enormous amounts of milk in a huge udder when fresh or due to previous teat damage. These cows need to have a teat sealant applied at the end of milking to prevent leaking in between milking. Collodian is an old-time sealant that is still available. Organic farmers should check with their certifier before using this routinely. Teat tape would be a reasonable alternative.

Teat Amputation

Sometimes a teat is "too far gone" as far as being smashed or cut that it may be more prudent to amputate (cut off) the teat. This should not be a first choice and should be reserved for truly needy situations. Some cuts leave a dangling end of a teat which will never be stitched back on (or heal correctly), and these are better snipped off. If half a teat remains, milk will flow unimpeded when milk letdown occurs. They may also become leakers. Obviously, this is not a good situation but may be better than having a non-healing teat to deal with every milking. These will also need calendula-echinacea ointment and wraps applied to help the cut surface progress to granulation tissue and to keep dirt from sticking to the end of the shortened teat.

Amputation is called for when a cow has gangrene mastitis. Once the cold, bluish teat that only releases squirts of gas upon stripping occurs, cutting off the teat is best, for it will allow free escape of noxious gangrenous gases. The cow herself will no doubt be systemically sick and need to be tended to, but removal of the dead teat should be done to help her locally.

Teat Fistula
(teat hole where it shouldn't be)

Once in a while a heifer comes fresh with a mini second teat opening somewhere along the side of the teat (or may even have a very small extra teat along the side of the regular teat). If it doesn't interfere with milking, don't worry about it. Or, use a wide bore teat cup and liner for that particular teat on that cow. A small fistula may be able to be surgically shut closed, but wait until the dry period to minimize mastitis risk.

A fistula also can be the result of a lacerated teat which wasn't tended to in time or didn't heal correctly even when stitched appropriately.

Floater/"Ball Valve"

Occasionally I am called by a farmer who cannot understand why the teat will not milk out. There won't be any mastitis, no tramped teat end, no problem that can be observed at all. The milk simply doesn't empty out of the quarter. This is almost always due to a "ball valve"/floater/teat spider blocking the milk canal. These are kind of fun to be called out to, even when the farmer knows what is wrong but cannot remove the obstacle. I remedy this by first finding where the floater is, then once it's in the teat, clamp off the teat with my left thumb and index finger while working the floater down the canal with my right thumb and index finger. Once it is lodged in the teat sphincter, using "gentle strength" I work it on outwards until it pops out to the ground somewhere nearby. Have someone tail jack the cow as she will kick reflexively as the little glob is worked out of the sphincter. It is usually a fatty glob that is brownish. No instruments should be needed to get these out. Sometimes in a really hyper/mean cow, a dose of xylazine will allow easier removal. They are sporadic and don't tend to recur.

Membranous Obstructions

Sometimes a first calf heifer will have a thin membrane in the teat blocking milk flow. These need to be aseptically pierced with a teat instrument. Let the vet do it, and be ready to use xylazine sedation as cows do not like to have their teats "opened." Some heifers will have very thin milk stream from all four quarters. Get the vet to open these up when the heifer is just fresh a day or two. Usually it takes only one insertion of the instrument into each teat for these types of problems.

Killing a Quarter

Sometimes it becomes necessary to kill a quarter, due to long-standing active infection of the mammary gland. The farmer gets tired of stripping out the bad secretion, but the quarter just won't die off on its own. It is possible to kill a quarter chemically with chlorhexidine, but only do so if there are only about

10-20 strips at most from the bad quarter at every milking. Use 60cc chlorhexidine infused into the quarter for 4 milkings in a row *without* stripping out the quarter at milking times. (Never use this method in goats, it will kill them.) There'll probably be some swelling of the quarter, but there probably already had been. Watch that the cow does not go off feed and check her temperature. Do not try to kill a functional quarter, like one that is milking a few pounds of milk each milking. Try some of the mastitis treatments which follow.

Increased Somatic Cell Counts (SCC)

Increased bulk tank somatic cell counts and addressing the problem is critical before being "shut-off" by the milk company. What exactly is meant by "somatic cell count"? This is the amount of white blood cells being secreted with the milk from the mammary gland. These cells are part of the cow's immune system and reflect the cow's natural attempts to fight off infection in the udder. So, in a sense, "somatic" or white blood cells are good, but having too many of them reduces milk quality and quantity. What causes the infection? Usually bacteria of different types find their way into the udder, and depending on the environment in the udder and the immune capabilities of the cow they take up residence and create problems. Typical bacteria we find are staph, strep or coliform. How do the bacteria ever get into the udder? Generally at milking times through improper teat preparation, improper milking technique, poor milking machine function (vacuum fluctuations at the teat ends during milking) and from the last cow milked to the next one as well as leaky teats. Generally a farm that consistently has a bulk tank count every month of 400,000-800,000 will have contagious staph or strep identified in milk cultures. Farms that have consistently low counts (less than 200,000) are more prone to flare-ups with coliforms in my opinion. This is because the cows' immune systems are not being challenged/turned on and this may allow an environmental coliform infection to sneak into the teat and gain a footing. Check your DHIA records for any cows with linear SCC of 5-9. Then run a California Mastitis Test (CMT) on those cows to determine which quarters are involved. (If you're not on DHIA, the only way to identify the cows is to CMT all cows in the herd.) Then culture the involved quarters to see which bacteria are involved. By knowing the bacteria, we know the factors that led up to the problem and can then control and prevent new infections.

A common finding in fresh cows is one or maybe two quarters that are high on the CMT paddle, yet show no (or few) flakes. Usually, I find this in cows which have not yet cleaned correctly or still have a retained placenta. Until the uterus is OK, there will be some high CMT, as well as a little extra udder swelling that doesn't seem to completely go away. Although veterinarians are not taught about a direct connection between the uterus and mammary gland in school, it doesn't take a rocket scientist to understand intuitively that a connection does indeed exist between these two organs in the female (especially right after birth!). Try it out for yourself — CMT cows that didn't clean right and check her later on DHIA. So, don't get too worked up by a mild infection in one quarter of a fresh cow that didn't clean. Instead, work on the uterus to get that corrected by 2-3 weeks fresh.

Here are some ways to improve bulk tank milk quality:

1) Strip cows with a strip cup before milking — this lets you visually check the milk and also gets the worst quality milk out before going into the pail or tank.

2) Use a "quarter milker" to keep out bad milk. These are excellent to keep bulk tank milk of good quality but do absolutely nothing for the cow with the mastitis.

3) Rinse milker unit out after a bad cow with a 1 part iodine to 9 parts hot water solution and change out solutions when it cools off. A study done a long time ago in Canada showed that by using 180° F water during milking times (to flush the milking units between cows), contagious bacteria were eliminated.

4) Wear latex gloves (these also reduce painful skin cracks in the winter).

5) Make sure the teats are dry and clean (teat opening itself is the most important).

6) Consider pre and post dipping.

7) Milk first calf heifers first — they are technically the cleanest milk.

8) Try to have the same person do the same things day after day so there no are variations in the milking routine (dairy cows are the pinnacle of the meaning "creatures of habit").

9) Immunoboost, 5cc IM/SQ one time can help to lower a cow's SCC for 1-2 months (thus its potential use also for the dry period). Unfortunately, it doesn't work on *Staph aureus* cases.

Mastitis

It is always best to culture milk to know which organism is present, if early enough in the disease. It is critical to be aggressive with natural treatments at the slightest hint of mastitis. One speck/flake should be enough to signal the beginning of treatment! A CMT plate is excellent at identifying sub-clinical infections. It is an extremely wise investment and should not be allowed to gather dust in the closet. The absolute best way to stay ahead of mastitis and to win the battle without antibiotics is to frequently strip out the affected quarters. No less than 4-5 times a day.

Milking hygiene is also critical to the prevention and during the treatment phase:

1) Clean and dry hands when stripping quarters.

2) Clean and dry teats before beginning to strip quarters.

3) Dip teats after milking; it can be as simple as saline-water or a calendula tincture in water all the way to the commercial products. Don't forget that glycerin has mild antiseptic properties as well as moisturizing properties.

(Macey, 2000)
Udder wash:
13 liters water
1 drop of pine oil
1 capful of hydrogen peroxide (35% — author)
1 oz. of clay

Teat dip:
4 liters water
5cc lavender oil
5cc pine oil
2cc eucalyptus oil
12cc cottonseed oil
5cc methylene blue (p.86)

For 3 milkings try this supportive therapy:
1) 10-20cc Vit. B in muscle.
2) 60cc Vit. C in different muscle.
3) 30cc IMPRO or Biocel CBT under the skin above the udder or at tail-head. Both IMPRO and the Biocel CBT can be given in the vein if needed for quicker action. If you are really good at using needles, an even more effective injection point would be acupuncture point K-10 — just where the udder meets the leg on the side of the mastitis (see Acupuncture Charts).
4) Choose an appropriate homeopathic remedy to give orally.
5) Some people like to give a bottle of vitamin C in the vein, especially if a coliform case.

Common presentation and types of mastitis

There are two basic types of mastitis — contagious or environmental. Only by culturing a mastitis cow can we identify exactly which kind of bacteria is present. And once that is known, appropriate management strategies can be undertaken. For instance, contagious germs usually require dipping cows with a germicide and dipping the milking units between cows in a disinfectant. It is also good to milk first calf heifers first and known infected cows last or separately as a whole group. With environmental germs, attention needs to be paid to cleaning the teats correctly and having clean and dry bedded areas for cows to lie down. Also, teat sphincter health needs to be evaluated, for a damaged teat end no doubt allows environmental germs to enter more easily. I have found that natural treatments (especially the colostrum-whey derivatives) work best on contagious mastitis.

Although it's always best to culture milk to identify for any organisms present, mastitis is such a common problem on dairy farms that a few pointers would be in order here.

1) If any milk has been cultured in the past, what was it? Often times the same kind of mastitis will recur in the same cow (although this is not always the case).

2) **Staph** mastitis cases can be contagious or environmental and tend to make milk be slightly thicker and slightly off-color when a flare up begins. Usually the mastitis comes and goes and is seen in older cows. Individual cow somatic cell counts will be chronically high *i.e.,* 5-9. Bulk tanks that are continually between 500,000-800,000 are usually contagious staph *(Staph aureus)* or contagious strep *(Strep agalactiae). Staph epidermidis* (Staph non-aureus) is an environmental bacteria that inhabits the skin of cattle and people. If this is cultured, it is usually easy to treat with natural treatments, but one needs to make sure the milk sample wasn't taken sloppily and contaminated by the hands of the person taking the sample. (It is always good to wear latex gloves when taking samples.) Occasionally, a severe form of staph occurs — blue bag. This is very much like clostridium mastitis.

3) **Strep** mastitis cases can be contagious or environmental and tend to keep milk white but has small specks/flakes in it at the beginning of stripping. It may seem slightly dilute but still white, although hot environmental strep cases can look exactly like a coliform mastitis case. Many streps are environmentally caused, but *Strep ag* is definitely contagious and now more recently *Strep uberis* can sometimes act as a contagious pathogen be passed from cow to cow. *Strep ag* infected milk can actually look quite normal, and this is a tip-off to me when the cow in question has a really high SCC (can be 2-4 million).

4) **Coliform** mastitis cases are environmental and the secretion will look watery or like lemonade with a hard quarter. Cows will initially have high fevers. Coliform can quickly involve the whole cow; it may be necessary to give hypertonic saline IV to get her to drink and help flush system; oxytocin can help to strip her out more effectively; if toxic, consider flunixin (allowed by NOSB for emergencies). Recent research has indicated that the coliform bacteria is found in the bloodstream as well, and this may be an indication to give an IV antibiotic (oxytetracycline usually) to help save the cow's life. If a cow is down and has diarrhea secondary to coliform mastitis, this gives a very poor prognosis as it indicates a toxic state. This occurs when coliform bacteria die and they release endotoxins (toxins from within their cell make up) into the bloodstream. The resulting reaction of the cow's system is what is so devastating to her (hence the use of steroids to suppress the reaction by the cow with

conventional treatment). There are many types of coliform mastitis — *E. coli, Klebsiella, Pseudomonas, Nocardia,* etc. Each type of bacteria can replicate itself rapidly, dictating aggressive and rapid intervention on the part of the farmer. Cows can be extremely sick within 12 hours. If you suspect a coliform infection, do not just wait and see if it will get better.

5) ***Arcanobacter/Actinomyces/Corynebacterium*** mastitis cases are environmental and are like pudding and have a sickeningly sweet odor. This kind of mastitis often occurs with tramped teats, especially if using dilators to try to keep the teat open. This is also the most common kind of mastitis a young non-lactating animal will develop if it was sucked on (breaks the natural seal) and/or had flies on its teat ends. This will usually cause a high fever (105°F) and be painful to strip out. This is also a common mastitis type in dry cows and it can really ravage them. The outcome for the calf may be abortion or birth of a dead calf at term. Fortunately this type of mastitis only occurs sporadically.

6) **Gangrene** mastitis cases are definitely environmental. This is fortunately a rare form of mastitis, but usually deadly. It is caused by clostridia bacteria which are found everywhere there is soil on earth. Some types of clostridia that are related and more familiar are the ones that cause tetanus (lockjaw) and botulism. Usually the cow will initially have a low grade fever (102.6-102.9° F) and maybe slightly thickened milk. This will rapidly become a port wine colored secretion and have a disgusting smell. Then the teat will begin to become cold, bluish and there will be no secretion, just gassy air being expressed when stripping out the cow (see Amputation). If the cow is down and is further along in age, she will nearly always die within about 4-8 hours. If, however, she is a young cow (1st calf heifer or 2nd calf cow) and standing and relatively bright, antibiotics (ampicillin in vein and penicillin in muscle) can save her life. However, the quarter will slough and be a real horrible looking mess for about a month and then dry up and be shrunken and leathery. Although it may be claimed that homeopathic lachesis can help these cows out without using antibiotics, I am not in a position to comment as I haven't seen it tried yet. Suffice it to say that fluid therapy will be needed as well as pumping the stomach. Since clostridia bacteria require anoxic conditions to thrive (no air), perhaps using liberal amounts of hydrogen peroxide as blasts of air may help. If the young cow seems to be strong enough to make it, once the teat is cold and blue, cut it off so the quarter can drain out whatever is left in there.

7) **Udder abscesses** — these usually indicate a *Staph aureus* infection (contagious). Usually these cows have high SCC for long periods, and then if the abscess releases its contents or it is somehow reabsorbed, the cow's SCC often will go down dramatically. Do not open these udder abscesses! They will eventually open on their own; once they do, then keep them clean as they begin to heal shut.

The following remedies are remedies for mastitis. Remember, it is absolutely mandatory to begin using the natural treatments as soon as you detect any hint that the cow is coming down with mastitis. If you jump on it, you'll probably see nice results, just as I did. If you're pretty sure which type of mastitis you have, the homeopathic nosode would be a wise choice in addition to other remedies. Although I have not found nosodes to work well as preventatives in general, I have found them to be valuable at the time of infection. The ideal nosode is a milk culture done on cows from your herd and then the culture is made into a nosode. However, the commercially available staph and strep nosodes do well. *Always, always check for fever if mastitis is present and the cow is not eating normally.*

1) Agri-Dynamics products — such as Biocel CBT inject 30cc under the skin, once daily for three days in a row, then one time again seven days later; can also be given IV (use brand new whole bottle). This product works really well on high somatic cell count cows.

2) IMPRO products — these products are from hyper immunized dry cows whose colostrum is then harvested and processed and becomes a source of passive antibodies. (The company will never tell you that due to FDA rules on making claims.) I think they are great products, especially if you know specifically what kind of mastitis you are dealing with. They have injectables specifically for strep or staph or coliform, etc., but also nice combination products. Although they are making more and more oral products, I still recommend the injectables as source of passive antibodies to act as a band-aid as well as immune support.

3) Natural, non-antibiotic udder infusion *(i.e., Phyto-Mast)* into udder for 4 milkings in a row. This is only available to veterinarians through Dr. Karreman.

4) Udder Mint, applied topically on a hot, hard quarter to cool and soothe the inflamed areas. Adding Belladonna Ø and Phytolacca Ø to the mint lotion would even help to further reduce the inflammation.

There are also ointments which can be used. For instance, use of the Phytolacca-Belladonna ointment would make sense for a hard quarter that is hot and painful. Peppermint ointment is commercially available and would be good for the same situation, though not as potent. Do not use the peppermint ointment longer than 2 days since it constricts the vessels, and it is not needed to continually constrict the vessels (but it is good to do so for a short time).

5) Coliform mastitis treatment — IV dextrose with 60cc tincture of garlic, echinacea, goldenseal, wild indigo, barberry, 100-200cc specific passive antibodies, 500cc vitamin C, 1000cc hypertonic saline, 1000-3000cc lactated ringers solution, 15cc flunixin; apply peppermint lotion on quarter and gives 20cc of above tincture 2-3 times daily.

6) *Apis* 30C could complement Coli nosode, as *Apis* is indicated for hard swellings (hard quarter). Also for "summer mastitis"; cows with good condition.

7) *Belladonna* homeochords for mastitis that comes on as the cow freshens; hard-quarter, off-feed; use especially if fever is seen; right-sided; once an hour for 4-6 hours.

8) *Phytolacca* homeochords for cows that fight stripping out; fever; stringy mastitis; chronic mastitis (thickened milk); left-sided.

9) *Aconite* homeochords for any mastitis that comes on fast; once every 30 minutes (can alternate while giving *Belladonna* given hourly).

10) *Pyrogen* 200C for fever, off-feed, possibly septic mastitis cows.

11) *SSC* a shot-gun approach *(Sulfur/Silica/Carbo* veg). Works nicely for sub-clinical mastitis (see only on CMT plate). Good for mid-lactation mastitis that comes and goes. Can help while trying to figure out the right remedy.

12) *Hepar sulph* 30C or 200C good for pudding-like discharge mixed in with milk; abscess somewhere on udder. Use low potency in beginning and increase to higher potency if seeing results.

13) *Pulsatilla* homeochords could complement Staph nosode nicely, as a "*Pulsatilla* discharge" would be thick and creamy; fresh cows.

14) *Bryonia* homeochords for fresh cows, especially Holsteins; cow lays on the effected side of hard quarter; fever; abscess somewhere on udder. Chronic, intermittent mastitis with swollen, firmness to udder (but not necessarily rock hard).

15) *Silica* 30C for hard quarter and scar tissue; abscess somewhere on udder; therapy for dry cows to help soften hardness in udder.

16) *Arsenicum* for chilly cows with watery mastitis that sip at water frequently.

17) *Gelsemium* 30C for generally tired-looking, weak animals. Works nicely in Guernseys.

18) *Calc carb* 30C for a slow, blocky cow. Has worked well with Staph nosode and also Strep nosode; thin white milk that perhaps borders towards watery.

19) *Lachesis* 30C would be indicated for any left-sided mastitis, especially for left-sided coliform with a thin bloody secretion; toxic cow.

Conventional: Antibiotics.

(Protocol Journal of Botanical Medicine, 1996) — Refer to Table 10 for veterinary dosages if dosage is not shown below:

Aconite root (Aconitum napellus) — for acute mastitis presenting pain, fever, chills and full hard pulse; every 2 hours.

Poke Root (Phytolacca americana) — for acute mastitis during or proceeding lactation, with enlarged lymph nodes, use as local ointment to reduce pain and swelling; 4 times daily.

Bryonia root (Bryonia alba) — for hard, pale, hot, very painful breasts; every 2-3 hours.

Echinacea root (Echinacea angustifolia) — acute infection, red and inflamed, with signs of septicemia; 1 ounce every couple hours.

Gelsemium root/herb (Gelsemium sempervirens) — for mastitis with pain and excessive nervous excitation/mania; every 2-3 hours.

Witch Hazel bark (Hamamelis virginiana) — as a local poultice when signs of venous engorgement.

Jaborandi (Pilocarpus jaborandi) — acute mastitis with suppression of milk, hot and dry mammary, dry mouth; every 2-3 hours.

Black Cohosh root (Cimicifuga racemosa) — inflammation with accompanying uterine tenderness and general muscular soreness; ¼ oz. every 2 hours.

Bearsfoot herb (Polymnia uvedalia) — "caked" with stagnant enlargement and impaired circulation with abscess; use as a local ointment or poultice.

Red Clover blossom (Trifolium pretense) — mastitis with lymphatic congestion, use with Phytolacca; 1 fluid oz. 4 times daily.

Figwort root/herb (Scrophularia nodosa) — hard, painful mammae with lymph involvement, use locally as a poultice or a salve; 3 times daily.

Blue violet (Viola spp.) — enlarged, painful mammae with headaches and constipation; ½ ounce 3 times daily (pg 71-72).

Combination tincture of:

Phytolacca, Hydrastis, Iris, Trifolium — ¼ oz. each to make 1 ounce. Then give ¼ ounce in 2 ounces water 4-6 times a day (pg. 66).

Echinacea angustifolia	400 drops
Hydrastis candensis	300 drops
Trifolium pretense	200 drops
Viola spp.	300 drops
Phytolacca americana	100 drops
Glycyrrhiza glabra	100 drops
3 times daily for 5-7 days	(pg. 78)

Herbal ointment:
Narrow-leaved Plantain Leaf, Goldenseal root, Chaparral, Cedar oil, Black walnut (bark, leaves), Burdock Root: compound equal parts and apply to infected area during acute inflammation as needed (pg. 79).

(Udall, 1922)

Milk frequently, restrict the diet, laxatives (salts, arecoline); hot compresses held in position with suspensory bandage.

Irrigate the teat canal with boric acid under strict aseptic precautions. Boric acid treatment: milk the udder completely dry; inject 120-180cc of a 3% sterile boric acid solution; milk out the udder in 3-4 hours until recovery; repeat the treatment in three days if necessary (p.123).

(Nuzzi, 1992)

Lymphatic formula: Burdock root (Arctium lappa), Cleavers (Galium spp.), Echinacea root (E. angustifolia), Ocotillo (Fouguieria splendens), Red Root (Ceanothus Americana), Wild Indigo Root (Baptisia) (p.64).

(Dadd, 1897)

"Garget" — use an infusion of elder and chamomile flowers, long application, while stripping out to collapse the over-distended vessels to their healthy diameter. Also give an aperient, if off-feed:

Extract of butternut	2 drams
Powdered capsicum	$1/_3$ tsp.
Thoroughwort tea	2 qts.

Give all in one dose (p.197)

(Dadd 1897)

If toxic:

Mayapple (Podophyllum peltatum)	1 tbsp
Sulfur	1 tsp
Cream of tartar	1 tsp
Hot water	2 qts.

Give all in one dose

And cold water to the head and rub spine and legs (below knees) with this counter-irritant:

Powdered bloodroot or cayenne	1 oz.
Powdered black pepper	½ oz.
Boiling vinegar	1 qt.

Rub in while hot with a piece of flannel

If trembling, use:

powdered ginger	½ tsp.
powdered cinnamon	½ tsp.
powdered goldenseal	½ tsp.
Catnip tea	½ gallon

Give mix as one dose (p.349)

Sore teats:

1st wash with castile soap and warm water, then apply equal parts lime water and linseed oil.

Chapped teats & chafed udder:

Same as above, but can substitute bayberry tallow, elder or marshmallow ointment.

If fever:

Lemon balm, Wandering milkweed, Thoroughwort, Lady slipper — 2 oz. of any into 2 qts. boiling water. When cool, strain and add a wine-glass of honey. If bad (septic — author) — add small quantity of capsicum and charcoal to the drink (p.178).

(Waterman, 1925)
Cattle: If fat & feverish —
Give laxative:

Epsom salts	1-2lbs.
Ginger	1 oz.
Water	2 qts.

Followed with ½ oz. doses of saltpeter, 2-3 times daily and in very bad cases, 15-20 drops fluidextract of aconite and 1 dram fluidextract of belladonna.

Apply to udder:

4 oz. Camphorated oil + 1oz. turpentine, then shake;

alternate with 3 oz. witch hazel + 3 oz. soap liniment + 2 oz. fluidextract of belladonna (p.166).

Into udder (if bacterial): 2 drams with 1 quart water. Shake and infuse 4-6 oz. into quarter, knead gently for 10 minutes, then milk out. Repeat 3 times a day (p.437).

(Alexander, 1929) cathartics to use when mastitis present:

Glauber or Epsom salt OR ½ grain eserine and 1 grain pilocarpine,

And: (for heifer) ½ oz. fluidextract pokeroot with 2 drams powdered saltpeter; (for cow) 6-8 drams fluidextract pokeroot with ½ oz. saltpeter.

Bathe udder in water and Epsom salt (100 F) (p.25).

Persistent swelling:

Rub in (2 times daily) 2 oz. of turpentine and 8 oz. lanolin or lard, after washing the udder clean with castile soap in warm water (p.27).

Fever with mastitis:

10 drops tincture aconite and 10 drops tincture belladonna leaves, alternately, every 2-3 hours until fever drops; then continue the belladonna every 4 hours and add in 2 drams fluidextract pokeroot to each dose. Also: ½ oz. saltpeter in water every 4 hours or twice daily after temperature falls (p.31).

Infuse weak solution of salt, boric acid or other mild antiseptic (p.33).

For hardening of the udder, stop other drugs and give 1 dram potassium iodide dissolved in water 3 times daily for 4 days and then twice dialy until iodinism. Can use iodine ointment if udder got hard — use twice daily (p.32).

20cc raw milk under the skin every 48 hours up to 10 days (5 treatments). Local reaction possible, painful hot edema and sometimes abscess. Usually no trace after 24 hours. No reaction usually means it will cure easily (p.42).

Sudden suppression of milk:

Fluidextract of nux vomica, 20-60 drops in water, 2-3 times daily.

Decrease of milk yield (not due to infection):

Usual steps in sickness — chilled, shivering, hair standing on end, loss of appetite, decreased rumination, then fever.

To break chill, rub strong liniment on throat, blanket the cow, then:

Every 2-3 hours give by mouth the following mix:

1-2 qts. warm tea or flaxseed tea with molasses

2-4 drams tincture of ginger

½-1 dram tincture of capsicum

20-30 drops nux vomica

potassium dichromate, 3-10 grains, every 3-4 hrs. to stimulate lymphatic circulation

Feed: thin oatmeal gruel or flaxseed tea, twice daily with 1 oz. powdered anise and 1 oz. fennel seeds.

Galactogogue (to mix into cow's feed daily): 1 tbsp. of the mix —

1 ½ parts each of powdered seed, caraway seed and juniper berries, ½ part flowers of sulfur, 1 part black sulfuret of antimony, and 5 parts salt.

Off-flavored milk:

Keep off problem paddock several hours before milking. Let young-stock or sheep graze wild onions, etc. Mix powdered wood charcoal in grain (or activated charcoal). 1 ½ lbs. to 100 lbs. grain mix.

(Alexander, 1929)
Rub in mix of:

Turpentine	1 part
Fluidextract pokeroot	1 part
Lard or sweet oil	8 parts

Give orally:
½ oz/ saltpeter + ½ oz. pokeroot (p. 71)

Poultice udder with antiphlogistine, applied hot and sprinkle spirits of camphor and turpentine on surface. Twice daily rub in a mix of equal parts carbolized oil, camphorated oil and compound soap liniment (p.72).

(Udall, 1943)
Decrease grain and milk out every 1-2 hours; within 1st 24 hrs., use ice packs; after 48 hrs., change to heat.
Acute flare-up of a chronic mastitis is usually better with heat.
Infusion of acriflavine (1:8000).
After acute heat and pain recede, ointment:

Iodine ointment	150 g.
Camphor	45 g.
Methyl salicylate	12 g.
Fat/lard	500 g. (p.64)

(Burkett, 1913)
Rub with hot camphorated oil twice daily. Orally give 8 tbsp. of hyposulphite of soda daily in feed or as drench (p.166).
Rub hot water 15-20 minutes, then dry and apply ointment:
3 tbsp. gum of camphor
4 tbsp. fluidextract belladonna
to 1 pt. clean, fresh lard
Apply 3 times daily

Infuse 3% hydrogen peroxide (p.219).

(M.R.C.V.S., 1914)
Rub in:

Fluidextract of belladonna	1 dram
Camphor	1 oz.

Spirit of wine   ½ oz.
Olive oil   3 oz.
Shake and apply 2-3 times daily

Orally: aloe, salts, nitre and ginger. If early, give with strong ale or whiskey.
chinosol   30 grains (= oxyquinoline; not allowed for organic)
glycerine   2 drams
warm water   1 gill
Infuse into quarter (p.130)

(Dun, 1910)
Boracic acid (3%) infusion, strip out 3 hrs. later.

Ointment: belladonna extract in Vaseline several times daily to decrease congestion, milk secretion and tenderness. When inflammation decreases, then switch to iodine ointment and iodine and salines internally. Decrease rich food to minimize milk production.

Infusion mix:
hydrogen peroxide (5%)
Boracic acid (5%)
Sodium fluoride (3%)
Sanitas (3%)*
*prepared by oxidation of oil of turpentine and containing camphoraceous bodies and hydrogen peroxide
(p.776)

(Winslow, 1919)
Infusion mix: warm 3% borax solution or 1% sodium fluoride. Infused, strip out in 15 minutes.

(de Bairicli Levy, 1984)
Fast 2 days, but give laxative drench daily:
laxative drench (during mastitis):
senna pods, large — 20 (1st soaked in ½ pint cold H2O 6 hrs)
ginger, ground — 1 tsp. (added to above)
3rd morning — 2 pints milk, ½ pt. tepid water + 10 heaped tbsp. molasses
Midday — steamed hay, softened by heating gently over hot water for 1 hr.
Add 2 lb. bran and 10 tbsp. molasses
Evening — repeat

Rx herbs, specific for cure of udder ailments — garlic and woodsage (woodsage or common [blueflower] sage):

Garlic — 2 whole roots grated into 1 pt. water; ½ pint AM & ½ pt. PM or 6-8 4-grain garlic tablets.

Blue Sage (as brew) — 2 handfuls, finely cut, into 1 ½ pts. Water with 2 tbsp. honey added. Give as a single drench in early AM. Can give herb finely cut in bran + molasses when fast is finished. Also: several handfuls of raspberry foliage and herb Robert, also southern wood, twice daily.

Convalescence — steamed or raw beetroot and comfrey.

Rub — 2 handfuls dock in 1 pt. water, give cold as a pack upon udder. (p.21)

The feeling of success is much greater when winning a battle by using natural treatments. The other nice thing is that, if needed, you can always reach for the antibiotics. But, by not relying on them you reduce the amount of drug residues you need to be concerned with, and you allow the cow to overcome the infection on her own. This will hopefully strengthen the cows in the long-run.

Acupuncture Charts

These may be used by farmers or veterinarians to help confirm a suspicion and also to massage a cow (by acupressure). By brushing cows regularly, the cows not only look nicer, but you will invigorate the meridians (energy paths) and probably enhance their well-being. Most acupuncture points on cows are about the size of a nickel and can be felt as a small depression at the points shown. It is perhaps easier by using the ring finger of the hand less commonly used to achieve a heightened sensitivity for the areas shown. I use acupuncture therapeutically on only a few conditions (down cows and reproduction) by injecting 5cc vitamin B12 into the site or using a T.E.N.S. electroacupuncture unit. Injecting homeochords could also be considered. Use a "tail jack" to avoid being kicked. Try to use a 20 or 22 gauge needle if employing injectible solutions. One very important point is located along the bottom junction of the nostrils. I have stimulated new born calves to breathe after a rough delivery by repeatedly "pecking" this spot with any needle deep enough to tap against the hard palate. (Charts by Mattison)

Kothbauer's Diagnostic Pain Points

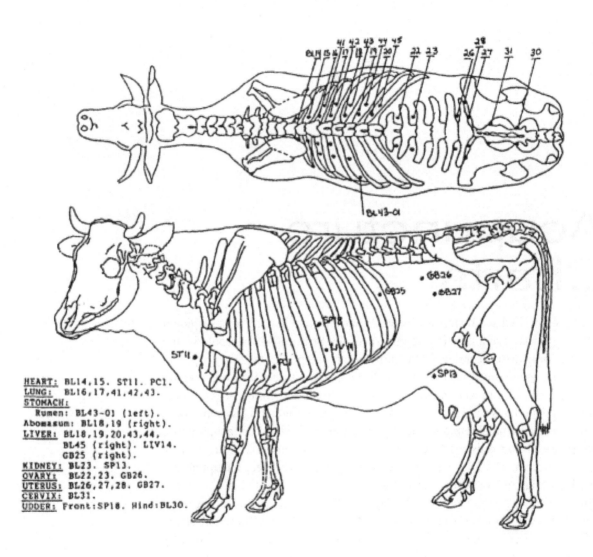

HEART: BL14,15. ST11. PC1.
LUNG: BL16,17,41,42,43.
STOMACH:
    Rumen: BL43-01 (left).
    Abomasum: BL18,19 (right).
LIVER: BL18,19,20,43,44,
       BL45 (right). LIV14.
       GB25 (right).
KIDNEY: BL23. SP13.
OVARY: BL22,23. GB26.
UTERUS: BL26,27,28. GB27.
CERVIX: BL31.
UDDER: Front:SP18. Hind:BL30.

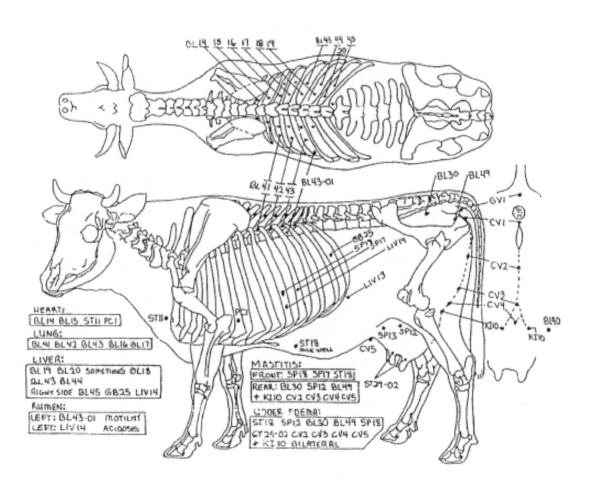

HEART:
BL14 BL15 ST11 PC1

LUNG:
BL41 BL42 BL43 BL16 BL17

LIVER:
BL19 BL20 SOMETIMES BL18
BL43 BL44
RIGHT SIDE BL45 GB25 LIV14

RUMEN:
LEFT: BL43-01 MOTILITY
LEFT: LIV14 ACIDOSIS

MASTITIS:
FRONT: SP18 SP17 ST19
REAR: BL30 SP12 BL49
+ KI10 CV1 CV3 CV4 CV5

UDDER 'DEMA:
ST12 SP12 BL30 BL49 SP18
ST24-02 CV1 CV3 CV4 CV5
+ KI10 BILATERAL

Calving Paralysis

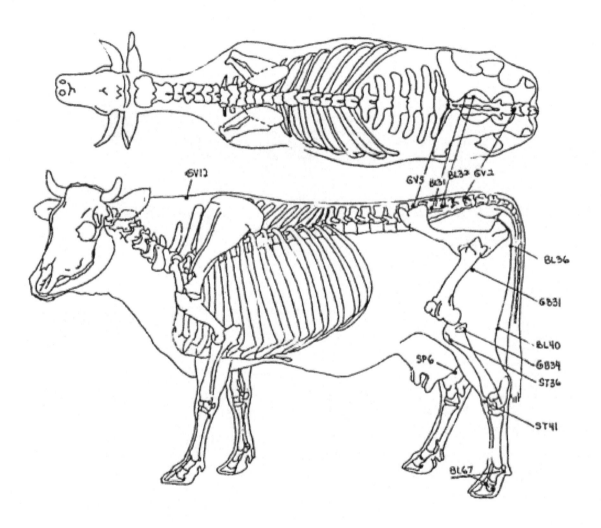

Reproductive Points

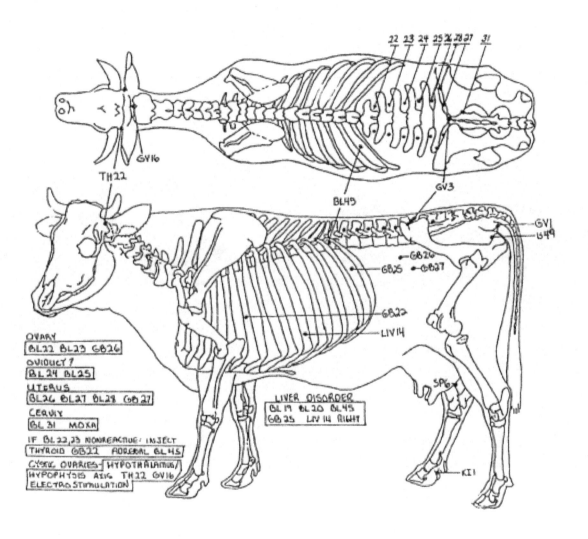

22 23 24 25 26 28 27 31

GV16

TH22

BL49

GV3

GV1
BL49

GB26
GB25 GB27

GB22

LIV14

OVARY
BL22 BL23 GB26

OVIDUCT ?
BL24 BL25

UTERUS
BL26 BL27 BL28 GB27

CERVIX
BL31 MOXA

IF BL22,23 NONREACTIVE: INJECT
THYROID GB22 ADRENAL BL45

CYSTIC OVARIES: HYPOTHALAMUS/
HYPOPHYSIS A216 TH22 GV16
ELECTRO STIMULATION

LIVER DISORDER
BL19 BL20 BL45
GB25 LIV14 RIGHT

SP6

KI1

Milk as Medicine

As an ending to this work, I would like to bring to the attention of the reader a topic which may be of interest — Certified Milk. Although it is historical in its content, there are many points of interest that a modern producer of milk, or a consumer of fresh pure milk bought directly from a farm, would do well to know. I include the topic of Certified Milk in this book only as an interesting aside, since it does not have to do directly with the natural treatment of dairy cows. Certainly some of the treatments for dairy cows as presented in this book (*i.e.*, the historical veterinary uses of plants) would have been used for cows during that period. Moreover, the topic of the production of Certified Milk is directly tied in with the ultimate end product — milk itself. Since there is a small but growing niche market in selling pure milk (raw milk) these days, it would be wise to see exactly how such a product was actually produced and regulated when it was in demand prior to the second World War. It should also be noted that Certified Milk was essentially a medicinal product. If it is produced in an extremely clean manner and is of superior quality, pure milk today can still be thought of as a medicinal product. In our present time, however, perhaps the correct term for fresh high quality pure milk would more properly be called a "nutraceutical" product. There are farmers who believe that whatever they offer for sale is the absolute best possible product — because it originates right from *their* farm. I can understand that a farmer is proud of his product, which is the result of many hours of hard labor. However, hard labor in producing crops from which to feed your milk cows does not necessarily translate into automatically the highest superior end product. There are a number of fine details that can either make or break the final product, some of which were mentioned early in the book regarding healthy dairy farms in general and some which are mentioned in the re-print on "The Production of Certified Milk."

Obviously, times have changed since back in the 1920's. Yet, there are some absolute truths regarding hygienic milking technique that still ring loud and clear today — especially for those farmers interested in direct marketing of their own non-pasteurized pure milk. As a veterinarian, I see a wide variety of hygiene levels among my farmers — some have extremely clean cows and excellent milk quality while others remain challenged as to keeping cows clean and their somatic cell count within reasonable limits. (There were definitely many reasons to pasteurize milk back in the old days, and there still remain valid reasons today.) The quality of pure milk products that farmers make right on the farm (especially raw butter) hinge directly on hygiene of barns, cows and milking technique. When consumers are willing to pay premiums for such products, I believe the consumer has a right to know that the products are only of the highest quality. I would promote visits by consumers any day during regular hours in order that they may see for themselves whether or not "their farmer" is worthy of the premiums received for their products. Obviously Certified Milk was of absolutely the highest quality, as can be seen by the 64 inspection items needed to be approved by the local medical Milk Commission. Hopefully those folks promoting fresh pure milk (raw milk) will keep in mind what doctors of the time had in mind when they thought of milk as medicine (*i.e.,* Certified Milk). There is no room for poor hygiene when it comes to making the best possible milk of the absolute highest quality. If the reader considers all the inspection steps, it becomes obvious that there were not many dairy farms which qualified to produce Certified Milk. It would probably be easier to get a farm USDA Certified Organic than it would be to get it to produce Certified Milk. But what a joy it would be today to drink Certified Organic Certified Milk!

"Pure Milk"**

The Certified Milk standards were obviously very difficult to accomplish and that was the goal. It was to allow only exceptionally well handled and cared for high quality milk to be called "Certified." These standards fell into disuse and there are no more Certified Milk dairies in existence. In their place have come a few "Certified Raw Milk" dairies in certain states which still allow unpasteurized milk to be sold. Pennsylvania is one of those states. Certified Raw Milk licenses are available if the farmer passes the regulations enacted regarding raw milk. There is quite a division between those who believe raw milk should be allowed and those that believe it should be outlawed.

In an effort to forestall and even reverse the trend to outlawing all raw milk, there is a grass roots effort being launched through the country and particularly in key battleground states to encourage the drinking of raw milk for its health benefits. Without fully elaborating on it, I will simply say that as a dairy practitioner who is out in the barn anytime of day or night, it takes an exceptional farmer to produce high quality pure milk over the long run. Simply diverting milk from the bulk tank that is normally destined for the milk truck and selling it to consumers directly is NOT in the best interests of consumers. There ARE legitimate reasons for pasteurization, and I won't go into all the details of why I think there are such reasons. However, IF a farmer is truly dedicated to producing only the highest quality milk possible, then marketing pure, un-pasteurized milk is a legitimate pursuit. Although the 64 checklist items needed to be considered in the old days for Certified Milk are no longer in force, a rather straight forward 16 point checklist (which I whole-heartedly endorse) was created by the respected holistic agricultural consultant, Jerry Brunetti. The 16 points, in their entirety, are below and should be the model from which farmers would be wise to follow if interested in making available healthy "nutraceutical" milk directly to consumers. The goals are not lofty or nearly impossible to meet (as were the Certified Milk checklist); they are worthy and attainable if one is truthfully trying to make the best product in the most honest way for the consumer. If attaining these goals, farmers are voluntarily going above and beyond the minimums set for raw milk licenses in Pennsylvania. To have consumers more involved with the farm, perhaps consider having a Cow Share program where the consumer legally owns a share of a cow and contracts the farmer to feed her, keep her clean and milk her for the consumer. If farmers are interested in this, they should do everything in writing. In addition, in states which require that all dairy products be from pasteurized milk, having personal contracts between farmer and consumer for the farmer to create the dairy product from the consumer's milk from a Cow Share program may keep everything legal.

"Pure Milk" — *term first used by Jerry Brunetti (Agri-Dynamics) to denote only the highest quality, truly healthful, non-pasteurized milk which is of a much higher standard than "Certified Raw Milk" (as permitted in Pa.)*

"Pure Milk Parameters"

1. Set Somatic Cell Count goals at less than 300,000-350,000.
2. Bulk Tank Samples (Bacterial).
   - a. Twice monthly SPC < 20,000; *E. coli* < 10.
   - b. Test every 3-4 months for *Salmonella, Listeria, Campylobacter*
3. Annual testing of TB/Brucellosis.
   - a. Annual testing of Leucosis & Johne's.
4. Farm Inspection by State inspector every 6 months.

5. Sanitarian Inspection every 3-6 months.

6. Water test every 6 months for coliform.

    a. Do annual water test for nitrates, heavy metals, chemicals, coliform, miscellaneous bacteria.

7. Sales on farm limited to 2 gallons/person/day in Pennsylvania

    a. Not so with a Cow Share program.

    b. If someone is re-selling, have his customers as shareholders.

8. Label on jug: Nutrition facts, expiration date (currently maximum of 14 days).

9. Before being permitted: current cow test for TB/Brucella; 3 milk tests for Somatic Cell Count — one per week for 3 weeks; water test; farm inspection (passing).

10. Promote Conjugated Linoleic Acids (CLA's)! Cows must be grass fed only; or less than 15% DMI as grain.

11. Pure Milk must be free of antibiotics, hormones, insecticides, wormers, fungicides, herbicides (even if not organic).

12. Pure Milk comes from mineralized soils growing diverse forages and herbs and balanced rations to maintain high levels of immunity.

13. Pure milk comes from clean cows, milked in clean stables/parlors; free of rodents, flies, mud holes, manure, and dead livestock.

14. Pure Milk is never from cows with mastitis, ketosis, retained placenta, milk fever. Always use a strip cup and a CMT paddle when in question.

15. Cow Share programs must have valid legal agreements with shareholders.

16. Customers transport milk products in refrigerated vessels.

The information on page 351 is provided for readers to see test milk results carried out on routine bulk tank milk samples.

"Certified Milk"

Here is the direct quote from Dr. A.S. Alexander's 1929 book *Udder Diseases of the Cow and Related Subjects*:

On June 25-26, 1923, The American Association of Medical Milk Commissioners adopted "Methods and Standards for the Production and Distributions of Certified Milk" and revised them on April 26, 1926. They were issued in pamphlet form by the Association, from the National Headquarters, 360 Park Place, Brooklyn, NY and are republished here (abridged), for study by all who are interested in the production of pure milk, as follows on pages 353-363.

Bulk Tank Sample	Bacteriology Results
Farm A	Coag. Neg, Staph sp. (VLG)
Farm B	Coag. Neg. Staph sp. (VLG), Klebsiella oxytoca (VLG) and Bacillus spp. (VLG)
Farm C	Gram (-) non-fermenting rod (VLG) and Micrococcus sp. (MG)
Farm D	Streptococcus bovis (HG) and Coag. Neg. Staph sp. (VLG)
Farm E	Alpha-hemolytic Streptococcus sp. (not Strep. agalactiae), Escherichia coli and coag. Neg. Staphylococcus sp. (VLG all)
Farm F	Klebsiella oxytoca, hemolytic coag. Neg. Staphylococcus sp. and coag neg. Staphylococcus sp. (VLG all)
Farm G	E. coli and gram neg non-fermenting rod (VLG of each)
Farm H	Alpha hemolytic Streptococcus sp. (not Strep. agalactiae) (LG) and coag neg. Staphylococcus sp. (VLG)
Farm I	Coag. Neg Staphylococcus sp. and alpha-hemolytic Streptococcus sp. (not Strep. agalactiae) (LG of each)
Farm J	Coag neg. Staphylococcus sp.(isolates #1 and #2, MG and HG respectively) and gram neg. non-fermenting rod (MG)
Farm K	E. coli (MG), gram neg. non-fermenting rod (HG), coag. Neg. Staphylococcus sp. (LG) and Micrococcus sp. (MG)
Farm L	E. coli (HG), gram neg. non-fermenting rod (HG), alpha-hemolytic Streptococcus sp. (not Strep. agalactiae) (MG), and coag. Neg. Staphylococcus sp. (MG)
Farm M	Strep. agalactiae (VLG), Streptococcus sp. (not Strep. agalactiae) (MG) and E. coli (VLG) Store sample Strep. bovis (LG) and Bacillus sp. (HG)

VLG = very light growth; LG = light growth; MG = medium growth; HG = high growth

Bulk tank sample	total CLA%	Butterfat %	Total CLA gm/liter
Farm A	1.19	3.49	0.37
Farm B	1.65	4.20	0.62
Farm C	1.81	3.44	0.56
Farm D	0.88	3.42	0.27
Farm E	1.22	3.70	0.41
Farm F	1.15	3.65	0.38
Farm G	1.53	3.70	0.51
Farm H	1.51	4.25	0.58
Farm I	1.11	3.43	0.34
Farm J	0.85	3.79	0.29
Farm K	1.49	4.84	0.65
Store sample	0.58	3.50	0.18
Average	1.28	3.78	0.43

Note: Bulk tank samples for bacteriology and conjugated linoleic acid determinations were collected in late May 2004 when pastures were lush.

Certified milk is the product of dairies operated in accordance with accepted rules and regulations formulated by authorized Medical Milk Commissions to insure its purity and adaptability not alone for infants but for children during the entire period of growth. It also has special qualifications to meet the requirements of the expectant mother, as a galactogogue during the period of lactation, and as a food for invalids.

The need for purer milk was experienced primarily by those engaged in the conservation of the life and health of infants. As a result there was formulated in 1892 a plan whereby Certified Milk would be produced by a dairyman under the control of a Medical Milk Commission designated by a representative medical society.

The first rules designed for this purpose were those contained in an agreement entered into by a Medical Milk Commission and the dairymen concerned. The rules contained in the original agreement mentioned, represented the essential requirements for the production of Certified Milk. Following this precedent, other Commissions were organized, which, in 1926, became federated into a national association known as the American Association of Medical Milk Commissions.

A fundamental object of this Association was to bring about the uniformity of standards and their perfection. This result has been approximated by the adoption from time to time of definite standards relating to the veterinary inspections of herds, the sanitary inspections of the farms and their equipment, the medical inspection of employees handling the milk, and the bacteriological and chemical examinations as to quality and purity. The Association recommends these methods and standards to component Commissions as ideal and to be as closely approximated as possible.

Duties of the Commission

After its organization, the Commission should designate a sanitary inspector, a veterinarian, a physician and an analyst to enforce its methods and standards. Any two of these positions may be filled by the same person if properly qualified. These officers shall be required to render regular reports of their inspections and examinations. An agreement written or otherwise may then be entered into with any dairyman who is desirous of undertaking the production of Certified Milk. Upon investigation of such plant and upon receipt of favorable reports from several experts, the dairyman should be authorized, in accordance with the terms of agreement, to employ the term Certified Milk. Such Commissions shall report promptly to the secretary of the American Association of Medical Milk Commissions in order that no delay be experienced in securing approved seals for the bottles.

The Commission's certificate shall continue as long as its standards and requirements are maintained. The Milk Commission reserves the right to change its standards within reasonable limits upon due notice being given the Producers and Distributors.

All specifications covering Certified Milk, except those referring to butterfat, shall also apply to Certified Cream.

Hygiene of the Dairy Buildings

1. Location of buildings. — Buildings in which Certified Milk is produced and handled shall be located as to insure proper shelter and good drainage, and at sufficient distance from other buildings, dusty roads, cultivated and dusty fields, and all other sources of contamination; provided, in the case of unavoidable proximity to dusty roads or fields, suitable arrangements shall be made to exclude dust.

2. Surroundings of buildings. — The surroundings of all buildings shall be kept clean and free from accumulations of dirt, rubbish, decayed vegetable or animal matter or animal waste, and the stable yard shall be well drained.

Milking Stables

3. Construction of milking stables. — The milking stables shall be constructed so as to facilitate the prompt and easy removal of waste products. The floors and platforms shall be made of cement or other nonabsorbent material only, and the gutters of cement only. The floors shall be properly graded and drained, and the manure gutters shall be from 6-8 inches deep and so placed in relation to the platform that manure will usually drop into them. The inside surface of the walls and all interior construction shall be smooth, with tight joints, and shall be capable of shedding water. The ceiling shall be smooth material and dust-tight. Horizontal and slanting surfaces which might harbor dust shall be avoided as far as possible. When cows are kept in a separate stable between milkings, this stable shall be so construct-

Clean cows in stable (above), filthy environment results in filthy cows and udders.

Tail tied to keep clean.

ed as to provide proper shelter. The floors shall be kept dry and reasonably clean.

4. Condition of milking stable. — The milking stable shall be kept scrupulously clean. Interior walls shall be light in color. The walls and ceilings shall be whitewashed often enough to keep them clean and white, at least twice a year, unless such walls and ceilings are of smooth cement finish or painted so that they can be washed frequently.

5. Drinking and feed troughs. — Drinking troughs or basins, and feed troughs and mixing floors shall be kept in a clean and sanitary condition at all times.

6. Stanchions. — Stanchions shall be constructed of iron or hardwood. When another method of fastening is used it shall be arranged so that the droppings will fall for the most part into the gutter. Provision shall be made to keep cows standing between cleaning and milking.

7. Ventilation. — The cow stables shall be provided with adequate ventilation either by means of some approved artificial device, or by the substitution of cheesecloth for glass in windows, each cow provided with a minimum of 600 cubic feet of air space.

8. Windows. — A sufficient number of windows shall be installed and so distributed as to provide satisfactory light and a maximum of sunshine; 4 square feet of window area to each 600 cubic feet of air space to represent the minimum. Windows shall be kept free from dust and dirt.

9. Exclusion of flies, etc. — All necessary measures should be taken to prevent the entrance of flies and other insects, rats and other vermin, into stables and dairy buildings, and proper methods for their destruction if they gain access.

10. Exclusion of children. — Children shall not be allowed in the dairy building nor in the stable during milking.

Milk Receiving Room

11. Milk receiving room. — The milk receiving room shall be held to mean any room or building established at or near the milking stables and used for the purposes of a central collecting room for milk as brought from the stables. Such room shall be provided and shall conform to the same rules as to construction, maintenance and cleanliness as applies to the milk bottling room in the bottling plant, and shall not be directly connected with the stable.

Dairy Building and Equipment

12. Dairy building. — A dairy building shall be provided which shall be located at a distance from the stables and dwelling prescribed by the local Commission and there shall be no hog-pen, privy, or manure pile at a higher level within 300 feet of it. The building shall contain separate rooms (suitably equipped) for handling and bottling Certified Milk only and for washing and sterilizing the utensils and bottles, together with a separate boiler room. Necessary cooling facilities shall be provided. The walls and ceilings of the dairy building shall be constructed of nonabsorbent material. It shall be kept clean and shall not be used for purposes other than the handling and storing of dairy products and milk utensils. It shall be provided with light and ventilation, and the floors shall be graded, concrete and watertight. The dairy building shall be well-lighted and properly screened, and drained through well-trapped pipes. No animals shall be allowed therein. No part of the dairy building shall be used for dwelling or lodging purposes, and the bottling room shall be used for no other purpose than to provide a place for clean milk utensils and for handling milk. During the bottling this room shall be entered only by persons employed therein or official inspectors. The bottling room shall be kept scrupulously clean and free from odors. Adequate lavatory facilities shall be provided.

13. Utensils. — All utensils shall be constructed as to be easily cleaned. The milk pails should preferably have an elliptical opening 5 x 7 inched in diameter. The cover of this pail should be so convex as to make the entire interior of the pail visible and accessible for cleaning. Pails shall be of heavy tin, and with seams flushed and made smooth with solder. All utensils used in handling milk shall be kept in good repair and as free as possible from rust.

14. Cleaning and sterilizing of bottles and utensils. — The dairy building shall be provided with approved apparatus for cleansing and sterilizing all bottles and utensils used in milk production. Thermometers shall be used on all sterilizers to record or indicate temperatures. All bottles and utensils shall be thoroughly cleaned by hot water and a strong solution of Sal soda or an equally effective cleansing agent, rinsed until the cleaning water is thoroughly removed, then sterilized by steam, dry heat or boiling water and kept inverted in a place free from dust and other contaminating materials until used. Milking machines when used shall be equally well cleaned and sterilized.

15. Water supply. — The entire water supply shall be absolutely free from contamination, and shall be sufficient for all dairy purposes. It shall be protected against flood or surface drainage, and shall be conveniently situated in relation to the milk house.

16. Nuisances. — Privies, pigpens, manure piles, and all other possible sources of contamination shall be so situated on the farm as to render impossible the contamination of the water supply, and shall be so protected by use of screens and other measures as to prevent their becoming breeding grounds for flies or sources of objectionable odors.

Herd Management & Stable Practices

17. Pastures and paddocks. — Pastures and paddocks to which cows have access shall be free from marshes or stagnant pools, crossed by no contaminated stream or one in direct danger of becoming contaminated. They must be sufficiently distant from all offensive conditions to suffer no bad effects from them. Pastures shall be free from infectious agents, and plants which might affect the milk deleteriously.

18. Exclusion of animals from the herd. — No horses, hogs, dogs, or other animals or fowls shall be allowed to come in contact with the Certified herd. All cows kept in the same barn with the milking cows shall receive the same supervision and care.

19. Bedding. — No moldy hay or straw, bedding from horse stalls or other unclean materials shall be used for bedding the cows. Only bedding which is clean, dry, absorbent and reasonably free from dust may be used.

20. Cleaning the milking stable and disposal of manure. — Soiled bedding and manure shall be removed from the milking stable at least twice daily, and the floors shall be swept and kept free from refuse. Such cleaning shall be done not less than one hour before the milking time. Manure, when removed from the stable, should be hauled directly to the field and spread, if possible. If stored temporarily near the stable, it should be placed on a cement floor and removed to the fields and spread at least every seven (7) days, or placed in a container screened against flies. Manure shall not be even temporarily stored within 300 feet of the barn or dairy building.

21. Clipping. — Long hairs shall be clipped from the udder and flanks of the cow, and from the tail above the switch. The switch should be cut to clear the ground or platform by at least four inches. If this is not desirable with pure bred animals then the switch shall be washed often enough to keep it clean and shall be brushed thoroughly before each milking.

22. Cleaning of cows. — Each cow in the milking stable shall be carefully groomed before each milking, all dry or wet manure, mud or other dirt being removed. The dry material shall be loosened with a curry comb and thoroughly brushed out with a stiff brush, and the wet material remaining after

brushing shall be washed off with a clean cloth and clean water. The brushing shall be completed a half hour before milking. Special attention shall be given to the udder and flanks. After the udders are brushed clean they shall be wiped with a clean, damp cloth, or they may be washed. If washed, they should be wiped with a dry clean cloth. It is recommended that a freshly-washed individual towel or the fresh surface of a folded towel be used for each cow.

23. Feeding. — All food stuffs shall be kept in an apartment separate from the cow stable. They shall be brought into the stable only immediately before the feeding hour, which shall follow milking. Only those foods shall be used which consist of clean, palatable, and nutritious materials, such as will not injure the health of the cows or unfavorably affect the taste or character of the milk. No dirty or moldy food, or food in a state of decomposition or putrefaction shall be fed. A well-balanced ration shall be used, and all changes of food shall be made slowly. The first few feedings of grass, alfalfa, ensilage, green corn, or other green feeds shall be given in small rations and increased gradually to full ration.

24. Exercise. — It is recommended that all dairy cows be turned out for exercise at least 2 hours in each 24 in suitable weather. Exercise yards shall be kept free from manure and other filth.

25. An inspection of the hygiene of a dairy shall be made at least once a month by the designated representative of the Milk Commission.

Veterinary Supervision of the Herd

26. Tuberculin tests. — When application is made for the certification of the milk of the herd, a tuberculin test of the herd shall be made and if any of the animals react, a second test shall be made within 60-90 days. Thereafter, the herd shall be tested semi-annually until it passes three successive tests without any reactions, after which it shall be tested annually, but when any reactions occur semi-annual tests shall be resumed. If no reactions are found on the first test, the second test shall follow in six months. If the entire herd has been subjected to a tuberculin test within 60 days prior to the receipt of application, this may, if satisfactory, be accepted as the primary test. The dates of all herd tests shall be definitely arranged with the Commission and a complete record of all tests shall be made by the veterinarian making the same and promptly reported by him to the secretary of the Commission. These reports shall be kept on file and a summary of all such tests shall be made available to the American Association of Medical Milk Commissions for statistical purposes.

27. The subcutaneous and intradermal methods of applying the tuberculin tests shall be used alternately in the tests prescribed in paragraph 26. The ophthalmic method shall be used as an adjunct to whatever method is used on the 60-90 day test and it is recommended that this method be used in conjunction

with the subcutaneous or intradermal methods on all tests. Each animal shall be subjected to physical examination when tuberculin tested. When the subcutaneous method is used all animals in the herd shall be included in each herd test except calves under 6 months of age and animals over this age which have been tuberculin tested within 60 days, and when the intradermal method is used all shall be included except animals which have been tested within 60 days. The latter shall be tested 60-90 days after the previous test if conditions require it. No new animals shall be brought on the farm without first having passed a satisfactory physical examination and tuberculin test within 60 days, and, unless from a herd with less than 10 per cent of reactors to the last annual or semi-annual test, they shall be kept segregated from the herd until they pass a second satisfactory examination and test 60 to 90 days later. All tests shall be made by a veterinarian approved by the Commission and shall be applied in accordance with the technique approved by the U.S. Bureau of Animal Industry or the State Livestock Sanitary authorities.

28. All reactors shall be removed from the herd immediately after discovery and the milk shall be discarded, and the stable and exercise yards used by them shall be cleaned and disinfected in a manner approved by the Commission.

29. If a herd is officially accredited or in the process of accreditation, the tuberculin tests made for this purpose may be accepted in place of the tests specified in paragraphs 26 and 27 when an official copy of such tests is presented to the Commission in satisfactory form.

30. Identification of cows. — Each dairy cow in each of the certified herds shall be sketched, labeled or tagged with a number or mark which will permanently identify her.

31. Herd record. — Each cow in the herd shall be registered in a herd record, which record shall be accurately kept so that her entrance and departure from the herd and her tuberculin testing can be identified. This record may contain other information useful to the owner or manager. This record shall be accessible to the Milk Commission or its representative and the producer shall be responsible for its accuracy.

32. Disposition of cows sick with diseases other than tuberculosis. — No cow known or suspected to be diseased shall be permitted to remain in the milking herd. Any cow showing signs of disease, or under suspicion of being diseased, shall be immediately removed from the milking stable to a separate compartment or building provided for such animals and the milk shall not be used. Any cow affected with mastitis (garget) shall be removed from the milking herd, even if only one quarter of the udder is affected, and none of the milk from any of the quarters shall be used. No cow separated on account of disease shall be returned to the milking herd without the approval of the veterinarian.

33. Milking and calving period. — Milk from all cows shall be excluded for a period of 45 days before and 7 days after parturition. Cows shall not be permitted to calve in the milking stable and shall not be returned to said stable

while the uterine discharges are putrid or purulent and under no condition before the seventh day following parturition.

34. Notification of veterinary inspector. — In the event of the occurrence of a disease which appears to be of a serious character between the visits of the veterinarian, or if a number of cows become sick at about the same time, it shall be the duty of the dairyman to withdraw such sickened cattle from the herd, to destroy their milk, and to notify the veterinary inspector by telegraph or telephone immediately.

35. Inspections. — A veterinary inspection of health of the herd shall be made at intervals of not more than one month. Each cow in milk shall be subjected to a careful physical examination at least once a month.

Employees
(remember that milking by hand was the standard method of milking cows in the 1920's)

36. Washing of hands. — Adequate and conveniently located facilities shall be provided for the milkers. The hands of the milkers shall be thoroughly washed and scrubbed, using soap (preferably liquid), warm water and brush, and carefully dried on a clean towel immediately before milking. The hands of the milkers shall be rinsed with clean water and carefully dried before milking each cow. The practice of moistening the hands with milk is forbidden.

37. Employees clothing. — Milkers and other persons handling the milk shall wear white suits and caps that shall be worn for no other purpose. They shall use not less than three freshly laundered suits and caps each week, and these when not in use shall be kept in a clean place, protected from dust and dirt.

38. Things to be avoided by milkers. — While engaged about the dairy or in handling the milk, employees shall not use tobacco or intoxicating liquors. They shall keep their fingers away from nose and mouth, and no milker shall permit hands, fingers, or mouth to come in contact with milk intended for sale. During milking, the milkers shall be careful not to touch anything but the clean top of the milking stool, the milk pail and the cow's teats. Milkers are forbidden to spit upon the floors and walls of the stables, or upon the floors and walls of the milk house, or into the water used for cooling or cleansing purposes. These precautions are necessary to avoid the possibility of any disease-producing bacteria getting into the milk.

39. Toilet rooms. — Toilet facilities for the milkers and others handling milk shall be provided and unless of the sanitary flush type shall be located outside of the stable or milk house. Flush closets shall not be in proximity to the bottling room or receiving room. All toilets shall be properly screened, shall be kept clean, and shall be accessible to wash basins, water, nail brush, soap and clean individual towels. Employees shall be required to wash and dry

their hands immediately after leaving the toilet room. Sanitary drinking fountains or individual drinking cups for the employees should be installed in convenient locations both in the stables and dairy houses.

Milking, Handling, Transportation

40. Fore milk. — The first streams from each teat shall be rejected. Such milk shall be milked through a very fine meshed strainer to help determine the presence or absence of garget and be collected in a separate vessel and shall not be milked upon the floors or into the gutters. If milk from any cow is bloody, stringy, or in anywise unnatural, the milk from that cow shall be rejected and the cow shall be removed from the milking herd and shall not be returned until approved by the veterinarian (see paragraph 32). If dirt gets into the pail, the milk shall be discarded and the pail washed and sterilized before it is used again. The pails shall be protected from extraneous contamination. The milking shall be done rapidly and quietly and the cows shall be treated kindly. Promptly after the milk is drawn it shall be removed from the stable and strained through strainers made of double layers of finely meshed cheese cloth or absorbent cotton or other approved material, thoroughly sterilized. Several containers shall be provided for each milking in order that they may be frequently changed.

41. Temperature of milk. — Proper cooling to reduce the temperature below 50 degrees F. shall be used. Coolers shall be so situated that they can be protected from flies, dust and odors. The milk shall be cooled immediately after being milked, and maintained at a temperature above freezing and below 50 degrees F. until delivered to the consumer.

42. Capping and sealing bottles. — Certified Milk is to be bottled and sealed upon the farm where it is produced. Milk, after being cooled and bottled, shall be immediately capped and sealed in a manner satisfactory to the Commission. The inner cap shall be applied only by a suitable capping machine. The seal shall include a sterile hood which completely covers the lip of the bottle and shall be marked on the exposed surface with the approved design of the American Association of Medical Milk Commissions, the name of the Commission, the name of the farm and such other information as state and local health authorities may require. Certified Milk may be transported and delivered in cans properly sealed and marked, to hospitals and other institutions under Medical supervision.

43. Transportation. — In transit the milk packages shall be kept reasonably free from dust and dirt. The wagons, trays, and crates shall be kept clean.

44. Certified Milk shall be delivered to the consumer in the shortest possible time after production as determined by each individual Milk Commission.

Distributors

45. Distributors must furnish a sufficient number of new or clean washed bottles to cover all shipments of Certified Milk.

46. Distributors must furnish deep covered boxes for shipping Certified Milk.

47. No bottles shall be collected from a house in which communicable disease exists except under conditions prescribed by the Health authorities.

48. Distributors handling Certified Milk must keep it at a temperature above the freezing point and below 50 degrees F. from the time it is received until delivered to the consumer. This means bottles shall be well packed in ice in warm weather, and not allowed to freeze in cold weather.

Employee Health & Hygiene

49. A medical officer, known as the attending dairy physician, shall be selected by the Commission, who should reside near the dairy producing Certified Milk. He shall be a physician in good standing and authorized by law to practice medicine; he shall be responsible to the Commission and subject to its direction. In case more than one dairy is under the control of the Commission and they are in different localities, a separate physician should be designated for the supervision of each dairy.

50. Every new employee for milking and handling milk or to be regularly employed by the farm shall be examined by the Medical Examiner before assuming his duties, except in cases of most extreme emergency, and all necessary laboratory tests shall be made when such are indicated.

51. The physician shall visit the farm weekly and examine all those who milk or handle the milk, and all regular employees, and shall report to the Milk Commission upon blank forms supplied for that purpose. This report shall be countersigned by the farm owner or manager.

52. No person suffering with any communicable disease, and no person recently exposed to such disease shall be employed in the production of Certified Milk. No person known to be a typhoid carrier shall be employed on a certified farm. No person known, or suspected, to have had typhoid fever shall be employed except by special permit of the Milk Commission. In the event of any illness likely to prove dangerous to the milk, the patient shall be immediately removed from the farm or properly isolated, and the Milk Commission notified. Satisfactory evidence of successful vaccination against smallpox within seven years shall be presented before employment. In case of an epidemic at or near a farm all employees shall be vaccinated.

53. Record of employees. — A record shall be kept on the dairy premises of each employee which shall show his name and address, date of employment,

his medical history, results of physical examination by physician, and the results of any examinations of cultures and their laboratory tests.

54. Dormitories. — When employees live upon the premises their dormitories shall be constructed and operated according to plans approved by the Commission. Sufficient and modern bathing and toilet facilities should be provided for all employees living on dairy premises.

STANDARDS

55. Certified Milk must be pure, clean, fresh, wholesome cows' milk in its natural state, not having been heated and without the addition of coloring matter or preservatives. Nothing must be added to the milk, and nothing must be taken away.

56. Examinations of milk are to be made with frequency and at the option of the Milk Commission according to the season and condition of the milk, and at least once a week. All samples required for examination shall be furnished by the Distributors and collected at the discretion of the Milk Commission. Reports upon examinations should be sent to the Producers and Distributors.

57. Collection of samples. — The samples to be examined shall be obtained from milk as offered for sale. The samples shall be received in the original packages, in properly iced containers, and they shall be so kept until examined, so as to limit as far as possible changes in their bacterial content.

58. Bacterial counts. — Certified Milk shall contain not over 100,000 bacteria per cubic centimeter. In case a count exceeding 100,000 bacteria per cubic centimeter is found, daily counts shall be made and if normal counts are not restored within 10 days the certification may be suspended. Bacterial examinations shall be made as soon as possible after collection and in no case shall the interval be over four hours. Immediately after the plates have been prepared the temperature may be taken.

59. The Methods and Techniques used in the Bacteriological Examinations of Certified Milk shall conform to the standard methods and technique of the American Public Health Association for bacteriological examinations of milk.

60. Determination of taste and odor of milk. — Immediately after the plates have been prepared, the taste and odor of the milk shall be determined. Chemical examinations may be made on samples previously examined for bacteria.

61. The Methods and Technique used in the Chemical Analysis shall conform to the methods and technique of the American Association of Official Agricultural Chemists.

62. Fat Standards. — Certified Milk shall contain an average of 4.0 percent of butterfat with a minimum of 3.5 percent. This average is to be based on a period of not over 90 days.

63. In case it is desired to maintain an average butterfat percent above or below the standard, the percent of butterfat shall be stated on the seal of the container. Such average shall be based on tests covering a period of 90 days or less.

64. The percentage of butterfat in Certified Cream shall either be stated on the cap or conform to legal standards.

Bibliography

The following bibliography shows the many works from which reference information was used in certain areas of this book. All these titles can be obtained either through bookstores or on-line booksellers of hard to find books (www.alibris.com). This list will give the reader an impression of how my views have been formed regarding veterinary medicine and the pharmacologic bases for treating dairy cattle.

Each entry has an annotation at the end of it to give the reader an idea of the genre to which the book generally belongs.

The key:
AG = Agricultural
BM = Botanical Medicine
CM = Conventional Medicine
CVM = Conventional Veterinary Medicine
EM = Eclectic Medicine
EP = Eclectic Pharmacy
EV = Eclectic Veterinary
LV = Lay Veterinary
MH = Medical Homeopathy
VA = Veterinary Acupuncture
VH = Veterinary Homeopathy

Albrechtsen, J., D.V.M. *The Sterility of Cows, Its Causes and Treatment.* Alexander Eger Publisher, Chicago, 1917. **(CVM)**

Alexander, A.S., D.V.M. *The Veterinary Adviser.* Orange Judd Pub. Co., New York, 1929. **(CVM)**

Alexander, A.S., D.V.M. *Udder Diseases of the Cow.* Richard G. Badger, Pub., Boston, 1929. **(CVM)**

Arthur, G., Noakes, D., and Pearson, H. *Veterinary Reproduction and Obstetrics.* 6th ed. Bailiere Tindall, London, 1983. (CVM)

Baker, E.T., D.V.M. *The Home Veterinarian's Handbook.* The MacMillan Company, New York, 1945. **(CVM)**

Boericke, W., M.D. *Materia Medica and Repertory.* 9th ed. B. Jain Publishers Pvt. Ltd., New Delhi, India, 1994. **(MH)**

Burkett, W., D.V.M. *The Farmer's Veterinarian.* Orange Judd Pub. Co., New York, 1913. **(CVM)**

Coleby, Pat. *Natural Cattle Care.* Acres U.S.A., Austin, Texas, 2001. **(LV)**

Chuan, Yu, D.V.M. *Traditional Chinese Veterinary Acupuncture and Moxibustion.* China Agricultural Press, Beijing, 1995. **(VA)**

Dadd, G.H. *American Cattle Doctor.* Orange Judd Pub. Co., 1897. **(EV)**

Day, C., M.R.C.V.S. *The Homeopathic Treatment of Beef and Dairy Cattle.* Beaconsfield Publishers Ltd., Bucks, England, 1995. **(VH)**

de Bairicli Levy, J. *The Complete Herbal Handbook for Farm and Stable.* Faber & Faber, Boston, 1984. **(LV)**

Der Marderosian, A. and Liberti, L. *Natural Product Medicine.* George F. Stickley Co., Philadelphia, 1988. **(BM)**

Duke, J. *The Green Pharmacy.* Rodale Press, Emmaus, PA, 1997. **(BM)**

Dun, Finlay, D.V.M. *Veterinary Medicines.* 12th ed. David Douglas, Edinburgh, 1910. **(CVM)**

Ellingwood, F., M.D. *American Materia Medica, Therapeutics and Pharmacognosy.* Eclectic Medical Publications, Sandy, OR, 1998. **(EM)**

Felter, H.W., M.D. *The Eclectic Materia Medica, Pharmacology and Therapeutics.* Eclectic Medical Publications, Sandy, OR, 2002. **(EM)**

Felter, H.W., M.D. and Lloyd, J.U. *King's American Dispensatory.* 18th ed. Vol. I & II. Eclectic Medical Publications, Sandy, OR, 1997. **(EP)**

Fish, Pierre A., D.V.M. *Veterinary Doses and Prescription Writing.* 6th ed. The Slingerland-Comstock Pub. Co., Ithaca, 1930. **(CVM)**

Foster, S. and Duke, J. *Peterson Field Guides Eastern/Central Medicinal Plants.* Houghton Mifflin Company, Boston, 1990. **(BM)**

Ganora, Lisa. "The Phytochemistry of Herbs." *HerbalChem.* www.herbalchem.net/GarlicIntroductory.htm. Nov. 2003. **(BM)**

Gibson, D., M.D. *Studies of Homeopathic Remedies.* Beaconsfield Publishers Ltd., Bucks, England, 1987. **(MH)**

Graczyk, Shiff, Karreman, Pautz. "Environmental and Geographical Factors Contributing to Watershed Contamination with *Cryptosporidium parvum* Oocysts." *Environmental Research* Vol. 82: 263-271. **(AG)**

Grainger, J. and Moore, C. *Natural Insect Repellents.* The Herb Bar, Austin, Texas, 1991. **(LV)**

Grosjean, N. *Veterinary Aromatherapy.* C.W. Daniel Co. Ltd., Essex, England, 1994. **(LV)**

Harrower, H., M.D. *Practical Organotherapy: The Internal Secretions in General Practice.* 2nd ed. The Harrower Laboratory, Glendale, CA, 1920. **(CM)**

Herbal Research Publications, ed. *Naturopathic Handbook of Herbal Formulas.* 4th ed. Herbal Research Publications, Inc., Ayer, MA, 1995. **(BM)**

Humphreys, F., M.D., D.V.M. *Manual of Veterinary Specific Homeopathy.* 3rd ed. Specific Homeopathic Medicine Co., New York, 1881. **(HV)**

Jones, Eli G., M.D. *Definite Medication.* The Therapeutic Publishing Company, Inc., Boston, 1911. **(EM)**

Jones, Eli G., M.D. *Reading the Eye, Pulse and Tongue for the Indicated Remedy.* Buckeye Naturopathic Press, East Palestine, OH, 1989. **(EM)**

Jousset, Pierre., M.D. *Practice of Medicine: Containing the Homeopathic Treatment of Diseases.* A.L. Catterton & Co., New York, 1901. **(MH)**

Karreman, H., V.M.D. "Can Cows on Organic Dairy Farms Compete?" *Hoard's Dairyman* Vol. 148 No. 12 (July 2003): 453. **(AG)**

Karreman, H., V.M.D. "Macroscopic Aspects of Dairy Farm Health." *Biodynamics* No. 234 (Mar./Apr. 2001): 3-8. **(AG)**

Karreman, H., V.M.D. "Snapshot of a Holistic Dairy Practice." *JAHVMA* Vol. 20 No. 4: 25-31. **(AG)**

Karreman, Wentink, Wensing. "Using Serum Amyloid A to Screen Dairy Cows for Sub-clinical Inflammation." *Veterinary Quarterly* Vol. 22 (2000): 175-8. **(CVM)**

Kothbauer, O. D.M.V. *Veterinary Acupuncture.* Zweimuhlen Verlag GmbH, Munchen, Germany, 1999. **(VA)**

Lewis, W. and Elvin-Lewis, M. *Medical Botany, Plants Affecting Man's Health.* John Wiley & Sons, New York, 1977. **(BM)**

Lloyd, J.U. *"A Treatise on Nux Vomica." Drug Treatise Number VIII.* Lloyd Brothers, Cincinnati, 1904. **(EP)**

Lloyd, J.U. *"A Treatise on Libradol: An External Remedy for Pain." Drug Treatise Number XVIII.* Lloyd Brothers, Cincinnati, 1907. **(EP)**

Lloyd, John Uri. *The Eclectic Medical Journal.* Lloyd Brothers, Cincinnati, March 1914. **(EP)**

Lust, John. *The Herb Book.* Benedict Lust Publications, New York, 2001. **(BM)**

Macey, Anne, ed. *Organic Livestock Handbook.* Canadian Organic Growers, Ottawa, Canada, 2000. **(AG)**

MacLeod, G., M.R.C.V.S. *The Treatment of Cattle by Homeopathy.* C.W. Daniel Co. Ltd., Essex, England, 1981. **(VH)**

Mattison, M., D.V.M. *"Bovine Acupuncture for Non-Acupuncturists."* Short course, Colorado State University, School of Veterinary Medicine. July 12, 2000. **(VA)**

Milks, H.J., D.V.M. *Practical Veterinary Pharmacology, Materia Medica and Therapeutics.* 4th ed. Bailliere, Tindall and Cox, London, 1940. **(CVM)**

Morrison, F. *Feeds and Feeding.* 21st ed. The Morrison Publishing Co., Ithaca, NY, 1949. **(AG)**

Mowry, D. *Next Generation Herbal Medicine.* 2nd ed. Keats Publishing, Inc., New Canaan, CT. 1990. **(BM)**

Mowry, D. *The Scientific Validation of Herbal Medicine.* Keats Publishing, Inc., New Canaan, CT. 1986. **(BM)**

M.R.C.V.S., An. *The Farm Vet.* MacDonald & Martin, London, 1914. **(CVM)**

Murphy, B. *Greener Pastures on Your Side of the Fence.* 4th ed. Arriba Publishing, Colchester, VT., 1998. **(AG)**

Muse, M., R.N. *Materia Medica, Pharmacology and Therapeutics.* W.B. Saunders Co., Philadelphia, 1937. **(CM)**

Nuzzi, Debra. *Pocket Herbal Reference Guide.* The Crossing Press, Freedom, CA.1992. **(BM)**

Osweiler, G., D.V.M. *Toxicology.* Lippincott Williams & Wilkins, Philadelphia, 1996. **(CVM)**

Potter, S., M.D. *Handbook of Materia Medica, Pharmacy, and Therapeutics.* P. Blakiston, Son & Co., Philadelphia, 1897. **(CM)**

Protocol Journal of Botanical Medicine, The. Vol. 1 No. 4 (Spring 1996). **(BM)**

Rebhun, W.C., D.V.M. *Diseases of Dairy Cattle*. *Lippincott* Williams & Wilkins, Baltimore, 1995. **(CVM)**

Richardson, Josiah. *The New-England Farrier, and Family Physician*. Josiah Richardson, Exeter, 1828. **(CVM)**

Roberts, D., D.V.M. *Practical Home Veterinarian*. 15th ed. Dr. David Roberts Veterinary Co., Waukesha, WI, 1913. **(CVM)**

Schoen, A. and Wynn, S. *Complementary and Alternative Veterinary Medicine, Principles and Practice*. Mosby, Philadelphia, 1998. **(BM)**

Scudder, J., M.D. *Specific Diagnosis: A Study of Disease with Special Reference to the Administration of Remedies*. Wilstach, Baldwin & Co., Printers, Cincinnati, 1874. **(EM)**

Scudder, J., M.D. *Specific Medication and Specific Medicines*. 10th ed. John M. Scudder & Sons Medical Publishers, Cincinnati, 1893. **(EM)**

Shaller, J., M.D. *A Therapeutic Guide to Alkaloidal Dosimetric Medication*. The Clinic Publishing Co., Chicago, 1905. **(EM/CM)**

Sheaffer, C.E., V.M.D. *Homeopathy for the Herd*. Acres U.S.A., Austin, Texas, 2003. **(VH)**

Stamm, G.W., D.V.M. *Veterinary Guide for Farmers*. Windsor Press, New York, 1950. **(CVM)**

Stedman, T. *Stedman's Medical Dictionary*. 25th ed. Williams & Wilkins, Baltimore, 1990. **(CM)**

Steffen, M., D.V.M. *"Special Veterinary Therapy." American Journal of Veterinary Medicine*. (1914). **(CVM)**

Steiner, R. *Agriculture*. Bio-Dynamic Agricultural Association, London, 1958. **(AG)**

Titus, Nelson N. *The American Eclectic Practice of Medicine as Applied to the Diseases of Domestic Animals*. N.N. Titus, Union, NY, 1865. **(EV)**

Transactions of the National Eclectic Medical Association. Vol. XVI (1889). **(EM)**

Transactions of the National Eclectic Medical Association. Vol. XXII (1895). **(EM)**

Transactions of the National Eclectic Medical Association. Vol. XXXIII (1905). **(EM)**

Tyler, M.L., M.D. *Homeopathic Drug Pictures*. Homeopathic Publishing Co., London, 1944. **(MH)**

Udall, D.H., D.V.M. *The Practice of Veterinary Medicine*. 4th ed. Ithaca, 1943. **(CVM)**

Udall, D.H., D.V.M. *Veterinarian's Handbook of Materia Medica and Therapeutics.* The MacMillan Company, New York, 1922. **(CVM)**

Varlo, Charles. *A New System of Husbandry.* Vol. 1. Philadelphia, 1785. **(CVM)**

Vitoulkas, G. *The Science of Homeopathy.* Grove Weidenfeld, New York, 1980. **(MH)**

Waterman, G.A., D.V.M. *The Practical Stock Doctor.* F.B. Dickerson Company, Lincoln, NE, 1925. **(CVM)**

West, C. *Australian Tea Tree Oil: First Aid for Animals.* Kali Press, Pagosa Springs, CO, 1998. **(LV)**

Winslow, K., D.V.M. *Veterinary Materia Medica and Therapeutics.* 8th ed. American Veterinary Publishing Co., Chicago, 1919. **(CVM)**

Conversions

Thanks to Richard H. Detweiler, V.M.D. for these conversions. They are from his class notes while at the University of Pennsylvania School of Veterinary Medicine, from which he graduated in 1948. The notes are from Professor Frank Lentz's classes in pharmacology. Dr. Detweiler's class was the last to have a course in "Materia Medica" at the veterinary school.

Apothecaries' Fluid Measure24

1 drop	=	1 minim
60 minims	=	1 fluid dram
8 fluid drams	=	1 fluid ounce
16 fluid ounces	=	1 pint
8 pints	=	1 gallon
480 minims	=	1 fluid ounce
7680 minims	=	1 quart

Apothecaries' (Troy) Weight

The grain (gr.) was derived by an act of Henry III of England in 1226. An English penny, called the sterling, round and without clipping, shall weigh 32 grains of wheat, well dried and gathered out of the middle of the ear.

20 grains	=	1 scruple
3 scruples	=	1 dram
8 drams	=	1 ounce
12 ounces	=	1 pound

Metric Conversions

Weight:

   15.43 grains = 1 gram

Measure:

   1 ounce = 30cc

   ppm = mg/l

Domestic measures were established by custom and estimated to be:

Tumblerful = 8 fluid ounces or 240cc

Teacupful = 4 fluid ounces or 120cc

Wineglass = 2 fluid ounces or 60cc

Tablespoonful (tbsp.) = 4 fluid drams or 16cc

Dessertspoonful = 2 fluid drams or 8cc

Teaspoonful (tsp.) = 1 fluid dram or 4cc

1 Drop (gtt.)* = 1 minim or 0.061cc

* the size of a drop is dependent upon the viscosity of the fluid.

Liquid Measure

4 gills = 1 pint

2 pints = 1 quart

4 quarts = 1 gallon

$31\frac{1}{2}$ gallons = 1 barrel

2 barrels = 1 hogshead

A U.S. gallon is the same as the English wine gallon and contains 231 cubic inches. A gallon of water weighs about $8\frac{1}{3}$ pounds.

Table of Approximately Equivalent Weights

1 mg. = $\frac{1}{64}$ gr.

1 gm. = $15\frac{1}{2}$ gr.

4 gm. = 1 dram

30 gm. = 1 ounce

500 gm. = 1 pound (av.)

Table of Approximately Equivalent Measures

16 minim	=	1cc
1 fluidram	=	4cc
1 fluidounce	=	30cc

Recommended Remedies

These are available from Agri-Dynamics
www.agri-dynamics.com 610-250-9280

Botanicals

1) Phyto-Mast (herbal udder infusion): By prescription only to licensed veterinarians: *Thymus vulgaris, Angelicaspp, Glycyrrhiza, Gaultheria,* Vitamin E Oil + Canola oil

2) Phyto-Biotic (herbal antibiotic tincture): *Allium sativum, Echinacea angustifolia, Hydrastis canadensis, Baptisia tinctoria, Berberis vulgaris*

3) Phyto-Gest (herbal digestive tincture): *Nux vomica, Zingiber officinalis, Gentiana lutea, Foeniculum vulgare*

4) Phytonic (herbal liver tincture): *Ceanothus americanus, Chelidonium majus, Silybum marianum, Taraxacum officinale, Mahonia aquifolium*

5) Phyto-Gesic (herbal pain formula tincture): *Corydalis yanhusuo* tuber, *Tenacetum parthenium* herb, *Filipendula ulmaria* herb, *Atractylodes macrocephala* rhizome

6) Heat Seek (capsules): Turnera diffusa var. aphrodisiaca, Dioscorea villosa, Mitchella repens, Actea racemosa, Viburnum opulus, Angelica sinensis, Oenothera biennis seed, Linum usitatissimum, vitamin B6

Individual concentrations of therapeutic botanicals

(see *Agri-Dynamics, Inc.* in Resource Contacts):

Aconitum napellus (Aconite), dried root (1:10)

Allium sativum (Garlic), fresh bulb (1:2)

Atropa belladonna (Belladonna), dried herb (1:10)

Baptisia tinctoria (Wild indigo), fresh root (1:3), dried root (1:2)

Berberis vulgaris (Barberry), dried root (1:4)

Bryonia dioica (White bryony), dried root (1:10)

Calendula officinalis (Wild Marigold), fresh flower (1:2)

Ceanothus americanus (Red root), fresh root (1:2)

Chelidonium majus (Celandine), fresh whole plant (1:2)

Convallaria majalis (Lilly of the Valley), dried herb (1:5)

Echinacea angustifolia (Cone flower), fresh root (1:3)

Foeniculum vulgare (Fennel), mature seed (1:2.5)

Gelsemium sempervirens (Yellow jasmine), dried root (1:10)

Gentiana lutea (Gentian, yellow), dried root (1:4)

Glycyrrhiza glabra (Licorice), dried root (1:1)

Mahonia aquifolium (Oregon grape), fresh root (1:2)

Mentha piperita (Peppermint), fresh leaf (1:2.5)

Phytolacca Americana (Poke weed), fresh root (1:2)

Silybum marianum (Milk thistle), mature seed (1:1.5)

Strychnos nux vomica (Quaker buttons), mature seed (1:10)

Taraxacum officinale (Dandelion), fresh root (1:2)

Veratrum viride (White hellebore), fresh rhizome (1:5)

Zingiber officinalis (Ginger), fresh rhizome (1:1)

Colostrum-whey products

1) Biocel CBT (can be given SQ or IV); 2) IMPRO injectible products (can be given SQ or IV)

Homeopathic remedies

(Top 30) (See *Washington Homeopathic Products* in Resource Contacts) Aconite, Apis, Antimonium, Arnica, Arsenicum, Belladonna, Bryonia, Calc carb, Calc phos, Carbo veg, Caulophyllum, Colocynthus, Folliculinum 6X,

Hepar sulph 10X, Hypericum, Iodium, Lycopodium, Merc corr, Nux vomica, Ovarian 5C, Phos, Phytolacca, Pulsatilla, Pyrogen, SSC (Sulfur-Silica-Carbo veg), Sepia, Sabina, Silica, Sulfur, Ustilago

Articles on
Natural Medicines

Abaineh, D. and Sintayehu, A. "Treatment Trial of Subclinical Mastitis with Herb Persicaria senegalense (Polygonaceae)." *Tropical Animal Health and Production* Vol. 33 (2001): 511-519.

Ademola, I.O. "Anthelmintic Activity of Extracts of Spondias mombin Against Gastrointestinal Nematodes of Sheep: Studies in vitro and in vivo." *Tropical Animal Health and Production* Vol. 37 No. 3 (Apr. 2005): 223-235.

Aly, A. "Licorice: A Possible Anti-Inflammatory and Anti-Ulcer Drug." *AAPS PharmSciTech* Vol. 6 No. 1 (2005): E74-E82.

Assis, L.M. "Ovicidal and Larvacidal Activity in vitro of Spigelia anthelmia Linn. Extracts on Haemonchus contortus." *Veterinary Parasitology* Vol. 117 No. 1-2 (3 Nov. 2003): 43-49.

Benie, T. and Thieulant, M.L. "Interaction of Some Traditional Plant Extracts with Uterine Oestrogen or Progestin Receptors." *Phytotherapy Research* Vol. 17 (2003): 756-760.

Biziulevichius, G.A. "In vivo Studies on Lysosubtilin: 1. Efficacy for Prophylaxis and Treatment of Gastrointestinal Disorders in Newborn Calves." *Veterinary Research* Vol. 28 (1997): 19-35.

Biziulevichius, G.A. "In vivo Studies on Lysosubtilin: 2. Efficacy for Treatment of Post-Partum Endometritis in Cows." *Veterinary Research* Vol. 29 (1998): 47-58.

Biziulevichius, G.A. "In vivo Studies on Lysosubtilin: 3. Efficacy for Treatment of Mastitis and Superficial Lesions of the Udder and Teats in Cows." *Veterinary Research* Vol. 29 (1998): 441-456.

Biziulevieius, G.A. "Comparative Antimicrobial Activity of Lysosubtilin and Its Acid-Resistant Derivative, Fermosorb." *International Journal of Antimicrobial Agents* Vol. 20 (2002): 65-68.

Burgos, R.A. "14-Deoxyandrographolide as a Platelet Activating Factor Antagonist in Bovine Neutrophils." *Planta Medica* Vol. 71 No. 7 (July 2005): 604-608.

Burris, T.P. "The Hypolipedemic Natural Product Guggulsterone Is a Promiscuous Steroid Receptor Ligand. *Molecular Pharmacology* Vol. 67 (2005): 948-954.

Choi, E.M. and Hwang, J.K. "Anti-Inflammatory, Analgesic and Antioxidant Activities of the Fruit of Foeniculum vulgare." *Fitoterapia* Vol. 75 No. 6 (Sept. 2004): 557-565.

Dadalioglu, I. "Chemical Compostitions and Antibacterial Effects of Essential Oils of Turkish Oregano (Origanum minutiflorum), Bay Laurel (Laurus nobilis), Spanish Lavender (Lavendula stoechas), and Fennel (Foeniculum vulgare) on Common Foodborne Pathogens." *J. Agric. Food Chem.* Vol. 52 (2004): 8255-8260.

DeBosscher, K. "A Fully Dissociated Compound of Plant Origin for Inflammatory Gene Repression." *PNAS* Vol. 102 No.44 (1 Nov. 2005): 15827-15832.

Delves-Broughton, J. "Applications of the Bacteriocin, Nisin." *Antonie Van Leeuwenhoek* Vol. 69 No. 2 (1996): 193-202.

Du, A. and Hu, S. "Effects of a Herbal Complex Against Eimeria tenella Infection in Chickens." *J. Vet. Med. B Infect. Dis. Vet. Public Health* Vol. 51 No. 4 (2004): 194-197.

Emmanuel, N. Ali. "Treatment of Bovine Dermatophilosis with Senna alata, Lantana camara and Mitracarpus scaber Leaf Extracts." *Journal of Ethnopharmacology* Vol. 86 (2003): 167-171.

Fang, H. "Modulation of Human Humoral Immune Response Through Orally Administered Bovine Colostrum." *FEMS Immunology and Medical Microbiology* Vol. 31 No. 2 (Aug. 2001): 93-96.

Fernandez, T.J. "The Potential of Tinospora rumphii as an Anthelmintic Against H. contortus in Goats." *Vetwork UK* Nov. 1997. www.vetwork .org. uk/pune13.htm#4.

Gilbert, B. "Synergy in Plant Medicines." *Current Medicinal Chemistry* Vol. 10 No. 1 (2003): 13-20.

Githiori, J.B. "Evaluation of Anthelmintic Properties of Some Plants Used as Livestock Dewormers Against Haemonchus contortus Infections in Sheep." *Parasitology* Vol. 129 Pt. 2 (Aug. 2004): 245-253.

Graczyk, T.K. "Therapeutic Efficacy of Hyperimmune Bovine Colostrum Treatment Against Clinical and Subclinical Cryptosporidium serpentis Infections in Captive Snakes." *Veterinary Parasitology* Vol. 74 No. 2-4 (31 Jan. 1998): 123-132.

Graefe, E. "Pharmacokinetics and Bioavailability of Quercitin Glycosides in Humans." *Journal of Clinical Pharmacology* Vol. 41 (2001): 492-499.

Greenspan, P. "Anti-Inflammatory Properties of the Muscadine Grape (Vitis rotundifolia)." *J. Agric. Food Chem.* Vol. 53 No. 22 (2 Nov. 2005): 8481-8484.

Gupta, M. "Evaluation of Antipyretic Potential of Veronia cineria Extract in Rats." *Phytotherapy Research* Vol. 17 (2003): 804-806.

Haridy, F.M. "Efficacy of Mirazid (Commiphora molmol) Against Fascioliasis in Egyption Sheep." *Journal of Egyptian Society of Parasitology* Vol. 33 No. 3 (Dec. 2003): 917-924.

Hemmatzadeh, F. "Therapeutic Effect of Fig Tree Latex on Bovine Papillomatosis." *J. Vet. Med. B Infect. Dis. Vet. Public Health* Vol. 50 No. 10 (Dec. 2003): 473-476.

Hogan, J.S. "Opsonic Activity of Bovine Serum and Mammary Secretion After Escherichia coli J5 Vaccination." *Journal of Dairy Science* Vol. 75 (1992): 72-77.

Hordegen, P. "The Anthelmintic Efficacy of Five Plant Products Against Gastrointestinal Thrichostrongylids in Artificially Infected Lambs." *Veterinary Parasitol* Vol. 117 No.1-2 (3 Nov. 2003): 51-60.

Hulse, E.C. and Lancaster, J.E. "The Treatment with Nisin of Chronic Bovine Mastitis Caused by Strep. uberis and Staphylococci." *Veterinary Record* Vol. 63 No.29 (21 July 1951): 477-480.

Husu, J. "Production of Hyperimmune Bovine Colostrum Against Campylobacter jejuni." *Journal of Applied Bacteriology* Vol. 74 No. 5 (May 1993): 564-569.

Jones, G.E. "Protection of Lambs Against Experimental Pneumonic Pasteurellosis by Transfer of Immune Serum." *Veterinary Microbiology* Vol. 20 (1989): 59-71.

Kohlert, C. "Bioavailability and Pharmacokinetics of Natural Volatile Terpenes in Animals and Man." *Planta Medica* Vol. 66 No. 6 (Aug. 2000): 495-505.

Kohlert, C. "Systemic Availability and Pharmacokinetics of Thymol in Humans." *Journal of Clinical Pharmacology* Vol. 42 No. 7 (July 2002): 731-737.

Koko, W.S. "Evaluation of Oral Therapy on Mansonial Schistosomiasis Using Single Dose of Balantines aegyptiaca Fruits and Praziquantel. *Fitoterapia* Vol. 76 No. 1 (2005): 30-34.

Koll, K. "Validation of Standardized High-Performance Thin-Layer Chromatographic Methods for Quality Control and Stability Testing of Herbals." *Journal of AOAC International* Vol. 86 No. 5 (Sep.-Oct. 2003): 909-915.

Larkins, N. "Potential Implications of Lactoferrin as a Therapeutic Agent." *American Journal of Veterinary Research* Vol. 66 No. 4 (Apr. 2005): 739-742.

Li, L.W. "Anti-Inflammatory Activity, Cytotoxicity and Active Compounds of Tinospora smilacina Benth." *Phytotherapy Research* Vol. 18 No. 1 (2004): 78-83.

Loew, D. "Approaching the Problem of Bioequivalence of Herbal Medicinal Products." *Phytotherapy Research* Vol. 16 (2002): 705-711.

Malinowski, E. "The Use of Some Immunomodulators and Non-Antibiotic Drugs in a Prophylaxis and Treatment of Mastitis." *Polish Journal of Veterinary Sciences* Vol. 5 No. 3 (2002): 197-202.

Marley, C.L. "The Effect of Birdsfoot Trefoil (Lotis corniculatus) and Chicory (Cichorium intybus) on Parasite Intensities and Performance of Lambs Naturally Infected with Helminth Parasites." *Veterinary Parasitology* Vol. 112 No. 1-2 (28 Feb. 2003): 147-155.

Massoud, A. "Treatment of Egyption Dicrocoeliasis in Man and Animals with Mirazid." *Journal of Egyptian Society of Parasitology* Vol. 33 No. 2 (Aug. 2003): 437-442.

Mbaria, J.M. "Comparative Effect of Pyrethrum Marc with Albendazole Against Sheep Gastrointestinal Nematodes." *Tropical Animal Health and Production* Vol. 30 No. 1 (Feb. 1998): 17-22.

McCurdy, C.R. "Analgesic Substances Derived From Natural Products (Nutraceuticals)." *Life Sciences* Vol. 78 No. 5 (22 Dec. 2005): 476-484.

Molan, A.L. "Effect of Condensed Tannins Extracted from Four Forages on the Viability the Larvae of Deer Lungworms and Gastrointestinal Nematodes." *Veterinary Record* Vol. 147 No. 2 (8 July 2000): 44-48.

Molan, A.L. "The Effect of Condensed Tannins from Seven Herbages on Trichostrongyle colubriformis Larval Migration in vitro." *Folia Parasitologica* Vol. 47 No. 1 (2000): 39-44.

Muruganandan, S. "Anti-Inflammatory Activity of Syzygium cumini Bark." *Fitoterapia* Vol. 72 No. 4 (May 2001): 369-375.

Naciri, M. "Treatment of Experimental Ovine Cryptosporidiosis with Ovine or Bovine Hyperimmune Colostrum." *Veterinary Parasitology* Vol. 53 No. 3-4 (June 1994): 173-190.

Naresh, R. "Clinical Management of SarcopticMange in Indian Buffalo Calves with a Botanical Ointment." *Veterinay Record* Vol. 156 No. 21(2005): 684-685.

Niezen, J.H. "The Effect of Feeding Sulla (Hedysarum coronarium) or Lucerne (Medicago sativa) on Lamb Parasite Burdens and Development of Immunity to Gastrointestinal Nematodes." *Veterinarian Parasitology* Vol. 105 No. 3 (2 May 2002): 229-245.

Olsen, E.J. "Effects of an Allicin-Based Product on Cryptosporidiosis in Neonatal Calves. *JAVMA* Vol. 212 No.7 (Apr. 1998).

Onyeyili, P.A. "Anthelmintic Activity of Crude Aqueous Extract of Nauclea latifolia Stem Bark Against Ovine Nematodes." *Fitoterapia* Vol 72 No. 1 (Jan. 2001): 12-21.

Ormrod, D.J. and Miller, T.E. "A Low Molecular Weight Component Derived from the Milk of Hyperimmunised Cows Suppresses Inflammation by Inhibiting Neutrophil Emigration." *Agents Action* Vol. 37 No. 1-2 (Sep. 1992): 70-79.

Paolini, V. "In vitro Effects of Three Woody Plant and Sainfoin Extracts on 3rd Stage Larvae and Adult Worms of Three Gastrointestinal Nematodes." *Parasitology* Vol. 129 Pt. 1 (Jul. 2004): 69-77.

Pessoa , L.M. "Anthelmintic Activity of Essential Oil of Ocimum gratissimum Linn. and Eugenol against Haemonchus contortus." *Veterinarian Parasitol* Vol. 109 No.1-2 (16 Oct. 2002): 59-63.

Quintus, J. "Urinary Excretion of Arbutin Metabolites after Oral Administration of Bearberry Leaf Extracts." *Planta Medica* Vol. 71 (2005): 147-152.

Reichling, J. "Topical Tea Tree Oil Effective in Canine Localized Pruritic Dermatitis – A Multi-Centre Randomized Double-Blind Controlled Clinical

Trial in the Veterinary Practice." *Dtsch. Tierartzl. Wochenschr.* Vol. 111 No. 10 (Oct. 2004): 408-414.

Robert, A.M. "Protection of Cornea Against Proteolytic Damage. Experimental of Procyanidolic Oligomers (PCO) on Bovine Cornea." *J. Fr. Ophtalmol.* Vol. 25 No. 4 (Apr. 2002): 351-355.

Rosengarten, F. "A Neglected Mayan Galactogogue – Ixbut (Euphorbia lancifolia)." *Journal of Ethnopharmacology* Vol. 5 No. 1 (Jan. 1982): 91-112.

Ross, R.P. "Developing Applications for Lactococcal Bacteriocins." *Antonie Van Leeuwenhoek* Vol. 76 (1999): 337-346.

Roth, J.A. "Enhancement of Neutrophil Function by Ultrafiltered Bovine Whey." *Journal of Dairy Science* Vol. 84 (2001): 824-829.

Saha, A. "Anti-Inflammatory, Analgesic and Diuretic Activity of Polygonum lanatum Roxb." *Pakistan Journal of Pharmaceutical Sciences* Vol. 18 No.4 (Oct. 2005): 13-17.

Sayyah, M., Hadidi, N., and Kamalinejad, M. "Analgesic and Anti-Inflammatory Activity of Lactuca sativa Seed Extract in Rats." *Journal of Ethnopharmacology* Vol. 92 No. 1-2 (June 2004): 325-329.

Sharma, S. "Pharmacokinetic Study of 11-Keto Beta-Bboswellic Acid." *Phytomedicine* Vol. 11 No. 2-3 (Feb. 2004): 255-260.

Sheikh, A."Passive Immunotherapy in Neonatal Calves – I. Safety and Potency of a J5 Escherichia coli Hyperimmune Plasma in Neonatal Calves." *Vaccine* Vol. 13 No. 15 (1995): 1449-1453.

Shen, C.L. "Comparative Effects of Ginger Root (Zingiber officinale Rosc.) on the Production of Inflammatory Mediators in Normal and Osteoarthritic Sow Chondrocytes." *J. Med Food.* Vol. 8 No. 2 (Sum. 2005): 149-153.

Sirtori, C. "Aescin: Pharmacology, Pharmacokinetics and Therapeutic Profile." *Phamacological Reasearch* Vol. 44 No. 3 (2001).

Slobodnikova, L. "Antimicrobial Activity of Mahonia aquifolium Crude Extract and Its Major Isolated Alkaloids." *Phytotherapy Research* Vol. 18 (2004): 674-676.

Stelwagen, K. and Ormrod, D.J. "An Anti-Inflammatory Component Derived from Milk of Hyperimmunised Cows Reduces Tight Junction Permeability in vitro." *Inflamm. Res.* Vol. 47 (1998): 384-388.

Tedesco, D. "Silymarin, a Possible Hepatoprotector in Dairy Cows: Biochemical and Histological Observations." *Journal of Veterinary Medicine Series A* Vol. 51 (2004): 85-89.

Tzamaloukas, O. "The Consequences of Short-Term Grazing of Bioactive Forages on Established Adult and Incoming Larvae Populations of Teladorsagia circumcinta in Lambs." *International Journal for Parasitology* Vol. 35 (2005): 329-335.

Uddin, Q. "Antifilarial Potential of the Fruits and Leaves Extracts of Pongamia pinnata on Cattle Filarial Parasite Setaria cervi." *Phytotherapy Research* Vol. 17 (2003): 1104-1107.

Yagi, A. "Antioxidant, Free Radical Scavenging and Anti-Inflammatory Effects of Aloesin Derivatives in Aloe vera." *Planta Medica* Vol. 68 (2002): 957-960.

Yong, L. "Enteric Disposition and Recycling of Flavanoids and Gingko Flavanoids." *Journal of Alternative and Complementary Medicine* Vol. 9 No. 5 (2003): 631-640.

Zhang, B. "Gaultherin, a Natural Salicylates Derivative from Gaultheria yunnanensis: Towards a Better Non-Steroidal Anti-Inflammatory Drug." *European Journal of Pharmacology* Vol. 530 No. 1-2 (13 Jan. 2006): 166-171.

Zhang, Y.H. "In vitro Inhibitory Effects of Bergenin and Norbergenin on Bovine Adrenal Tyrosine Hydroxylase. *Phytotherapy Research* Vol. 17 No. 8 (Sep. 2003): 967-969.

Resource Contacts

Acres U.S.A.
P.O. Box 91299
Austin, TX 78709
T- 800-355-5313
F- 512-892-4448
www.acresusa.com

Agri-Dynamics/Jerry Brunetti
P.O. Box 267
Martins Creek, PA, 18063
www.agri-dynamics.com
T- 610-250-9280
T- 877-393-4484

American Farmland Trust
1200 18th St. NW, Suite 800
Washington, DC 20036
T- 800-431-1499
F- 202-659-8339
www.farmland.org

**Appropriate Technology Transfer
for Rural Areas (ATTRA)**
P.O. Box 3657
Fayetteville, AR 72702
T- 800-346-9140
www.attra.org

American Botanical Council
www.herbalgram.org

**American Holistic Veterinary
Medical Association**
www.ahvma.org

**Biodynamic Farming and
  Gardening Association, Inc.**
25844 Butler Road
Junction City, OR 97448
T- 541-998-0105
T- 888-516-7797
F- 541-998-0106
biodynamic@aol.com
www.biodynamics.com

Botanical.com
www.botanical.com/botanical/
  mgmh/mgmh.html

Canadian Organic Growers
323 Chapel Street
Ottawa, ON K1N 7Z2
Canada
T- 613-216-0741
F- 613-236-0743
www.cog.ca

CAVM Community
CAVM Discussion List for DVM's,
  VMD's, MD's, PhD's in the health
  sciences
http://cavm.net

Eclectic Institute, Inc.
www.eclecticherb.com

Eclectic Medical Publications
www.eclecticherb.com/emp/

**European Agency for the
  Evaluation of Medicinal
  Products**
www.emea.eu.int/htms/vet/mrls/
  a-zmrl.htm

Farming
http://farmingmagazine.net

Fetrell
Jeff Mattocks
P.O. Box 265
Bainbridge, PA 17502
T- 717-367-1566
www.fetrell.com

**FDA Approved Animal
  Drug Products**
http://dil.vetmed.vt.edu/NADA/
  NADA.cfm

Forage Systems Research Center
http://aes.missouri.edu/fsrc/

Gaia Herbs
108 Island Ford Rd.
Brevard, NC 28712
T- 888-917-8269
T- 828-884-4242
F- 828-883-5960
www.gaiaherbs.com

Graze
Joel McNair, Editor/Publisher
P.O. Box 48
Belleville, WI 53508
T- 608-455-3311
F- 608-455-2402
graze@ticon.net
www.grazeonline.com

**Health Sciences Libraries
  Consortium**
www.hslc.org

HEEL Biotherapeutics
P.O. Box 11280
Albuquerque, NM 87192
T- 800-621-7644
F- 505-275-1672
www.heelusa.com

Henriette's Herbal Homepage
www.henriettesherbal.com

Herbalist and Alchemist Books
P.O. Box 553
Broadway, NJ 08808
T- 908-835-0822
F- 908-835-0824
www.herbaltherapeutics.net

Herbal Vitality, Inc.
2085 Contractors Rd. #8
Sedona, AZ 86336
T- 888-204-6455
herbvitality@sedona.net

HerbMed
http://herbmed.org

Homeopathic & Flower Remedies
Product/Resources Guide
www.lightparty.com

**Homeopathic Medical Society
of the State of Pennsylvania**
P.O. Box 353
Palmyra, PA 17078-0353
T- 717-838-9563
F- 717-838-0377

Horizon Organic Dairy
P.O. Box 17577
Boulder, Co 80308
T- 888-494-3020
www.horizonorganic.com

IMPRO Products
3 Allamakee St.
Waukon, IA 52172
T- 800-626-5536
F- 563-568-4259

K.O.W. Consulting Associates
Weaver Feeding & Management,
   LLC
25800 Valley View Rd.
Cuba City, WI 53807
T- 608-762-6948
F- 608-762-6949
tweaver@mhtc.net
www.kowconsulting.com

Northeast Grazing Guide
www.umaine.edu/grazingguide

**Northeast Organic
   Farming Association**
NOFA-VT
PO Box 697
Richmond, VT 05477
T- 802-434-4122
F- 802-434-4154
www.nofavt.org

**Northeast Organic Dairy
   Producers Alliance**
www.nodpa.com

Organic Ag Info
www.organicaginfo.org

Old/Antiquarian Books:
www.alibris.com

OrganicValley Dairy Products
One Organic Way
La Farge, WI 54639
T- 888-444-6455
F- 608-625-3025
www.organicvalley.com

Organic Materials Review Institute
P.O. Box 11558
Eugene, OR 97440-3758
T- 541-343-7600
F- 541-343-8971
info@omri.org
www.omri.org

Penn Dutch Cow Care
1272 Mt. Pleasant Rd.
Quarryville, PA 17566
T- 717-529-0155
www.penndutchcowcare.org

Penn State Grazing Research and Education Center
www.agronomy.psu.edu/Research/
graze.htm

Pennsylvania Association for Sustainable Agriculture
114 West Main Street
P.O. Box 419
Millheim, PA 16854-0419
T- 814-349-9856
F- 814-349-9840
www.pasafarming.org

Pennsylvania Certified Organic
406 S. Pennsylvania Ave.
Center Hall, PA 16828
T- 814-364-1344
F- 814-364-4431
info@paorganic.org

www.paorganic.org
Phytochemical and Ethnobotanical Databases
www.ars-grin.gov/duke/

Rarebooks
www.planetherbs.com

Regional Farm & Food Project
P.O. Box 339
Chatham, NY 12037
T- 518-271-0744
www.farmandfood.org

Ruminant Nutrition
www.asft.ttu.edu/ansc5311/

Seven Stars Farm, Inc.
501 W. Seven Stars Rd.
Phoenixville, PA 19460
T- 610-935-1949
F- 610-935-8292
www.sevenstarsfarm.com/farm.htm

Southwest School of Botanical Medicine
(excellent site for Eclectic literature)
www.swsbm.com/homepage/

PlantMedicine
www.phytotherapy.info

The Phytochemistry of Herbs
www.herbalchem.net

University of Pennsylvania School of Veterinary Medicine
www.upenn.edu

University of Wisconsin Forage and Extension Links
www.uwex.edu/ces/forage/links.htm

University of Wisconsin — Pasture Management and Grazing
www.uwrf.edu/grazing/

USDA-ARS Pasture Systems and Watershed Management Research Unit
http://pswru.arsup.psu.edu/

USDA National Organic Program (NOP)
1400 Independence Avenue, SW
Room 4008, South Building
Washington, DC 20250
www.ams.usda.gov/nop/indexIE.htm

USDA Natural Resources Conservation Service
307 B Airport Rd.
Smoketown, PA 17576-0211
717-396-9423

Vanbeek Scientific
3689 460th Street
Orange City, IA 51041
T- 800-346-5311
www.vanbeekscientific.com

Veterinary Botanical Medical Association
ww.vbma.org

Washington Homeopathic Products
33 Fairfax St
Berkeley Springs, WV 25411
T- 800-336-1695
F- 304-258-6335
www.homeopathyworks.com

Glossary

Therapeutic terms (Fish, 1930; p. 42-56)

Abortifacient — An agent causing premature birth of young. (ergot)

Alkaloid — An organic base containing nitrogen and form salts with acids. (atropine)

Alterative — A medicine used to modify nutrition so as to overcome morbid processes. (sassafras)

Analgesic — A medicine used to alleviate pain. (opium)

Anesthetic — An agent used to produce insensibility to pain. (ether)

Anodyne — An agent which diminishes sensibility to pain. (compound spirit of ether)

Anthelmintic — A remedy for destroying or expelling worms or to prevent their development. (santonin)

Antidote — A substance to counteract poisons. (belladonna)

Anti-emetic — An agent which allays vomiting. (bismuth subnitrate)

Anti-febrile — An agent for the reduction of fever. (acetanilid)

Anti-parasitic — A substance that destroys or drives away insects. (essential oils)

Anti-pyretic — A medicine to reduce body temperatures in fevers. (salicylic acid)

Anti-septic — An agent antagonizing sepsis or putrefaction. (phenol)

Anti-spasmodic — A medicine for preventing or relieving spasms. (valerian)

Astringent — A medicine causing contraction or constriction of tissues. (tannin)

Bitter — A medicine with a bitter taste stimulating the gastro-intestinal mucosa without materially affecting the general system. (gentian)

Blister — An agent, which when applied to the skin, causes a local inflammatory exudation of serum under the epidermis. (cantharides)

Cachexia — A term used to designate any morbid tendency or depraved condition of general nutrition.

Carminative — A remedy which helps allay pain by causing the expulsion of flatus from the alimentary canal. (asafetida)

Cathartic — A medicine which quickens or increases evacuations from the intestines. (castor oil)

Cathartic, cholagogue — An agent stimulating the stool and flow of bile at the same time. (podophyllin)

Cathartic, drastic — A medicine producing violent action of the bowels with griping pain. (jalap)

Cathartic, hydrogogue — A remedy which causes copious watery stools. (elaterium)

Cathartic, simple — A substance which causes one or two actions of the bowels. (senna)

Caustic — An agent used to kill living tissue. (silver nitrate)

Cholagogue — A drug provoking the flow of bile. (podophyllum)

Demulcent — A mucilaginous or oily substance to soothe and protect irritated mucous membranes. (ulmus)

Deodorant — A substance to conceal or destroy foul odors. (phenol)

Diaphoretic — A medicine to produce sweating. (pilocarpine)

Digestive — A tonic which promotes digestive processes. (quassia)

Disinfectant — A substance with the power of destroying disease germs or the noxious properties of decaying organic matter. (hydrogen peroxide)

Diuretic — A drug to increase the secretion of urine. (buchu)

Ecbolic — A drug to produce abortion. (ergot)

Emetic — A medicine to produce vomiting. (ipecac)

Emmenagogue — A drug to stimulate menstruation. (savin)

Emollient — A substance used externally to mechanically soften and protect tissues. (flaxseed poultice, oils)

Expectorant — A medicine to act upon the pulmonary mucous membrane to increase or alter its secretions. (balsams)

Febrifuge — An agent to decrease fever. (aconite)

Galactogogue — An agent to increase the secretion of milk. (pilocarpine)

Germicide — An agent to destroy parasites. (carbolic acid)

GRAS — "Generally recognized as safe."

Hematinic — A tonic for the blood. (iron preparations)

Hydragogue — An agent causing full watery discharges from the bowels. (gamboge)

Hydrotic — An agent to produce perspiration. (diaphoretics)

Hyperesthetic — Increasing sensitiveness of the skin.

Insecticide — A remedy to destroy insects. (pyrethrum)

Lactogogue — An agent to increase the secretion of milk. (malt)

Laxative — A medicine acting mildly in opening or loosening the bowels. (sulfur)

Local anesthetic — A medicine to destroy sensation, when applied locally. (cocaine)

Lubricant — An agent to soothe irritation in the throat, fauces, etc. (olive oil, honey)

Mechanical — an agent acting on a physical basis. (slippery elm)

Medicament — An agent used for curing diseases or wounds. (belladonna)

Mydriatic — an agent causing dilatation of the pupil. (atropine)

Myotic — A drug causing contraction of the pupil. (morphine)

Narcotic — A powerful remedy causing stupor. (opium)

Nephritic — Medicine used in renal diseases. (uva ursi)

Nervine — Medicine to calm the nervous system. (bromides)

Odontalgic — An agent for the relief of toothache. (oil of cloves)

Opiate — A medicine causing sleep. (opium)

Oxytocic — An agent to aid or produce parturition. (ergot; cotton root)

Oxyuricide — An agent destructive to parasitic (oxyuris) worms. (santonin)

Palliative — A remedy for the relief but not necessarily the cure of the disease. (morphine)

Panacea — A remedy pretending to cure all diseases. (some Patent Medicines)

Parasiticide — A remedy for the destruction of parasites. (calcium sulphide)

Parturient or parturifacient — A medicine to aid in the birth of the young. (ustilago)

Peristaltic — A drug increasing the movement or contraction of the intestines (strychnine)

Placebo — An inert substance given to satisfy a patient.

Protective — An agent to protect the part to which it is applied. (collodion)

Purgative — A medicine to produce increased discharges from the bowels. (aloe)

Refrigerant — An agent which produces the sensation of coolness. (alcohol)

Restorative — A medicine for causing a return of body vigor. (tonics)

Rubefacient — An agent causing irritation and redness of the skin. (mustard)

Septic — An agent that promotes putrefaction. (bacteria)

Simple bitter — A drug with a bitter taste. (calumba; quassia)

Sorbefacient — A medicine causing abortion. (ergot)

Specific — A remedy supposed to exert a special action in the prevention or cure of certain diseases. (quinine in malaria, potassium iodide in actinomycosis)

Stimulant — A medicine to increase or quicken functional activity. (strychnine)

Stomachic — A drug to stimulate functional activity of the stomach. (gentian)

Styptic — Agents causing contraction of blood vessels to check bleeding. (alum)

Sudorific — A medicine or agent causing increased sweating. (jaborandi)

Synergist — A drug which cooperates or assists the action of another.

Taenicide — A remedy for destroying tape worms. (male fern)

Taenifuge — An agent to expel tape worms. (areca nut)

Tonic — a medicine promoting nutrition and giving tone to the system. (arsenic)

Uterine — An agent acting upon the uterus (ustilago)

Vermicide — An agent to destroy parasitic worms. (creosote)

Vermifuge — An agent to expel parasitic worms. (arecoline hydrobromide; purgatives)

Vulnerary — Any remedy or agent for healing wounds. (ointments, etc.)

Zoiatrica — veterinary medicines.

Medical Terminology (Stedman's Medical Dictionary, 1990)

Alimentary — Relating to food or nutrition.

Amenorrhea — Absence or abnormal cessation of the menses.

Anasarca — A generalized infiltration of edema fluid into the subcutaneous connective tissue.

Anorexia — Diminished appetite; aversion to food.

Aphonia — loss of the voice as a result of disease or injury of the organ of speech.

Ascites — Accumulation of serous fluid in the peritoneal cavity.

Asthenia (asthenic) — weakness, adynamia; weakness or debility.

Atonic — Relaxed; without normal tone or tension.

Balanitis — Inflammation of the glans penis or the glans clitoridis.

Borborygmy — Rumbling or gurgling noises produced by movement of gas in the alimentary canal, and audible at a distance.

Buccal — Pertaining to, adjacent to, or in the direction of the cheeks.

Carbuncle — Deep-seated pyogenic infection of the skin and subcutaneous tissues, usually arising in several contiguous hair follicles, with the formation of connecting sinuses; often preceded by fever or malaise.

Catarrh — Simple inflammation of a mucous membrane.

Cellulitis — Inflammation of cellular or connective tissue

Cicatrization — The process of scar formation; the healing of a wound otherwise than by first intention.

Clonic — Marked by alternate contraction and relaxation of muscle.

Colic — Spasmodic pains in the abdomen; relating to the colon.

Congestion — Presence of an abnormal amount of fluid in the vessels or passages of a part or organ

Cutaneous — Relating to the skin.

Cystitis — Inflammation of the urinary bladder.

Dropsy — Old term for edema.

Dyspepsia — Gastric indigestion; impaired gastric function.

Dyspnea — Shortness of breath, a subjective difficulty or distress in breathing, usually associated with disease of the heart or lungs.

Emphysema — Presence of air in the interstices of the connective tissue of a part.

Endometritis — Inflammation of the endometrium.

Endophyte — A plant parasite living within another organism.

Exudate — Any fluid that has exuded out of a tissue or its capillaries, more specifically because of injury or inflammation.

Febrile — Feverish; denoting or relating to fever.

Flatulence — Presence of an excessive amount of gas in the stomach or intestines.

Goiter — A chronic enlargement of the thyroid gland, not due to a neoplasm.

Hematoma — A localized mass of blood that is relatively or completely confined within an organ or tissue, a space, or potential space; it is usually clotted (or partly clotted).

Hematuria — Any condition in which the urine contains blood or red blood cells.

Hyperemia — The presence of an increased amount of blood in a part or organ.

Hypertrophy — General increase in bulk of a part or organ, not due to tumor formation.

Inertia — Denoting inactivity or lack of force, lack of physical vigor, or sluggishness of action.

Inflammation — A fundamental pathologic process consisting of a dynamic complex of reactions that occur in the affected blood vessels and adjacent tissues in response to an injury or abnormal stimulation caused by a physical, chemical or biologic agent.

Keratitis — Inflammation of the cornea.

Laceration — A torn or jagged wound, or an accidental cut wound.

Laminitis — Founder; a painful inflammation of the sensitive lamina to which the hoof is attached.

Leucorrhea —Discharge from the vagina of a white or yellowish viscid fluid containing mucus and pus cells.

Menorrhagia — Hypermenorrhea.

Metritis — Inflammation of the uterus.

Mucopurulent — Pertaining to an exudates that is chiefly purulent (pus), but containing relatively conspicuous proportions of mucous material.

Neuralgia — Pain of a severe, throbbing, or stabbing character in the course or distribution of a nerve.

Orchitis — Inflammation of the testis.

Otorrhea — A discharge from the ear.

Periosteal — Relating to the periosteum.

Pharyngitis — Inflammation of the mucous membrane and underlying parts of the pharynx.

Post-partum — after having given birth.

Puerperal — Relating to the period after birth.

Purulent — Containing, consisting of, or forming pus.

Pyelitis — Inflammation of the renal pelvis.

Sepsis — The presence of various pus forming and other pathogenic organisms, or their toxins, in the blood or tissues; septicemia is a common form of sepsis.

Sinusitis — Inflammation of the mucous membranes lining the sinuses.

Sthenic — Strong; active; said of a fever with a bounding pulse, high temperature and active delirium.

Stomatitis — Inflammation of the mucous membranes of the mouth.

Subinvolution — Arrest of the normal involution of the uterus following birth with the organ remaining abnormally large.

Suppuration — The formation of pus.

Tenesmus — A painful spasm of the anal sphincter with an urgent desire to evacuate the bowel or bladder, involuntary straining, and the passage of but little fecal matter or urine.

Tonic — In a state of continuous unremitting action; denoting especially a muscular contraction.

Varicose — Relating to dilated veins or a state of dilated veins.

Vasomotor — Causing dilation or constriction of the blood vessels; denoting the nerves which have this action.

Index

immunoglobulin, 253
immunoglobulin IgA, 103
impacted rumen, 217-219
impactions, 153
infertility, 286-287
inflammation, 128-129
inflammatory states, 125
input-substitution, 190
insect powder, 156
insect stings, 272
intestinal irritation, 147
intranasal IBR/PI3, 243
intravenous (IV) fluids, 197
iodine, 152
iodine deficiency, 287
Iodium, 182
ipecac, 152, 182
ipecachuahana (ipecac), 152
irritants, 162

J-5, 301
jaborandi/pilocarpus, 152-153
jaw abscess, 234
Johne's disease, 228, 315
Jousset M.D., Pierre, 111
jugular vein, 197-199
juniper oil, 153

K.O.W. Consulting Associates, 23-24
kamala, 153
kava kava, 153
kelp, 283
ketosis, 214-215
ketosis diagram, 215
ketosis, nervous, 215-216
killing a quarter, 325-326
kino, 153
kousso, 154
krameria, 154

lacerations (teat cuts), 323-324
Lachesis, 182
lactucarium, 154
lameness, forelimb, 273
laminitis, 224
lard, 305-306, 310
laudanum, 154
Law's dip, 162
laxative, 145, 159, 168
Ledum, 182
leg abscesses, 273-274
leprosy, 145
Lepto, 298
Lepto hardjo bovis, 298-299
Leptospirosis, 297, 298
leucosis, 247
leukemia, 247
lice, 175
licorice root, 150
life-threatening conditions,
        62-63, 64
Lillium tigrinum, 182
liver problem, 214
Lloyd, John Uri, 116-117
lobelia, 154
lockjaw, 272, 330
low concentration medicinals, 118
lumpy jaw, 174, 272-273
lungworms, 174
Lycopodium, 134-135, 182
lymph nodes, 102-103, 248
lymphatic diseases, 152

Mad Cow Disease, 250-251
Mag phos, 182
maggots, 269
male fern, 154
male generative organs, 172
malignant edema, 276
malnutrition, 146
mammary gland, infection, 325-326
manganese, 251

SSC, 183

staph, 329

Staph nosode, 184

Staph/Strep nosode, 184

Stedman's medical terminology, 296-399

stillingia, 160

stimulant, 143, 147, 149, 150, 154, 156

stomach worms, 36-37

stomachic, 129, 144, 145, 151, 157, 161, 168

stomachic tonic, 149

stomachs, calves, 315

stramonium, 160

streambank fencing, 37

strep, 329

Strep nosode, 184

strongyles, 174

strophanthus, 160

strychnine, antidote, 154

styptics, 173

subacute rumen acidosis (SARA), 13, 224-225

Sulfur, 184

Sulfur, constitutional, 188

sumbul, 160-161

suppuration, 143

swollen joints, 275-276

Symphytum, 184

synchronization programs, 16

synchronization protocols, 294

synthetic substances, 60

tails, docked, 55

tanacetum, 161

tapeworm, 150, 153, 154, 174

taraxacum, 161

teat amputation, 324

teat blisters, 321

teat cuts, 323-324

teat fistula, 325

teat leakage, 324

teat spider, 325

teat, stepped on, 322

tetanus, 128, 272, 330

therapeutic terms (Fish), 393-396

thiosinamin, 161

thrush, 174

Thuja, 184

thymol, 161-162

thyroid glands, 152

tiglii, 162

tobacco, 162

tonga, 162

tonics, 173

tonics, general, 167

Total Mixed Ration (TMR), 12

toxemia, 205

triticum, 162

turpentine, 162-163

turpentine (oil of), 163

twins, 305

twisted cecum, 222

twisted stomach, 220-222

udder abscesses, 330

udder edema, 319-321

udder swelling, 319-321

ulcers in calves, 236-237

ultraviolet light, 40

umbilical abscess, 235-236

umbilical hernia, 235-236

underfeeding, 287

United States Pharmacopeia (USP), 123

urethra inflammation, 149

urinary antiseptics, 172

urinary, irritation, 162

urinary organs, debilitated, 146

urinary sedatives, 172

urinary troubles, 164

urine, excreted, 38

urine pooling, 295

Urtica urens, 184

Duchess

1987-2003

Traveling companion and confidante for 16 years.

The Moo News

The *Moo News* is the newsletter that I send out monthly to my local clients. It usually discusses clinical problems that I have seen or foresee in that time period. It occasionally also has some advertisements by my farmers selling cows or other dairy related items. The *Moo News* is accessible on my website at *www.penndutch cowcare.org* and then clicking *Moo News.* If the reader does not have access to a computer, please either call me at 1-717-529-0155 (collect calls will not be accepted) or write:

Hubert J. Karreman, VMD
1272 Mt. Pleasant Rd.
Quarryville, PA 17566
USA

IN THIS GROUNDBREAKING WORK, Dr. Karreman invites us to journey into the world of dairy cows from a truly holistic perspective. With much thought and direct experience, he builds a foundation from which to view dairy cows as animals that occupy a unique agro-ecological niche in our world. From within that niche, he describes how cows can be treated for a wide variety of problems with plant-derived and biological medicines. Drawing upon veterinary treatments from the days before synthetic pharmaceuticals, and tempering them with modern knowledge and clinical experience, Dr. Karreman bridges the world of natural treatments with life in the barn in a rational and easy-to-understand way. In describing treatments for common dairy cow diseases, he covers practical aspects of biologics, botanical medicines, homeopathic remedies, acupuncture and conventional medicine. This book should serve as a useful reference for years to come.

Praise for *Treating Dairy Cows Naturally*

This comprehensive text provides a broad overview of sustainable agriculture, basics of grazing and soil characteristics, physiology, and Complementary and Alternative Veterinary Medicine and includes mechanics of the system and the philosophies from which it is derived. . . . It is especially useful for farmers considering the transition to organic production or those who just wish to use a more sustainable method for farming and to reduce synthetic inputs on their farms.

– *Journal of the American Veterinary Medical Association*

Here at long last is a comprehensive resource book for farmers and veterinarians alike on the treatment of health problems in dairy cattle, from an organic perspective. Three thumbs up for *Treating Dairy Cows Naturally.*

– *The Natural Farmer* magazine

Dr. Karreman's medical hands-on experience on livestock farms is self-evident. He has effectively connected "all the dots" that make for a sustainable ecological farming unit. . . . No practitioner or stockman dare be without this wonderful, well-researched book.

– Jerry Brunetti, Agri-Dynamics

This is the book the organic dairy community has been waiting for, a must read for every bovine lover. You don't need to be a farmer or a vet to love *Treating Dairy Cows Naturally.*

– Kelly Shea, Horizon Organic Dairy

I have found *Treating Dairy Cows Naturally* to be an invaluable resource. Homeopathic, natural and the rediscovered Eclectic remedies are presented in a clear and farmer-friendly manner. It may sound cliche but . . . for the busy farmer, student, or vet . . . if you're only going to buy one book on organic dairy cattle management, this is the one!

– David Griffiths, Seven Stars Farm

DR. HUBERT KARREMAN is uniquely qualified to present this all-encompassing work on ecological dairy farming. A dairy practitioner who tends to cows any time of day or night, he is also an active member in many organizations that promote the health and welfare of animals. Since graduating from the University of Pennsylvania School of Veterinary Medicine in 1995, Dr. Karreman has been continuously active in the American Veterinary Medical Association, American Association of Bovine Practitioners, American Holistic Veterinary Medical Association, and Veterinary Botanical Medical Association. He has been an invited speaker at the AVMA and AHVMA 2006 annual meetings, the Universities of Connecticut, Maryland, Minnesota and New Hampshire as well as Cornell, Tufts and the University of Pennsylvania veterinary schools. He has also lectured at the New York State Veterinary Medical Society and the Maine Veterinary Society. In 1999 he was nominated by the AABP and subsequently appointed by the AVMA to its Task Force on Complementary and Alternative Veterinary Medicine. In 2005 Dr. Karreman was appointed to the USDA National Organic Standards Board for a five-year term.

ISBN 978-1-60173-000-8

EAN
9 781601 730008
54000

WHATEVER IT TAKES, DOC!

ACRES U.S.A.   w w w . a c r e s u s a . c o m

Made in United States
Orlando, FL
12 November 2024

53804681R00239